THE LASER LITERATURE

An Annotated Guide

THE
LASER
LITERATURE

An Annotated Guide

by Kiyo Tomiyasu
Consulting Engineer
General Electric Research and Development Center
Schenectady, New York

Ⓟ Springer Science+Business Media, LLC 1968

The author and the publisher gratefully acknowledge the permission of The Institute of Electrical and Electronics Engineers, Inc., to reproduce the five bibliographies which appeared in the June 1965, August 1965, June 1966, November 1966, and July 1967 issues of the *IEEE Journal of Quantum Electronics.*

Library of Congress Catalog Card Number 68-21474

ISBN 978-1-4899-6165-5   ISBN 978-1-4899-6321-5 (eBook)
DOI 10.1007/978-1-4899-6321-5

Originally published by Plenum Press in 1968.

Softcover reprint of the hardcover 1st edition 1968

PREFACE

Coherent properties of light have been known for many years; however, a method of generating spatially and temporally coherent light with a useful amount of output power was not available until recently. In 1958 Drs. C. H. Townes and A. L. Schawlow proposed a coherent light generator called an optical maser, and in 1960, Dr. T. H. Maiman made the astonishing announcement that he generated coherent red light by using optically pumped ruby. Immediately thereafter widespread activity began on the development of the optical maser, or laser, as well as on the applications of lasers. Concurrently the publication of papers describing these developments increased at a very rapid rate. In order to be aware of all of the laser developments, the author had to keep abreast of the literature and store the references in an easily retrievable form. Thus a laser bibliography arranged in subject categories was started.

The first collection of about 600 laser references was compiled between May 1963 and May 1964; this appeared in a company report and was not published in the open literature. The second collection was compiled between June 1964 and December 1964. In early 1965 the Institute of Electrical and Electronics Engineers (IEEE) started publishing the IEEE Journal of Quantum Electronics, and the second collection of references appeared in the June 1965 issue. Subsequently four more laser bibliographies were prepared, and these were published in the following issues of the IEEE Journal of Quantum Electronics: August 1965, June 1966, November 1966, and July 1967. This book is based on these five published bibliographies and the first collection of about 600 references.

The book contains 3990 references with an index of 3335 authors, and the references were compiled during the period from May 1963 through December 1966. Most of the references were taken from journals readily accessible to the author; hence the journals cited are primarily those published in the United States. Some journals published outside the United States in English or which have been translated into the English language are also included.

The references are listed in chronological order within each subject category. A few of the references are dated prior to the compilation period. Some of the references are listed under more than one subject category because of the contents of the paper. Brief annotations are added to most of the references. Where possible, errata and comments are listed with the original reference.

An explanation of some of the 27 subject categories may be helpful. In Category 4, all references are listed that concern neodymium ion lasers regardless of the host material ($CaWO_4$, YAG, glass, etc.). All gaseous ion, metal vapor, and molecular gas lasers are listed in Category 7; notable among these are the argon ion and carbon dioxide lasers. Chelate lasers are listed in Category 8. The general field of nonlinear optics has been divided and the references are listed under Categories 19, 20, and 23. References concerned with self-constriction or self-focusing of optical beams are found in Category 23. References on ring lasers are listed in Category 26, and a few references on laser instrumentation are given in Category 27.

In preparing the author index some difficulty arose since a few authors were not consistent in the use of first names and initials. An effort has been made to identify the proper authors, and it is hoped that all problems were resolved correctly.

The author wishes to acknowledge the discussions held with Dr. K. Frank Tittel, who collaborated in establishing the subject categories of this bibliography. My secretary, Miss Joan Smida, aided immeasurably in compiling the list of authors. Finally, a great deal of credit is due to my wife, Eiko, who gave constant encouragement, provided continual assistance, prepared the author index, and made numerous valuable suggestions.

KIYO TOMIYASU

Schenectady, New York
May 1968

CONTENTS

1. BOOKS AND GENERAL LASER PAPERS

1.1 Bibliography of the Open Literature on Lasers
Edward V. Ashburn (Naval Ordnance Test Station)
J. Opt. Soc. Am., Vol. 53, pp. 647-652, May 1963 (208 references)

1.2 Advances in Optical Masers
Arthur L. Schawlow (Stanford U.)
Scientific American, Vol. 209, No. 1, pp. 34–45, July 1963

1.3 *Lasers and Applications*
Edited by W. S. C. Chang
Ohio State U., Columbus, Ohio, 267 p., 1963

1.4 *Proceedings of the Symposium on Optical Masers*
Edited by Jerome Fox
Microwave Res. Inst. Symposia Series, Vol. XIII
Polytechnic Institute of Brooklyn, Brooklyn, N. Y., 652 p., 1963

1.5 *Proceedings of the Third International Conference on Quantum Electronics*
Edited by P. Grivet (Paris U.) and N. Bloembergen (Harvard U.) Columbia U. Press, N. Y., Vol. 1, pp. 1-966; Vol. 2, pp. 967-1923, 1964

1.6 Review of Solid State Lasers
C. G. B. Garrett (BTL)
Proc. Third Int. Conf. Quantum Electronics, Columbia U. Press, N. Y., Vol. 2, pp. 971-984, 1964

1.7 *Laser Abstracts*
A. K. Kamal (Purdue U.)
Plenum Press, N. Y., Vol. 1, 177 p., 1964

1.8 *Masers and Lasers*
M. Brotherton
McGraw-Hill Book Co., N. Y., 207 p., 1964

1.9 *Proceedings of the International School of Physics "Enrico Fermi", Course XXXI—Quantum Electronics and Coherent Light*
Edited by P. A. Miles
Academic Press, N. Y., 371 p., 1964 (Course Director, C. H. Townes; Varenna, Italy; August 19-31, 1963, 22 papers)

1.10 *Optical Masers*
G. Birnbaum (North American Aviation)
Suppl. 2, Advances in Electronics and Electron Physics, Academic Press, N. Y., 306 p., 1964.

1.11 *Optical Masers*
O. S. Heavens (U. of London, England)
John Wiley & Sons, New York, N. Y. (Methuen & Co., London), 103 p., 1964.

1.12 Bibliography of the Open Literature on Lasers. II
Edward V. Ashburn (N.O.T.S.), Bela A. Lengyel (Hughes Res.), and Raymond W. Merry (Lockheed)
J. Opt. Soc. Am., Vol. 54, pp. 135-142, January 1964 (390 references, paper classification)

1.13 P–N Junction Lasers
G. Burns and M. I. Nathan (IBM)
Proc. IEEE, Vol. 52, pp. 770-794, July 1964 (extensive review, over 162 references)

1.14 *Nonlinear Optics*
N. Bloembergen (Harvard U.)
W. A. Benjamin, Inc., New York, N. Y., 222 p., 1965 (general treatment, experimental results)

1.15 Nonlinear Optical Phenomena
Paul N. Butcher
Ohio State U., Engineering Experiment Station, Bulletin 200, 143 p., 1965 (Ohio State U. Seminar lecture notes)

1.16 Lasers and Their Applications
Conference sponsored by I.E.E. Electronics and Science Divisions, I.E.E.E. (United Kingdom and Eire Section), and I.E.R.E. Held in London, England, September 29–October 1, 1964.
(gas lasers; solid state lasers; injection lasers; modulation-demodulation-detection; propagation and communications; measurement techniques and radiation hazards; ranging, navigation and metrology; and machining and welding, uses in physics and medicine; 59 abstracts and papers)

1.17 *The Commercial Development and Application of Laser Technology*
Robert Saltonstall, Jr., et al.
Hobbs, Dorman & Co., New York, 143 p., 1965 (general review, several applications)

1.18 *Optical Physics*
Max Garbuny (Westinghouse)
Academic Press, New York, 466 p., 1965 (includes lasers, nonlinear optics, detection)

1.19 *Quantum Optics and Electronics*
Edited by C. DeWitt, A. Blandin, and C. Cohen-Tannoudji
Gordon and Breach, Science Publishers, New York, 621 p., 1965 (lectures delivered at Les Houches in 1964, University of Grenoble, Grenoble, France, in English or French, 8 lecturers, quantum theory, coherence, pumping, optical masers, nonlinear optics)

1.20 Inversion Mechanisms in Gas Lasers
W. R. Bennett, Jr. (Yale U.)
Appl. Optics, Suppl. 2: Chemical Lasers, pp. 3–33, 1965 (comprehensive, known gas laser transitions tabulated, gases and molecules)

1.21 Proceedings of the International Symposium on Laser-Physics and Applications. Bern, Switzerland; October 12-15, 1964
Zeitschrift fur angewandte Mathematik und Physik, Vol. 16, No. 1, pp. 1–184, January 25, 1965

1.22 Optics in Laser Research
H. de Lang (Philips Res. Labs., The Netherlands)
Z. angewandte Mathematik und Physik, Vol. 16, No. 1, pp. 7–14, January 25, 1965

1.23 Crystalline Solid State Lasers
D. W. Goodwin (Royal Radar Estab., Gt. Malvern, England)
Z. angewandte Mathematik und Physik, Vol. 16, No. 1, pp. 35–48, January 25, 1965 (general considerations)

1.24   Solid State Lasers with CW Emission
       K. Gürs (Siemens and Halske, Germany)
       Z. angewandte Mathematik und Physik, Vol. 16, No. 1, pp. 49–
       62, January 25, 1965

1.25   Gas Lasers
       J. Haisma (Philips Res. Labs., Eindhoven, The Netherlands)
       Z. angewandte Mathematik und Physik, Vol. 16, No. 1, pp. 74–
       84, January 25, 1965

1.26   Omphaloskepsis on Current Trends in Optics
       L. M. Biberman, A. Glass, and W. D. Montgomery (IDA)
       Appl. Optics, Vol. 4, pp. 259–262, March 1965 (general com-
       ments on lasers taken from October 1964 APS Meeting)

1.27   **Parametric Principles in Optics**
       I. P. Kaminow (BTL)
       IEEE Spectrum, Vol. 2, No. 4, pp. 35-43, April 1965
       **(nonlinear optics, interaction, mixing, modulation, Raman
       effect)**

1.28   Coherence Properties of Optical Fields
       L. Mandel and E. Wolf (U. of Rochester)
       Rev. Mod. Phys., Vol. 37, No. 2, pp. 231–287, April 1965 (ex-
       tensive review, coherence effects, superposition effects, numer-
       ous references)

1.29   The Laser
       Edited by H. E. Whipple
       Annals of the New York Academy of Sciences, Vol. 122, Art. 2,
       pp. 571–834, May 28, 1965 (Conference Cochairmen, Leon Gold-
       man and Joseph Weber; May 4–5, 1964, New York, N. Y.;
       29 papers. Review, tutorial, devices, applications, biological
       effects)

1.30   Laser Emission Lines and Materials
       A. J. Bevolo (Washington U.) and W. A. Barker (U. of Santa
       Clara)
       Appl. Optics, Vol. 4, pp. 531–543, May 1965 (table of 357
       wavelengths from 65 laser systems, performance data)

1.31   Comments on the Paper by A. J. Bevolo and W. A. Barker
       A. Lempicki and H. Samelson (G.T. & E.)
       Appl. Optics, Vol. 5, p. 168, January 1966 (Appl. Optics, Vol. 4,
       pp. 531–543, May 1965; errors and omissions)

1.32   Visible and UV Laser Oscillation at 118 Wavelengths in Ionized
       Neon, Argon, Krypton, Xenon, Oxygen and Other Gases
       W. B. Bridges and A. N. Chester (Hughes)
       Appl. Optics, Vol. 4, pp. 573–580, May 1965 (2677 to 7993 Å
       range; singly, doubly and triply ionized atoms; pulsed dc dis-
       charge)

1.33   Spectroscopy of Ion Lasers
       William B. Bridges and Arthur N. Chester (Hughes)
       IEEE J. Quantum Electronics, Vol. QE-1, pp. 66–84, May 1965
       (review, tabulation of wavelengths)

1.34   Masers and Lasers
       Solid State Abstracts, Vol. 6, pp. 337–408, April-May 1965 (744
       references, a few references published in 1966, subject categories)

1.35   Laser Bibliography
       Compiled by K. Tomiyasu (GE)
       IEEE J. Quantum Electronics, Vol. QE-1, pp. 133–156, June
       1965 (compiled during June-December 1964, over 650 references,
       24 subject categories)

1.36   Bibliography of the Open Literature on Lasers. III
       E. V. Ashburn (Lockheed)
       J. Opt. Soc. Am., Vol. 55, pp. 752–766, June 1965 (compilation
       period from July 1, 1963 to December 10, 1964, 818 references
       listed alphabetically by authors, subject classification)

1.37   **Progress in Semiconductor Lasers**
       Benjamin Lax (MIT)
       IEEE Spectrum, Vol. 2, No. 7, pp. 62–75, July 1965

(background of recent developments, magnetic effects, applications)

1.38   Laser Bibliography II
       Compiled by K. Tomiyasu (GE)
       IEEE J. Quantum Electronics, Vol. QE-1, pp. 199–219, August
       1965 (compiled during January-June 1965, 561 references, 25
       subject categories)

1.39   1964 Nobel Lecture—Production of Coherent Radiation by
       Atoms and Molecules
       Charles H. Townes (MIT)
       IEEE Spectrum, Vol. 2, No. 8, pp. 30–43, August 1965 (his-
       torical review of maser and laser)

1.40   Bibliography of the Open Literature on Lasers IV
       Edward V. Ashburn (Lockheed)
       J. Opt. Soc. Am., Vol. 55, pp. 1040–1045, August 1965 (closing
       date of bibliography is December 31, 1964, 262 references listed
       alphabetically by authors, subject classification)

1.41   Wavefront Reconstruction Photography
       Emmett Leith and Juris Upatnieks (U. of Michigan)
       Physics Today, Vol. 18, No. 8, pp. 26–30, August 1965 (review,
       general principles, 3-D)

1.42   P-N Junction Electroluminescence and Diode Lasers
       Henry T. Minden (Sperry Rand)
       IEEE Trans. on Parts, Materials and Packaging, Vol. PMP-1,
       pp. 40–47, September 1965 (review, tutorial, 0.64 to 8.5 μ wave-
       lengths)

1.43   Interaction of Light and Microwave Sound
       C. F. Quate, C. D. W. Wilkinson, and D. K. Winslow
       (Stanford U.)
       Proc. IEEE, Vol. 53, pp. 1604–1623, October 1965 (compre-
       hensive, parametric equations, diffraction patterns, experiments)

1.44   Absorption, Spontaneous Emission, Stimulated Emission, and
       Maxwell's Equations
       A. I. Mahan (Johns Hopkins U.)
       J. Opt. Soc. Am., Vol. 55, pp. 1611–1616, December 1965 (nega-
       tive or positive conductivity, sink- or source-type electric field,
       complex dielectric constant, complex permeability)

1.45   *Proceedings of the Physics of Quantum Electronics Conference*
       San Juan, Puerto Rico, June 28–30, 1965
       Edited by P. L. Kelley, B. Lax, and P. E. Tannenwald (M.I.T.)
       McGraw-Hill Book Co., New York, 861 p., 1966

1.46   Quantum Oscillator and Amplifier Investigations
       N. G. Basov (Lebedev Institute of Physics, Academy of Sci-
       ences of the USSR, Moscow, USSR)
       Proc. Physics of Quantum Electronics Conf., McGraw-Hill Book
       Co., New York, pp. 411–423, 1966

1.47   *An Introduction to Coherent Optics and Holography*
       George W. Stroke (U. of Michigan)
       Academic Press, New York, 270 p., 1966

1.48   *Introduction to Laser Physics*
       Bela A. Lengyel (San Fernando Valley State College)
       John Wiley & Sons, New York, 311 p., 1966

1.49   Nonlinear Optics
       P. S. Pershan (B.T.L.)
       Progress in Optics, Vol. 5, edited by E. Wolf, John Wiley &
       Sons, New York, pp. 83–144, 1966

1.50   Continuous Operation is Near for Uncooled Diode Lasers
       Michael F. Lamorte (RCA)

Electronics, Vol. 39, No. 1, pp. 95–99, January 10, 1966 (basic principles, temperature, power)

1.51 Bibliography of the Open Literature on Lasers. V
Edward V. Ashburn (Lockheed)
J. Opt. Soc. Am., Vol. 56, pp. 263–267, February 1966 (186 additional references, period of 1958–1964, classification headings)

1.52 Recent Applications of Lasers
O. S. Heavens (U. of York, England)
British J. Appl. Phys., Vol. 17, pp. 287–309, March 1966 (general review)

1.53 1966 International Quantum Electronics Conference—Digest of Technical Papers
IEEE J. Quantum Electronics, Vol. QE-2, No. 4, April 1966 (pp. i–lxxi following p. 85)

1.54 Bibliography on Holograms
R. P. Chambers and J. S. Courtney-Pratt (B.T.L.)
J. Soc. Motion Picture and Television Engineers, Vol. 75, pp. 373–435, April 1966 (over 180 papers, 1948 to 1966, many abstracts)

1.55 Laser Bibliography III
K. Tomiyasu (G.E.)
IEEE J. Quantum Electronics, Vol. QE-2, pp. 124–151, June 1966 (compiled during July-December 1965, 644 references, 25 subject categories)

1.56 Progress in Ionized-Argon Lasers
Roy A. Paananen (Raytheon)
IEEE Spectrum, Vol. 3, No. 6, pp. 88–99, June 1966

1.57 *The Laser*
William V. Smith and Peter P. Sorokin (IBM)
McGraw-Hill Book Co., New York, 498 p., 1966 (basic principles, pumping, Q-switching, Raman lasers, injection lasers, applications)

1.58 *Laser—Lichtverstärker und—Oszillatoren*
Dieter Röss
Akademische Verlagsgesellschaft, Frankfurt am Main, Germany, 722 p., 1966 (in German, comprehensive, principles, devices, applications, laser materials, emission wavelengths, 3141 references)

1.59 *Laser Receivers; Devices, Techniques, Systems*
Monte Ross (McDonnell Aircraft)
John Wiley & Sons, Inc., New York, 405 p., 1966 (noise, information theory, detectors, modulators, transmission, components and systems)

1.60 *Lasers; A Series of Advances*
Edited by Albert K. Levine (GT & E Labs.)
Marcel Dekker, Inc., New York, 365 p., 1966 (7 contributors, crystal lasers, organic lasers, Q-switching, resonators)

1.61 *Proceedings of the 8th Annual Electron and Laser Beam Symposium, April 6–8, 1966*
Edited by G. I. Haddad
Sponsored by U. of Michigan and IEEE, 520 p.

1.62 Wavefront Reconstruction or "Holography"
Dennis Gabor (U. of London, England)
Proc. 8th Annual Electron and Laser Beam Symposium, April 6–8, 1966. Sponsored by U. of Michigan and IEEE, pp. 1–19 (fundamentals, applications)

1.63 Opening Remarks: Fourth International Quantum Electronics Conference
N. G. Basov (Lebedev Phys. Inst., Acad. of Sci., USSR)
IEEE J. Quantum Electronics, Vol. QE-2, pp. 354–357, September 1966

1.64 Crystalline Solid Lasers
Z. J. Kiss and R. J. Pressley (RCA)
Appl. Optics, Vol. 5, pp. 1474–1486, October 1966 (survey, host materials, characteristics of devices, applications)

1.65 Glass Lasers
E. Snitzer (American Optical)
Appl. Optics, Vol. 5, pp. 1487–1499, October 1966 (summary, Nd, Yb, and Er lasers, thermal effects, Faraday rotation glasses)

1.66 Gas Lasers
A. L. Bloom (Spectra-Physics)
Appl. Optics, Vol. 5, pp. 1500–1514, October 1966 (review, several gases, applications)

1.67 Semiconductor Lasers
Marshall I. Nathan (IBM Watson)
Appl. Optics, Vol. 5, pp. 1514–1528, October 1966.
(review, list of materials, characteristics, applications)

1.68 Optical Transmission Research
S. E. Miller and L. C. Tillotson (BTL)
Appl. Optics, Vol. 5, pp. 1538–1549, October 1966 (summary, modulation, atmospheric effects, guided propagation, detection)

1.69 Nonlinear Optics
R. W. Minck, R. W. Terhune, and C. C. Wang (Ford)
Appl. Optics, Vol. 5, pp. 1595–1612, October 1966 (review, parametric oscillation, self-trapping of beams, stimulated scattering)

1.70 Crystalline Solid Lasers
Z. J. Kiss and R. J. Pressley (RCA)
Proc. IEEE, Vol. 54, pp. 1236–1248, October 1966 (survey, host materials, characteristics of devices, applications)

1.71 Glass Lasers
E. Snitzer (American Optical)
Proc. IEEE, Vol. 54, pp. 1249–1261, October 1966 (summary, Nd, Yb, and Er lasers, thermal effects, Faraday rotation glasses)

1.72 Gas Lasers
A. L. Bloom (Spectra-Physics)
Proc. IEEE, Vol. 54, pp. 1262–1276, October 1966 (review, several gases, applications)

1.73 Semiconductor Lasers
Marshall I. Nathan (IBM Watson)
Proc. IEEE, Vol. 54, pp. 1276–1290, October 1966 (review, list of materials, characteristics, applications)

1.74 Optical Transmission Research
S. E. Miller and L. C. Tillotson (BTL)
Proc. IEEE, Vol. 54, pp. 1300–1311, October 1966 (summary, modulation, atmospheric effects, guided propagation, detection).

1.75 Nonlinear Optics
R. W. Minck, R. W. Terhune, and C. C. Wang (Ford)
Proc. IEEE, Vol. 54, pp. 1357–1374, October 1966 (review, parametric oscillation, self-trapping of beams, stimulated scattering)

1.76 Electroluminescence and Semiconductor Lasers
Henry F. Ivey (Westinghouse)
IEEE J. Quantum Electronics, Vol. QE-2, pp. 713–726, November 1966 (review, principles, all semiconductor lasers, 189 references)

1.77 Laser Bibliography IV
  K. Tomiyasu (GE)
  IEEE J. Quantum Electronics, Vol. QE-2, pp. 726–755, November 1966 (compiled during January through June 1966,

734 references, 27 subject categories)

1.78 The Hologram—Properties and Applications
  E. G. Ramberg
  RCA Rev., Vol. 27, pp. 467–499, December 1966 (review)

2. LASER ANALYSES

2.1 Non-equilibrium Distributions of Molecular Vibrational States
  M. W. Muller, A. Sher, R. Solomon, and D. G. Dow (Varian)
  Appl. Phys. Letters, Vol. 2, pp. 86–88, February 15, 1963

2.2 Maser Action without Population Inversion
  Dietrich Marcuse (BTL)
  Proc. IEEE, Vol. 51, pp. 849–850, May 1963

2.3 Energy Density Distribution in a Polished Cylinder of Laser Material
  W. R. Sooy and M. L. Stitch (Hughes)
  J. Appl. Phys., Vol. 34, pp. 1719–1723, June 1963

2.4 On an Active Interference Filter as an Optical Maser Amplifier
  H. Jacobs, R. A. Bowden, and L. Hatkin (U.S. Army)
  Proc. IEEE, Vol. 51, p. 933, June 1963

2.5 Analysis of Transients and Stability in an Idealized Two-Level Laser System
  J. E. Ludman (AF CRL)
  Appl. Optics, Vol. 2, pp. 862–863, August 1963

2.6 Effects of Output Coupling on Optical Masers
  P. A. Miles (MIT) and I. Goldstein (Raytheon)
  IEEE Trans. on Electron Devices, Vol. ED-10, pp. 314–318, September 1963

2.7 Maximum Gain for Forward- and Backward-Wave Optical Maser Amplifiers
  H. Jacobs, D. A. Holmes, L. Hatkin, and F. A. Brand (U.S. Army)
  J. Appl. Phys., Vol. 34, pp. 2617–2624, September 1963

2.8 On Maser Rate Equations and Transient Oscillations
  C. L. Tang (Raytheon)
  J. Appl. Phys., Vol. 34, pp. 2935–2940, October 1963

2.9 Effect of a Small Stokes Shift on the Operation of Three-Level Masers
  Peter J. Warter, Jr. (Princeton U.)
  J. Appl. Phys., Vol. 34, pp. 2966–2972, October 1963

2.10 Linewidth and Conditions for Steady Oscillation in Single and Multiple Element Lasers
  J. A. Fleck, Jr. (U. of California)
  J. Appl. Phys., Vol. 34, pp. 2997–3003, October 1963

2.11 Steady-State Output Power of a Laser as a Function of the Single-Pass Gain
  Irwin Tobias (Rutgers U.)
  J. Appl. Phys., Vol. 34, pp. 3200–3204, November 1963

2.12 Super Radiant Narrowing in Fluorescence Radiation of Inverted Populations
  Ammon Yariv and R. C. C. Leite (BTL)
  J. Appl. Phys., Vol. 34, pp. 3410–3411, November 1963

2.13 Criterion for Continuous Amplitude Oscillations of Optical Masers
  Jerome I. Kaplan (MIT)
  J. Appl. Phys., Vol. 34, pp. 3411–3412, November 1963

2.14 Simple Method of Quantum-Efficiency Measurement
  N. C. Chang (G. T. & E. Labs.)
  J. Opt. Soc. Am., Vol. 53, pp. 1315–1317, November 1963

2.15 Designing Lasers with Pump-Power Charts
  Robert A. Kaplan (Wheeler Labs.)
  Electronics, Vol. 36, No. 52, pp. 23–28, December 27, 1963

2.16 Energy and Power Considerations in Injection and Optically Pumped Lasers
  Amnon Yariv (Watkins-Johnson)
  Proc. IEEE, Vol. 51, pp. 1723–1731, December 1963

2.17 A Method for Determining the Threshold of Laser Materials
  A. Lempicki and R. L. Martin
  Proc. IEEE, Vol. 51, pp. 1778–1779, December 1963

2.18 The Coherence Brightened Laser
  R. H. Dicke (Princeton U.)
  Proc. Third Int. Conf. Quantum Electronics, Columbia U. Press, N. Y., Vol. 1, pp. 35–54, 1964

2.19 Unified Theory of Steady-State Cavity Masers
  D. E. McCumber (BTL)
  Proc. Third Int. Conf. Quantum Electronics, Columbia U. Press, N. Y., Vol. 1, pp. 95–100, 1964

2.20 Relaxation Mechanisms, Dissociative Excitation Transfer and Mode Pulling Effects in Gas Lasers
  W. R. Bennett, Jr. (Yale U.)
  Proc. Third Int. Conf. Quantum Electronics, Columbia U. Press, N. Y., Vol. 1, pp. 441–458, 1964

2.21 Steady State Population Distributions in Quantum Mechanical Systems
  W. A. Barker and J. D. Keating (McDonnell Aircraft)
  Proc. Third Int. Conf. Quantum Electronics, Columbia U. Press, N. Y., Vol. 1, pp. 741–749, 1964

2.22 Theory of Power Output and Optimum Coupling in Laser Oscillators
  A. Yariv (BTL)
  Proc. Third Int. Conf. Quantum Electronics, Columbia U. Press, N. Y., Vol. 2, pp. 1055–1064, 1964

2.23 Theory of Relaxation Spikes in Two-Level Laser Amplifiers
  H. A. Trenchard (Westinghouse)
  Proc. Third Int. Conf. Quantum Electronics, Columbia U. Press, N. Y., Vol. 2, pp. 1089–1096, 1964

2.24 Dynamic Behavior of Quantum Mechanical Oscillators
  G. Makhov and O. Risgin (U. of Michigan)
  Proc. Third Int. Conf. Quantum Electronics, Columbia U. Press, N. Y., Vol. 2, pp. 1121–1129, 1964

2.25 A Novel Type of Maser Using Stimulated Emission of Bremsstrahlung
  D. Marcuse (BTL)
  Proc. Third Int. Conf. Quantum Electronics, Columbia U. Press, N. Y., Vol. 2, pp. 1161–1167, 1964

2.26 Amplitude and Frequency Control in Solid State Optical Masers
  J. M. Burch and D. L. Wood (Natl. Physical Lab., England)
  Proc. Third Int. Conf. Quantum Electronics, Columbia U. Press, N. Y., Vol. 2, pp. 1339–1346, 1964

2.57 Atomic Beam Lasers
      N. G. Basov and V. S. Letokhov (Lebedev Physics Inst., Academy of Sciences, U.S.S.R.)
      JETP Letters, Vol. 2, pp. 3–5, July 1, 1965 (atomic beam right angle to laser beam)

2.58 On the Intensity Distribution for a Two-Mode Laser Beam
      G. J. Troup (Monash U., Australia)
      Physics Letters, Vol. 17, pp. 264–265, July 15, 1965

2.59 Theory of Optical Maser Amplifiers
      F. T. Arecchi and R. Bonifacio (U. of Milan, Italy)
      IEEE J. Quantum Electronics, Vol. 1, pp. 169–178, July 1965 (electromagnetic wave and medium interaction)

2.60 Generalized Solutions for Optical Maser Amplifiers
      N. Kumagai and H. Yamamoto (Osaka U., Japan)
      IEEE Trans. on Microwave Theory and Techniques, Vol. MTT-13, pp. 445–451, July 1965 (Laplace transforms, transient terms, nonlinear above threshold)

2.61 Stability of Coupled-Mode Laser Equations
      J. A. Fleck, Jr., and R. E. Kidder (Lawrence Rad. Lab., U. of California)
      J. Appl. Phys., Vol. 36, pp. 2327–2328, July 1965 (corrected equation, predict irregular and regular spiking)

2.62 Theory of a Traveling-Wave Optical Maser
      Frederick Aronowitz (Honeywell)
      Phys. Rev., Vol. 139, pp. A635–A646, August 2, 1965 (both standing and traveling waves)

2.63 Saturation Effects in High-Gain Lasers
      W. W. Rigrod (BTL)
      J. Appl. Phys., Vol. 36, pp. 2487–2490, August 1965 (high loss in cavity)

2.64 Quantum Statistical Dynamics of Laser Amplifiers
      A. E. Glassgold (New York U.) and Dennis Holliday (RAND)
      Phys. Rev., Vol. 139, pp. A1717–A1734, September 13, 1965 (Heisenberg picture, density matrix, damping)

2.65 Mode Confinement and Gain in Junction Lasers
      W. W. Anderson (Stanford U.)
      IEEE J. Quantum Electronics, Vol. QE-1, pp. 228–236, September 1965 (three-layer media, gain-phase relations)

2.66 Theory of FM Laser Oscillation
      S. E. Harris and O. P. McDuff (Stanford U.)
      IEEE J. Quantum Electronics, Vol. QE-1, pp. 245–262, September 1965

2.67 Correction: Theory of FM Laser Oscillation
      S. E. Harris and O. P. McDuff
      IEEE J. Quantum Electronics, Vol. QE-2, p. 49, February 1966 (IEEE J. Quantum Electronics, Vol. QE-1, pp. 245–262, September 1965)

2.68 Theory of Laser Cascades
      H. Haken (Technische Hochschule, Stuttgart, Germany) and R. Der Agobian and M. Pauthier (Laboratoire Central de Télécommunications, Paris, France)
      Phys. Rev., Vol. 140, pp. A437–A447, October 18, 1965 (gas and solid-state lasers, homogeneously and inhomogeneously broadened line)

2.69 The Role of Multiphoton Processes in Establishing the Limiting Power of Quantum Oscillators
      F. V. Bunkin and A. M. Prokhorov (Lebedev Physics Inst., Academy of Sciences, U.S.S.R.)

      Soviet Physics JETP, Vol. 21, pp. 725–726, October 1965 (10^4 W from GaAs, 2×10^{11} W from ruby)

2.70 Interaction of Laser Modes
      L. A. Ostrovskiĭ (Gorkiĭ State U., U.S.S.R.)
      Soviet Physics JETP, Vol. 21, pp. 727–732, October 1965 (steady-state and transient processes in 2-mode laser)

2.71 Coherent Division of Quanta
      L. I. Gudzenko and G. M. Guro (Lebedev Physics Inst., Academy of Sciences, U.S.S.R.)
      Soviet Physics JETP, Vol. 21, pp. 756–760, October 1965

2.72 Oscillations in a System Comprising Two-Level Molecules and a Radiation Field
      V. F. Chel'tsov
      Soviet Physics JETP, Vol. 21, pp. 761–764, October 1965

2.73 Theory of Magnetic Effects in Optical Maser Amplifiers and Oscillators
      C. V. Heer and R. D. Graft (Ohio State U.)
      Phys. Rev., Vol. 140, pp. A1088–A1104, November 15, 1965

2.74 On the Regular and Irregular Spiking Behaviour of Solid-State Lasers
      Helmut K. V. Lotsch (Stanford U.)
      International J. of Electronics, Vol. 19, pp. 453–467, November 1965 (parabolic cylinder functions in Fabry-Perot cavity, filaments, ruby laser experiments, up to 20 meter long cavity, aperture limited)

2.75 Higher Order Calculation of the Lamb Dip in the Output of an Optical Maser
      Kiyoji Uehara and Koichi Shimoda (U. of Tokyo, Tokyo, Japan)
      Japanese J. Appl. Phys., Vol. 4, pp. 921–927, November 1965 (shallower dip by including fifth order, Doppler limit)

2.76 Breadth of Decay Quanta in Gas Lasers
      J. A. White (NBS)
      J. Opt. Soc. Am., Vol. 55, pp. 1436–1442, November 1965 (Doppler bifurcations, low and high power, large and small quanta decay)

2.77 On the Theory of Gas Lasers
      A. K. Popov (Physics Inst., Siberian Div., Academy of Sciences, U.S.S.R.)
      Soviet Physics JETP, Vol. 21, pp. 856–858, November 1965 (threshold and power calculations)

2.78 Pressure Effects in the Output of a Gas Laser
      B. L. Gyorffy and W. E. Lamb, Jr. (Yale U.)
      Proc. Physics of Quantum Electronics Conf., McGraw-Hill Book Co., New York, pp. 602–610, 1966

2.79 Multiple Quantum Processes in Magnetic Field-Tuned Optical Masers
      R. L. Fork and M. Sargent III (B.T.L.)
      Proc. Physics of Quantum Electronics Conf., McGraw-Hill Book Co., New York, pp. 611–619, 1966

2.80 Thermally Pumped Infrared Masers
      Koichi Shimoda (U. of Tokyo, Japan)
      Proc. Physics of Quantum Electronics Conf., McGraw-Hill Book Co., New York, pp. 635–642, 1966

2.81 Intensity Fluctuations in the Output of cw Laser Oscillators. I
      D. E. McCumber (B.T.L.)
      Phys. Rev., Vol. 141, pp. 306–322, January 1966

2.82 Theory of Double Resonance in Gaseous Lasers
      W. Culshaw (Lockheed)

Phys. Rev., Vol. 142, pp. 204–216, February 4, 1966 (RF perturbation between Zeeman sublevels)

2.83 Analysis of the Uniform Rate Equation Model of Laser Dynamics
Thomas J. Menne (McDonnell Aircraft)
IEEE J. Quantum Electronics, Vol. QE-2, pp. 38–44, February 1966

2.84 Effect of Spatial Dependence in the Single-Mode Laser Rate Equations
T. J. Menne and F. J. Rosenbaum (McDonnell Aircraft)
IEEE J. Quantum Electronics, Vol. QE-2, pp. 47–49, February 1966

2.85 Nonlinear Mode Interaction in Lasers
N. G. Basov, V. N. Morozov and A. N. Oraevskiĭ (Lebedev Physics Institute, U.S.S.R.)
Soviet Phys. JETP, Vol. 22, pp. 622–628, March 1966

2.86 Erratum: Spectral Output and Spiking Behavior of Solid-State Lasers
C. L. Tang, H. Statz, and G. deMars (Raytheon)
J. Appl. Phys., Vol. 37, p. 2203, April 1966 (J. Appl. Phys., Vol. 34, p. 2289, 1963)

2.87 Effects of Transverse and Axial Magnetic Fields on Gaseous Lasers
W. Culshaw and J. Kannelaud (Lockheed)
Phys. Rev., Vol. 145, pp. 257–267, May 6, 1966 (some 1.15 μ He–Ne laser experiments)

2.88 Quantum Theory of an Optical Maser
M. Scully and W. E. Lamb, Jr. (Yale U.)
Phys. Rev. Letters, Vol. 16, pp. 853–855, May 9, 1966

2.89 Effect of Inhomogeneities on the Operation Regime of Solid-State Lasers
A. F. Suchkov (Lebedev Physics Institute, Academy of Sciences, U.S.S.R.)
Soviet Physics JETP, Vol. 22, pp. 1026–1031, May 1966

2.90 Automodulation of the Radiation from a Laser with a Two-Mode Resonator
L. A. Ostrovskiĭ (Gor'kiĭ State University, U.S.S.R.)
Soviet Physics JETP, Vol. 22, pp. 1053–1058, May 1966

2.91 Etude de la largeur spectrale et des variations de longueur d'onde de la lumière emise par un laser déclenché
J. Ch. Viénot, A. Orszag, J. Pasteur, R. Saron, et J. Bulabois (Université de Besançon, France)
Appl. Optics, Vol. 5, pp. 1003–1007, June 1966

2.92 Oscillation Conditions for Superradiant and Feed Back Amplifier Lasers
Thomas R. Carver (Princeton U.)
Appl. Optics, Vol. 5, pp. 1090–1091, June 1966

2.93 Theoretical Assessment of a High Power Continuous-Wave 4-Level Solid Laser
A. C. Selden (Atomic Weapons Research Estab., England)
British J. Appl. Phys., Vol. 17, pp. 729–736, June 1966 (tens of watts predicted from Nd doped glass)

2.94 Spectral Properties of Stimulated Emission in a Broad Pumping Range
B. L. Livshitz and V. N. Tsikunov (Acad. of Sci., USSR)
Soviet Phys. JETP, Vol. 22, pp. 1260–1263, June 1966 (saturation of number of axial modes)

2.95 Quantum Theory of a Gas Laser
C. R. Willis (Boston U.)
Physics Letters, Vol. 21, pp. 634–635, July 1, 1966

2.96 Frequency and Polarization Locking Phenomena in a Laser with Axial Magnetic Field
H. Pelikan (Tech. Hochschule Stuttgart, Germany)
Physics Letters, Vol. 21, pp. 652–653, July 1, 1966

2.97 Effects of Atomic Degeneracy and Cavity Anisotropy on the Behavior of a Gas Laser
Walter M. Doyle and Matthew B. White (Aeronutronic)
Phys. Rev., Vol. 147, pp. 359–367, July 8, 1966

2.98 Giant-Pulse Laser Activity in Neodymium-Doped Silicate Glass: The Energy Conversion Process
J. H. Wenzel (GE)
J. Appl. Phys., Vol. 37, pp. 3100–3110, July 1966 (standing waves in resonator, radial variation in gain)

2.99 Theory of Steady Multimode Oscillation of a Solid-State Laser
L. Ronchi (Centro Microonde, Italy)
Nuovo Cimento, Vol. 44B, pp. 372–386, August 11, 1966

2.100 Interaction of Linearly and Circularly Polarized Fields in a Laser Amplifier with an Axial Magnetic Field
Andrew Dienes (California Inst. of Tech.)
Appl. Phys. Letters, Vol. 9, 142–145, August 15, 1966

2.101 Governing Influence of Atomic Degeneracy on Mode Interactions in a Gas Laser
W. M. Doyle and M. B. White (Aeronutronic)
Phys. Rev. Letters, Vol. 17, pp. 467–470, August 29, 1966 (gas laser experiments)

2.102 Radiation Interactions between Laser Oscillators with Different Active Elements and Different Frequency
H. Inaba, Y. Isawa, and N. Suda (Tohoku U., Japan)
IEEE J. Quantum Electronics, Vol. QE-2, pp. 222–229, August 1966 (Q-sw ruby interacting with Nd-glass laser, double photon absorption)

2.103 Measurement of Saturation Induced Optical Nonreciprocity in a Ring Laser Plasma
P. H. Lee and J. G. Atwood (Perkin-Elmer)
IEEE J. Quantum Electronics, Vol. QE-2, pp. 235–243, August 1966 (0.63-μ He–Ne laser, nonreciprocal index due to unequal intensity laser beams)

2.104 Effect of Spectral Hole-Burning and Cross Relaxation on the Gain Saturation of Laser Amplifiers
Amado Y. Cabezas and Richard P. Treat (Hughes Aircraft)
J. Appl. Phys., Vol. 37, pp. 3556–3563, August 1966 (Nd laser experiments)

2.105 Pulsed Stimulated Emission in a Hydrogen-Atom Beam Laser
G. M. Strakhovskiĭ and A. V. Uspenskiĭ (Lebedev Phys. Inst., Acad. of Sci., USSR)
Soviet Phys. JETP, Vol. 23, pp. 247–249, August 1966 (analytical, two relaxation times)

2.106 Dependence of the Radiation Intensity of a Gas Laser on the Magnetic Field
M. I. D'Yakonov and V. I. Perel' (Ioffe Phy. Tech. Inst., USSR)
Soviet Phys. JETP, Vol. 23, pp. 298–303, August 1966

2.107 Derivation of the Relation between Two Weakly Coupled Nonlinear Optical Oscillators
H. de Lang (Philips, Netherlands)
Appl. Phys. Letters, Vol. 9, pp. 205–207, September 1, 1966 (ring laser example)

2.108 Quantum Theory of Laser Radiation. I. Many-Atom Effects
J. A. Fleck, Jr. (Lawrence Rad. Lab., U. of California)
Phys. Rev., Vol. 149, pp. 309–321, September 9, 1966

2.109 Quantum Theory of Laser Radiation. II. Statistical Aspects of Laser Light
J. A. Fleck, Jr. (Lawrence Rad. Lab., U. of California)
Phys. Rev., Vol. 149, pp. 322–329, September 9, 1966

3. RUBY LASER

J. Appl. Phys., Vol. 35, pp. 3062–3063, October 1964 (up to 2 per cent Cr, 3–4 ms R-line lifetime at 297°K, max. of 12 ms at 0.2 per cent and 77°K)

3.85 Spectral Properties of a Single-Mode Ruby Laser: Evidence of Homogeneous Broadening of the Zero-Phonon Lines in Solids
C. L. Tang, H. Statz, G. A. deMars, and D. T. Wilson (Raytheon)
Phys. Rev., Vol. 136, pp. A1–A8, October 5, 1964 (unidirectional traveling wave ring laser, emission with less than 0.005 cm^{-1} (150 Mc), regular spiking, thermal shift in frequency)

3.86 Polarization of the Light Output from a Ruby Optical Maser
J. H. Brunton (BTL)
Appl. Optics, Vol. 3, pp. 1241–1246, November 1964 (seven rubies, extensive work, polarization varies with pumping energy)

3.87 Ruby Laser Action at the R_2 Wavelength
C. J. Hubbard and E. W. Fisher (AIL)
Appl. Optics, Vol. 3, pp. 1499–1500, December 1964 (used Lyot-Ohmann birefringent filter to suppress R_1 oscillation)

3.88 Measurements of Cavity Loss in a Pulsed Ruby Laser
D. Chen (Honeywell)
Nature, Vol. 205, pp. 271–272, January 16, 1965 (lower threshold yields narrower R_1 line width)

3.89 Many-Element Lasers
R. Pratesi and G. Toraldo Di Francia (Universita di Firence, Italy) and L. Ronchi (Centro Microonde del C.N.R., Florence, Italy)
Z. angewandte Mathematik und Physik, Vol. 16, No. 1, pp. 68–71, January 25, 1965 (7 elements, regular spiking in ruby)

3.90 Undamped Regular Spiking of High Energy Lasers
D. V. Keller, B. I. Davis, and M. E. Graham (Northrop)
Z. angewandte Mathematik und Physik, Vol. 16, No. 1, p. 71, January 25, 1965 (100 j output ruby laser, rod immersed in glycerin, product of spike frequency and spike width is constant)

3.91 Optical Properties and Laser Thresholds of Thirty-nine Ruby Laser Crystals
G. W. Dueker, C. M. Kellington, M. Katzmann, and J. G. Atwood (Perkin-Elmer and USA ERDL)
Appl. Optics, Vol. 4, pp. 109–118, January 1965 (extensive investigation of Linde rubies, laser threshold not a good measure of optical quality)

3.92 The Effect of Optical Pump Pulse Shape on Ruby Inversion
P. N. Mace and G. McCall (Los Alamos)
Proc. IEEE, Vol. 53, p. 74, January 1965 (little effect calculated)

3.93 Continuous Operation of Ruby Laser at Room Temperature
V. Evtuhov and J. K. Neeland (Hughes)
Appl. Phys. Letters, Vol. 6, pp. 75–76, February 15, 1965 (water cooled, elliptical pump cylinder, 2 kw mercury arc lamp, 800 w threshold, 70 mw output, spikes)

3.94 Relation between Absorption and Emission in the Region of the R Lines of Ruby
D. F. Nelson and M. D. Sturge (BTL)
Phys. Rev., Vol. 137, pp. A1117–A1129, February 15, 1965 (function of polarization, 20–373° K, fluorescent lifetime, laser implications given)

3.95 Amplification in a Thick Ruby Lens
E. R. Lanczi (Mitre)
Appl. Optics, Vol. 4, p. 255, February 1965 (gain of 2, also image amplified)

3.96 Power-Dependent Frequency Shifts in Ruby Lasers at 77°K
M. Birnbaum and T. L. Stocker (Aerospace)
J. Appl. Phys., Vol. 36, pp. 396–402, February 1965 (R_1 doublet splitting decreases with power level)

3.97 **Search for a Stokes' Shift in the R_1 Lines of Ruby**
P. J. Warter, Jr., Ramon U. Martinelli, and James W. Brault (Princeton U.)

J. Appl. Phys., Vol. 36, pp. 468–470, February 1965 (no shift observed)

3.98 Flaws in Ruby Laser Crystals
K. Janowski and H. Conrad (Aerospace)
J. Appl. Phys., Vol. 36, pp. 663–664, February 1965 (streaks)

3.99 Effects of Gamma Irradiation on the Energy Output of Ruby Laser Crystals
W. R. Davis, A. C. Menius, Jr., M. K. Moss, and C. R. Philbrick (U. of North Carolina)
J. Appl. Phys., Vol. 36, pp. 670–672, February 1965 (increased output with moderate irradiation from ^{60}Co)

3.100 Regular Emission from a Many-Element Laser During the Pumping Pulse
R. Pratesi and G. Toraldo di Francia (U. di Firenze, Italy)
Proc. IEEE, Vol. 53, pp. 196–197, February 1965 (ten-element ruby laser, regular spiking)

3.101 Ruby Laser Oscillations Modulated by Ultrasonic Vibration
Y. Sakai (Matsushita, Japan)
Proc. IEEE, Vol. 53, pp. 204–205, February 1965 (nickel transducer, 24 kc, superimposed modulation)

3.102 Saturation of the Cathodoluminescence of Ruby
M. W. Levine and M. Subramanian (Purdue U.)
Appl. Phys. Letters, Vol. 6, pp. 87–89, March 1, 1965 (77°K, 6 ma/cm^2 saturates R_1 fluorescence, 100 ms pulses, up to 20 kv)

3.103 Polarization of Phonon-Assisted Fluorescence in Ruby
E. W. Fisher and Z. H. Heller (AIL)
J. Appl. Phys., Vol. 36, p. 870, Part 1, March 1965

3.104 The Effect of Spatial Modulation of Pump Light on the Longitudinal-Mode Spectra of Ruby Lasers
V. Evtuhov (Hughes)
Appl. Phys. Letters, Vol. 6, pp. 141–142, April 1, 1965 (shielding bands on rod changes output spectra)

3.105 Study of the Output Spectra of Ruby Lasers
V. Evtuhov and J. K. Neeland (Hughes)
IEEE J. Quantum Electronics, Vol. QE-1, pp. 7–12, April 1965 (frequency separation between transverse modes, mode hopping to lower frequencies, mode selection)

3.106 Descriptive Theory of Spiking Pulses in Optically Pumped Lasers
E. L. Steele (Autonetics)
IEEE J. Quantum Electronics, Vol. QE-1, pp. 42–49, April 1965 (3-level laser, rate equations, inversion time, dump time)

3.107 Effective Fluorescent Lifetimes in Ruby Laser Rods
J. A. Fleck, Jr. (U. of California)
J. Appl. Phys., Vol. 36, pp. 1301–1306, April 1965 (Brewster-angle ends, roughened side, water immersed)

3.108 Infrared Absorption Spectra from Excited States in Ruby
T. Kushida (Tokyo Shibaura Electric Co., Japan)
J. Phys. Soc. Japan, Vol. 20, p. 619, April 1965 (scanned between 4000 and 14000 cm^{-1}, induced absorption between 6000 and 8000 cm^{-1}, strongest at 6600 cm^{-1})

3.109 Ein Kontinuierlicher Wassergekühlter Rubinlaser
K. Gürs (Siemens and Halske, Germany)
Physics Letters, Vol. 16, pp. 125–127, May 15, 1965 (0.03% doping, 1800 watt threshold)

3.110 Interdependence of Threshold, Filament Position and Efficiency in a Linearly Pumped Ruby Rod
E. W. Sucov (Westinghouse)
Appl. Optics, Vol. 4, pp. 593–596, May 1965 (sharp focal region not at expected location, 40% improvement in conversion efficiency)

3.111 Immersion Liquids for Ruby Lasers
M. E. Graham, B. I. Davis, and D. V. Keller (Northrop)
Appl. Optics, Vol. 4, pp. 613–615, May 1965 (high index liquids, photostability consideration, optically transparent, viscosity)

3.112   Importance of Ruby Absorption Bands for Laser Action
L. T. Long (U. S. NOL)
Appl. Optics, Vol. 4, p. 626, May 1965 (Corning 0-52 and 3-71 filters)

3.113   Square Law Behavior of Photocathodes at High Light Intensities and High Frequencies
A. M. Johnson (BTL)
IEEE J. Quantum Electronics, Vol. QE-1, pp. 99-101, May 1965 (beats greater than ruby laser emission linewidth)

3.114   Problem of Spike Elimination in Lasers
H. Statz, G. A. DeMars, and D. T. Wilson (Raytheon), and C. L. Tang (Cornell U.)
J. Appl. Phys., Vol. 36, pp. 1510-1514, May 1965 (Kerr cell in traveling-wave ring resonator, fast response feedback control)

3.115   Visible-Spectrum Absorption Cross Section of Cr^{3+} in the ^2E State of Pink Ruby
Y. C. Kiang and F. C. Unterleitner (General Dynamics)
Bull. Am. Phys. Soc., Series II, Vol. 10, No. 5, p. 608-EF9, June 1965 (new broad bands between 5100 and 6500 Å, new narrow band at 4540 Å)

3.116   Dependence of Spectral Composition of Stimulated Emission on the Velocity of Motion of the Crystal
B. L. Livshitz, V. P. Nazarov, L. K. Sidorenko, and V. N. Tsikunov (Inst. of General and Inorganic Chemistry, Academy of Sciences, U.S.S.R.)
JETP Letters, Vol. 1, pp. 136-138, June 1, 1965 (ruby moving at 35 cm/s, fewer modes, higher spectral density)

3.117   Effect of a Focused Ruby-Laser Beam on the Ruby
T. P. Belikova and E. A. Sviridenkov (Lebedev Physics Inst., Academy of Sciences, U.S.S.R.)
JETP Letters, Vol. 1, pp. 171-172, June 15, 1965 (internal fractures accompanied by visible flash, blue and orange glow, perhaps due to excitation from ^2E level)

3.118   Simultaneous Laser Oscillation at R_1 and R_2 Wavelengths in Ruby
J. A. Calviello, E. W. Fisher, and Z. H. Heller (A.I.L.)
IEEE J. Quantum Electronics, Vol. QE-1, p. 132, June 1965 (normal and Q-sw operation)

3.119   A Ruby Laser with External Mirrors of Large Spacing
Tadao Shimizu (Inst. of Physical and Chemical Research, Tokyo, Japan) and Fujio Shimizu, Minato Kawaguti, and Koichi Shimoda (U. of Tokyo, Japan)
Japanese J. Appl. Phys., Vol. 4, pp. 445-451, June 1965 (up to 24.5 meters, low and high Q modes)

3.120   Study on Saturation Process by Anomalous Dispersion of Ruby Laser
Pil Hyon Kim and Susumu Namba (Inst. of Physical and Chemical Research, Tokyo, Japan)
Japanese J. Appl. Phys., Vol. 4, pp. 469-470, June 1965 (kryptocyanine bleachable dye, laser wavelength depends on concentration and intensity)

3.121   Indication of an Energy Extremum in a Ruby Laser from Spiking Data
J. A. Detrio and E. J. Evans (Picatinny Arsenal)
Physics Letters, Vol. 17, pp. 118-119, July 1, 1965 (150 kc/s spikes near threshold)

3.122 $\bar{E}$-Level Population of Ruby vs. Pumping
V. Daneu, C. A. Sacchi, and O. Svelto (Istituto di Fisica del Politecnico, Milan, Italy)
Appl. Optics, Vol. 4, pp. 863-866, July 1965 (spiral flashtube, axial fluorescence at rod center and periphery, rod losses)

3.123   Excitation of Modes and Oscillation Kinetics in a Ruby Laser with a Concentric Resonator
V. V. Korobkin, A. M. Leontovich, and M. N. Smirnova (Lebedev Physics Inst., Academy of Sciences, U.S.S.R.)
Soviet Physics JETP, Vol. 21, pp. 53-58, July 1965 (regular and irregular spiking, emission spectra, explanations)

3.124   Changes in the Resonator of a Ruby Laser when Heated by Pumping
A. P. Veduta, A. M. Leontovich, and V. N. Smorchkóv (Lebedev Physics Inst., Academy of Sciences, U.S.S.R.)
Soviet Physics JETP, Vol. 21, pp. 59-63, July 1965 (0.63-μ laser interferometer probe, 4.8°C temp rise, nonuniform across rod diameter, distortion during 3.5 ms pump pulse)

3.125   Output Power and Energy in a Q-Switched Ruby Laser with a Saturable Absorber
G. Potenza and A. Sona (Laboratori C.I.S.E., Milan, Italy)
Nuovo Cimento, Vol. 38, pp. 1438-1440, August 1, 1965 (multiple spikes, higher peak power and lower energy output with increasing concentration of phthalocyanine in nitrobenzene, absorber loss of 10^{-4})

3.126   Mode-Locking Effects in an Internally Modulated Ruby Laser
Thomas Deutsch (Raytheon)
Appl. Phys. Letters, Vol. 7, pp. 80-82, August 15, 1965 (KDP, 50 and 148 Mc/s, varied cavity length)

3.127   Stimulated Emission from 4.3% Abundant Cr^{50} Ions in Ruby
L. W. Riley, M. Bass, and E. L. Hahn (U. of California)
Appl. Phys. Letters, Vol. 7, pp. 88-90, August 15, 1965 (0.05 percent concentration, Q-sw, about 70°K, weak 6934.38 Å output)

3.128   Thermal Effects in Optically Pumped Laser Rods
R. L. Townsend, C. M. Stickley, and A. D. Maio (AFCRL)
Appl. Phys. Letters, Vol. 7, pp. 94-96, August 15, 1965 (distortion during and after pumping, ruby and Nd glass, beam divergence)

3.129   Optical Distortion in Ruby Lasers During Pumping
D. White and D. Gregg (Lawrence Rad. Lab., U. of California)
Appl. Optics, Vol. 4, p. 1034, August 1965 (divergence of beam emerging from amplifier increased from 1 to 5 mrads, double ellipse pump housing, two linear flash lamps, nonsymmetrical beam cross section)

3.130   Polarization Effects in a Roof-Top Ruby Laser
W. E. K. Gibbs and R. E. Whitcher (Defense Standards Labs., Australia)
Appl. Optics, Vol. 4, pp. 1034-1035, August 1965 (subsidiary output beams)

3.131   Change of Optical Path Length in Laser Rods Within the Pumping Period
H. Welling, C. J. Bickart, and H. G. Andresen (U. S. Army Electronics Command Labs.)
IEEE J. Quantum Electronics, Vol. QE-1, pp. 223-224, August 1965 (interferometric method, ruby and Nd glass, distortion during and after pump pulse)

3.132   Oscillating Modes in Ruby Lasers with Nonuniform Pumping Energy Distribution
Tingye Li and J. G. Skinner (BTL)
J. Appl. Phys., Vol. 36, pp. 2595-2596, August 1965

3.133   Negative Dispersion in the R_1 Line of Ruby
N. K. Bel'skii and A. M. Leontovich (Lebedev Physics Inst., Academy of Sciences, U.S.S.R.)
Soviet Physics JETP, Vol. 21, pp. 497-499, August 1965 (index change of 3.7×10^{-8} when pumped)

3.134 Effect of Trapped Light on the Output of a Ruby Laser
J. Linn and J. Free (Korad)
Appl. Optics, Vol. 4, pp. 1099–1101, September 1965 (mode competition, spiking, near-field patterns, immersion)

3.135 Evaluation of Specially Grown Ruby Laser Rods
E. W. Sucov (Westinghouse)
Appl. Optics, Vol. 4, pp. 1107–1112, September 1965 (laser performance not predictable from passive tests, discussion)

3.136 Alignment of Cr^{3+} in Ruby
G. F. Hull, Jr., J. T. Smith, and A. F. Quesada (Baird-Atomic)
Appl. Optics, Vol. 4, pp. 1117–1120, September 1965 (pump with circularly polarized 6943 Å, 10^{-7} second spin-lattice relaxation time)

3.137 Visible Spectrum Absorption Cross Section of Cr^{3+} in the ^2E State of Pink Ruby
Y. C. Kiang, J. F. Stephany, and F. C. Unterleitner (General Dynamics/Electronics)
IEEE J. Quantum Electronics, Vol. QE-1, pp. 295–298, October 1965 (several absorption lines from excited state)

3.138 Traveling-Wave Ruby Laser with a Passive Optical Isolator
M. Hercher, M. Young, and C. B. Smoyer (U. of Rochester)
J. Appl. Phys., Vol. 36, p. 3351, October 1965 (ring resonator, first spike is pure traveling wave)

3.139 Cutoff of Ruby Laser Emission by Pulsed Irradiation
D. M. J. Compton, R. A. Cesena, J. F. Bryant, and B. L. Gehman (General Dynamics)
Proc. IEEE, Vol. 53, pp. 1668–1669, October 1965 (30 MeV Linac pulses of 200 mA, similar results with Nd laser)

3.140 Moden des kontinuierlichen Rubinlasers mit Fabry-Perot-Resonator
Dieter Röss (Siemens & Halske, Germany)
Z. Naturforschung, Vol. 20a, pp. 1348–1354, October 1965 (ellipsoidal pump cavity, mercury lamp)

3.141 Mode Competition and Self-Locking Effects in a Q-Switched Ruby Laser
Hans W. Mocker (Honeywell) and R. J. Collins (U. of Minnesota)
Appl. Phys. Letters, Vol. 7, pp. 270–273, November 15, 1965 (passive Q-switch, varied cavity, few axial modes)

3.142 On the Regular and Irregular Spiking Behaviour of Solid-State Lasers
Helmut K. V. Lotsch (Stanford U.)
International J. of Electronics, Vol. 19, pp. 453–467, November 1965 (parabolic cylinder functions in Fabry-Perot cavity, filaments, ruby laser experiments, up to 20 meter long cavity, aperture limited)

3.143 Perfection of Ruby Laser Crystals
L. S. Birks, J. W. Hurley, and W. E. Sweeney (U. S. Naval Research Lab.)
J. Appl. Phys., Vol. 36, pp. 3562–3565, November 1965 (Czochralski best, flux grown intermediate, flamefusion least perfect, no laser tests)

3.144 Self-Reversal of R Lines in Red Ruby
Walter Jekeli (Clarkson College of Tech.)
J. Opt. Soc. Am., Vol. 55, pp. 1442–1445, November 1965 (1 percent Cr_2O_3, 298 and 77°K, blue pump light)

3.145 The Change in the Emission Spectrum of a Ruby Laser during Generation
A. M. Kubarev and V. I. Piskarev (Radio Physics Inst., Gorkiĭ State U., U.S.S.R.)
Soviet Physics JETP, Vol. 21, pp. 823–825, November 1965 (near 100°K, 30°K increase reduces wave number by 1 cm^{-1}, shift during pulse depends on mirror reflectivity and emission wavelength)

3.146 Effect of Ground State ESR Saturation on Ruby Laser Output at 90°K
A. Szabo and T. Igarashi (National Research Council, Canada)
Appl. Phys. Letters, Vol. 7, pp. 289–290, December 1, 1965 (11.49 Gc/s saturating microwaves)

3.147 Erratum: Effect of Ground State ESR Saturation on Ruby Laser Output at 90°K
A. Szabo and T. Igaraski (National Research Council, Canada)
Appl. Phys. Letters, Vol. 8, p. 102, February 15, 1966 (Appl. Phys. Letters, Vol. 7, p. 289, 1965)

3.148 Phosphorescence and Band Structure of Ruby
Z. L. Morgenshtern and V. B. Neustruev (Lebedev Physics Institute, Academy of Sciences, U.S.S.R.)
JETP Letters, Vol. 2, pp. 316–318, December 1, 1965 (2–3 days prolonged phosphorescence after high-power optical excitation)

3.149 Beam Characteristics of Ruby Optical Masers
T. S. Jaseja, M. K. Dheer, and D. Madhavan (Indian Inst. of Tech., Kampur, India)
Appl. Optics, Vol. 4, pp. 1643–1647, December 1965 (2.4×10^{-4} rad divergence, beat freq line width of 500–700 kc/s)

3.150 Beam Divergence Measurement for Q-Switched Ruby Lasers
R. W. Waynant, J. H. Cullom, I. T. Basil, and G. D. Baldwin (Westinghouse)
Appl. Optics, Vol. 4, pp. 1648–1651, December 1965 (beam splitter, 136-cm focal length camera, intensity contour, 0.79 mrad beam)

3.151 Pump Power Dependence of Ruby Laser Starting and Stopping Time
A. E. Siegman and J. W. Allen (Stanford U.)
IEEE J. Quantum Electronics, Vol. QE-1, pp. 386–393, December 1965 (also "bouncing-ball" modes with polished side walls at high pump power)

3.152 Noise Properties of Pulsed Ruby Laser Amplifiers
I. J. D'Haenens and C. R. Giuliano (Hughes)
IEEE J. Quantum Electronics, Vol. QE-1, pp. 393–397, December 1965 (spontaneous fluorescence has broader beam and wider bandwidth than laser beam)

3.153 Spiking Behavior of a Multimode Ruby Laser
C. A. Sacchi and O. Svelto (Istituto di Fisica del Politecnico, Milano, Italy)
IEEE J. Quantum Electronics, Vol. QE-1, pp. 398–400, December 1965 (spherical resonator, regular spiking, aperture effects, near field patterns, Fabry-Perot spectrum)

3.154 Phase Locking of Modes in Lasers
H. Statz and C. L. Tang (Raytheon)
J. Appl. Phys., Vol. 36, pp. 3923–3927, December 1965 (important with multimode lasers, some data using ruby, analyses)

3.155 Optical Double Resonance Experiments in Ruby
Takashi Kushida and Pravin Parikh (Toshiba Central Res. Lab., Japan)
J. Phys. Soc. Japan, Vol. 20, pp. 2312–2313, December 1965 (ruby laser pump induces excited state 2610 Å absorption)

3.156 The Pulsed Ruby Maser as a Light Amplifier
D. A. Berkowitz (Mitre Corp.) and W. S. Boyle (BTL)
Proc. IEEE, Vol. 53, p. 2165, December 1965 (correction to

P. P. Kisliuk and W. S. Boyle, Proc. IRE, Vol. 49, pp. 1635–1639, November 1961)

3.157 On the Filamentary Nature of Laser Action
Helmuth K. V. Lotsch (Northrop)
Appl. Sci. Res., Vol. 12, Sect. B, No. 5–6, pp. 451–469, 1965–1966 (large Fresnel number, single and multiple filaments, ruby laser results)

3.158 Mechanisms of Optical Emission from Ruby Excited by Short Pulses of Relativistic Electrons
D. M. J. Compton, J. F. Bryant, and R. A. Cesena (General Dynamics Corp.)
Proc. Physics of Quantum Electronics Conf., McGraw-Hill Book Co., New York, pp. 305–314, 1966

3.159 Effects of Excited-State Absorption on a Ruby Light Amplifier
C. S. Naiman and B. DiBartolo (Mithras), and A. Linz (M.I.T.)
Proc. Physics of Quantum Electronics Conf., McGraw-Hill Book Co., New York, pp. 315–321, 1966

3.160 Quasicontinuous Ruby Giant Pulse Laser using a Saturable Absorber as a Q Switch
Dieter Roess and Günter Zeidler (Siemens & Halske A. G., Munich, Germany)
Appl. Phys. Letters, Vol. 8, pp. 10–12, January 1, 1966 (ellipsoidal pump cavity, room temp., very low saturation level in methylene blue absorber)

3.161 Measurement of Ground-State Population in Ruby under Optical Pumping
Toshizo Nakaya (Konan U., Japan)
Japanese J. Appl. Phys., Vol. 5, pp. 79–85, January 1966 (Hg probe light orthogonally through rod)

3.162 Beats between the Modes of a Ruby Laser
V. V. Korobkin and A. M. Leontovich (Lebedev Physics Institute, Academy of Sciences, U.S.S.R.)
Soviet Physics JETP, Vol. 22, pp. 6–10, January 1966 (radiation patterns, streak photographs, spikes)

3.163 $^2E \rightarrow {}^2T_2$ Absorption Spectrum of Ruby
G. K. Klauminzer, P. L. Scott, and H. W. Moos (Stanford U.)
Phys. Rev., Vol. 142, pp. 248–250, February 4, 1966 (near 1.5 μ, 95°K)

3.164 Effect of Ultraviolet Pumping on Ruby Laser Output
R. L. Greene, J. L. Emmett, and A. L. Schawlow (Stanford U.)
Appl. Optics, Vol. 5, pp. 350–351, February 1966 (detrimental, highest output with Pyrex filter)

3.165 Studies of Ruby Superfluorescence and Population Inversion
Petras V. Avizonis and William R. Willoughby (Air Force Weapons Laboratory)
J. Appl. Phys., Vol. 37, pp. 682–687, February 1966

3.166 Growth of Broad Linewidth Ruby Crystals
I. Adams, J. W. Nielsen, and M. S. Story (Litton)
J. Appl. Phys., Vol. 37, pp. 832–836, February 1966

3.167 Regenerative Ruby Laser Amplifiers
H. Jacobs, J. Castro, F. A. Brand, C. LoCascio, S. Weitz, and G. Novick (U. S. Army Electronics Command)
J. Opt. Soc. Am., Vol. 56, pp. 149–156, February 1966

3.168 Spatial Distribution of the Electric Field Produced by Focusing the Output of a Ruby Laser
T. M. Barkhudarova, G. S. Voronov, V. M. Gorbunkov, and N. B. Delone (Lebedev Physics Inst., Academy of Sciences, U.S.S.R.)
Soviet Physics JETP, Vol. 22, pp. 269–271, February 1966 (Q-sw ruby, 45 and 120 mm focal length lenses)

3.169 Single-Mode Operation of a Room-Temperature CW Ruby Laser
Dieter Roess (Siemens & Halske A. G., Germany)
Appl. Phys. Letters, Vol. 8, pp. 109–110, March 1, 1966 (1200 W threshold)

3.170 Laser with Nonresonant Feedback
R. V. Ambartsumyan, N. G. Basov, P. G. Kryukov, and V. S. Letokhov (Lebedev Physics Institute, U.S.S.R.)
JETP Letters, Vol. 3, pp. 167–169, March 15, 1966 (diffuse and specular mirrors in laser cavity, two ruby amplifiers)

3.171 Thermal Dependence of Ruby Laser Emission
Jerald R. Izatt, Robert C. Mitchell, and Harold A. Daw (New Mexico State U.)
J. Appl. Phys., Vol. 37, pp. 1558–1562, March 15, 1966 (0.046 Å/°K at 200°K, less at lower temperatures)

3.172 Ruby Whisker Growth and Characteristics
D. M. Schuster, M. C. Teich, and E. Scala (Cornell U.)
J. Appl. Phys., Vol. 37, pp. 1621–1623, March 15, 1966 (vapor phase reaction, up to 4% Cr)

3.173 Laser Operation in a Wedge Shaped Ruby
Takao Tanaka, Chuhei Suzuki, Takasuke Fukui, and Kakuo Futami (KDD Res. Lab., Japan)
Japanese J. Appl. Phys., Vol. 5, pp. 258–259, March 1966 (wedge cross section, regular spiking)

3.174 Features of the Time Behavior of the Generation in a Laser with Moving Ruby Crystal
B. L. Livshitz, V. P. Nazarov, L. K. Sidorenko, A. T. Tursunov, and V. N. Tsikunov (USSR Academy of Sciences, USSR)
JETP Letters, Vol. 3, pp. 179–181, April 1, 1966 (tends to smear spikes into continuous output, 1 mm diaphragm improves continuity)

3.175 Laser Properties of a Vapor-Grown Ruby
J. R. O'Connor (M.I.T.), P. S. Schaffer (Lexington Labs.) and R. A. Bradbury (A. F. Cambridge Res. Labs.)
Appl. Phys. Letters, Vol. 8, pp. 336–337, June 15, 1966 (77°K, 25 J threshold, low spiking level)

3.176 Pulsed Transmission Mode Operation in the Case of a Mode Locking of the Modes of a Non Q-Spoiled Ruby Laser
M. Michon, J. Ernest, and R. Auffret (CGE, France)
Physics Letters, Vol. 21, pp. 514–515, June 15, 1966 (Kerr cell within laser cavity, self modulation, train of pulses)

3.177 Synchronization of Giant Pulse Lasers
H. Opower and W. Kaiser (Tech. Hochschule München, Germany)
Physics Letters, Vol. 21, pp. 638–640, July 1, 1966 (within 3 ns, cryptocyanine, ruby, 3 rods, over 500 MW total output)

3.178 Comment on the Effect of Trapped Light on the Output of a Ruby Laser
J. McKenna and J. G. Skinner (BTL)
Appl. Optics, Vol. 5, pp. 1241–1242, July 1966

3.179 Relaxation Time for the $^4F_1 \rightarrow {}^2E$ Transition in Ruby
S. A. Pollak (TRW/Systems)
Bull. Am. Phys. Soc., Vol. 11, p. 776, abstract FH1, July 1966 (1.2 × 10⁻⁸ s at room temperature)

3.180 High Power Non-Spiking Operation of Ruby Laser
D. V. Keller (Defense Research Corp.), and B. I. Davis (Northrop)
IEEE J. Quantum Electronics, Vol. QE-2, pp. 179–181, July 1966 (immersed ruby, feedback circuit, virtually ripple free, 10 kW during pulse)

3.181 90° Rotation Between Near Field and Far Field of Ruby Lasers
Dieter Roess (Siemens and Halske, Germany)

IEEE J. Quantum Electronics, Vol. QE-2, pp. 181–182, July 1966 (caused by asymmetrical pumping, thermal curvatures of Fabry-Perot cavity surfaces)

3.182   Liquid-Nitrogen Cooling of a Ruby Rod
D. L. Mickey (Princeton U.)
J. Appl. Phys., Vol. 37, pp. 2963–2964, July 1966 (rod ends cooled)

3.183   Direct $^2T_1$-^2E Phonon Relaxation in Ruby and Its Effect Upon R-Line Breadth
Joseph A. Calviello, Edward W. Fisher, and Zindel H. Heller (AIL)
J. Appl. Phys., Vol. 37, pp. 3156–3160, July 1966

3.184   Absorption Spectrum of Optically Pumped Ruby. I. Experimental Studies of Spectrum in Excited States
Takashi Kushida (Tokyo Shibaura, Japan)
J. Phys. Soc. Japan, Vol. 21, pp. 1331–1341, July 1966 (between 5,500 to 45 000 cm^{-1})

3.185   Absorption Spectrum of Optically Pumped Ruby. II. Theoretical Analyses
Masaki Shinada, Satoro Sugano (U. of Tokyo, Japan), and Takashi Kushida (Tokyo Shibaura, Japan)
J. Phys. Soc. Japan, Vol. 21, pp. 1342–1352, July 1966

3.186   Direct Spectroscopic Detection of Ruby Laser Giant Pulse Off-Axial Mode Structure
Daniel J. Bradley, Malcolm S. Engwell, and A. W. McCullough (U. of London, England), and George Magyar and Martin C. Richardson (UKAEA, England)
Appl. Phys. Letters, Vol. 9, pp. 150–152, August 15, 1966

3.187   Radiative Coupling Between Two Different Solid-State Lasers
H. Inaba, Y. Isawa, and N. Suda (Tohoku U., Japan)
Physics Letters, Vol. 22, pp. 293–295, August 15, 1966 (Q-sw ruby laser beam into Nd-glass laser, two-photon excitation by ruby, enhanced Nd-laser output)

3.188   Analysis of Room Temperature CW Ruby Lasers
Dieter Roess (Siemens and Halske, Germany)
IEEE J. Quantum Electronics, Vol. QE-2, pp. 208–214, August 1966 (ellipsoidal pump cavity, Hg lamp, also pulsed at 120 pps, thermal lens effect in ruby)

3.189   Single Transverse and Longitudinal Mode Q-Switched Ruby Laser
V. Daneu, C. A. Sacchi, and O. Svelto (Milano Politecnico, Italy)
IEEE J. Quantum Electronics, Vol. QE-2, pp. 290–293, August 1966 (spherical mirrors, vanadium phathalocyanine saturable absorber)

3.190   Mode Coupling in a Ruby Laser
R. H. Pantell and R. L. Kohn (Stanford U.)
IEEE J. Quantum Electronics, Vol. QE-2, pp. 306–310, August 1966 (modulator within laser cavity)

3.191   Measurement of Fractional Metastable-State Population of Cr Ions in Ruby under Q-Switching Operation
Toshizo Nakaya (Konan U., Japan)
Japanese J. Appl. Phys., Vol. 5, pp. 689–695, August 1966 (Hg 5461 Å probe light through ruby)

3.192   Correlation between Laser Performance and Crystal Homogeneities of Ruby Laser Rods
Katsuhiko Nishida (Nippon Electric Co., Japan)
Japanese J. Appl. Phys., Vol. 5, p. 727, August 1966 (degree of coherence, fringe count, output energy)

3.193   Analysis of a Room-Temperature CW Ruby Laser of 10-mm Resonator Length: The Ruby Laser as a Thermal Lens

Dieter Roess (Siemens and Halske, Germany)
J. Appl. Phys., Vol. 37, pp. 3587–3594, August 1966 (axial and transverse modes)

3.194   Coupling Effects in a Passive Q-Switched Ruby Laser
V. Degiorgio and M. Giglio (Lab. CISE, Italy)
Nuovo Cimento, Vol. 45B, pp. 69–71, September 11, 1966 (vanadium phthalocyanine, two halves of beam decouples with 1.5 mm diameter blocking wire)

3.195   Internal Self-Damage in a 25 MW Ruby Laser Oscillator
D. J. Bradley, A. W. McCullough, and P. D. Smith (U. of London, England)
British J. Appl. Phys., Vol. 17, pp. 1221–1222, September 1966 (numerous 200 μm diameter bubbles)

3.196   Ruby Laser Loss Measurement by Comparison of R_1, R_2 Thresholds
D. C. Hanna, W. A. Gambling, and R. C. Smith (Southampton U., England)
IEEE J. Quantum Electronics, Vol. QE-2, pp. 507–510, September 1966

3.197   Characteristics of a Traveling-Wave Ruby Single-Mode Laser as a Laser Radar Transmitter
I. Goldstein and A. Chabot (Raytheon)
IEEE J. Quantum Electronics, Vol. QE-2, pp. 519–523, September 1966 (20 kW, 1 μs, 1 pps master oscillator, 24 dB gain power amplifier)

3.198   Absorption and Emission Properties of Optically Pumped Ruby
Takashi Kushida (Tokyo Shibaura, Japan)
IEEE J. Quantum Electronics, Vol. QE-2, pp. 524–531, September 1966 (metastable state absorption from 5500 to 45,000 cm^{-1}, emission near R lines)

3.199   Energy Transfer in Ruby
G. F. Imbusch (BTL)
IEEE J. Quantum Electronics, Vol. QE-2, pp. 532–537, September 1966

3.200   Output Spectra of Nd:YAG and Ruby Lasers and Implications for Laser Linewidth Determining Mechanisms
W. A. Specht, Jr., J. K. Neeland, and V. Evtuhov (Hughes Research)
IEEE J. Quantum Electronics, Vol. QE-2, pp. 537–541, September 1966 (inhomogeneous broadening, spatial relaxation effects)

3.201   Feedback Control of a Q-Switched Ruby Laser
C. H. Thomas and E. V. Price (E.G. & G.)
IEEE J. Quantum Electronics, Vol. QE-2, pp. 617–623, September 1966 (goal of 10 kW and 5 to 10 μs output pulses, experiments, theory)

3.202   Laser Operation with Liquid Semiconductor Mirrors
M. Birnbaum and T. L. Stocker (Aerospace)
IEEE J. Quantum Electronics, Vol. QE-2, pp. 632–635, September 1966 (ruby laser, one liquid selenium mirror at 500°C, reflectivity measurements)

3.203   Quenching Effects in Coupled Lasers
P. W. Pheneger and R. H. Pantell (Stanford U.)
IEEE J. Quantum Electronics, Vol. QE-2, pp. 644–648, September 1966 (ruby lasers, end coupled)

3.204   Simultaneous Giant Pulses from Five Ruby Laser Oscillators
David W. Gregg and Scott J. Thomas (Lawrence Rad. Lab., U. of California)
J. Appl. Phys., Vol. 37, pp. 3750–3753, September 1966 (oscillators were phase locked, three systems tried, low beam divergence)

4. NEODYMIUM LASER

4.21 Laser Oscillation at 1.06 μ in the Series $Na_{0.5}Gd_{0.5-x}Nd_xWO_4$
G. E. Peterson and P. M. Bridenbaugh (BTL)
Appl. Phys. Letters, Vol. 4, pp. 173–175, May 15, 1964 (N_2 temp., up to 50% Nd, 6-180 μs lifetime, 15 j min. threshold)

4.22 Erratum: Laser Oscillation at 1.06μ in the Series $Na_5Gd_{5-x}Nd_x$ WO_4
G. E. Peterson and P. M. Bridenbaugh
Appl. Phys. Letters, Vol. 5, p. 39, July 15, 1964 (Appl. Phys. Letters, Vol. 4, p. 173; 1964)

4.23 **Laser Oscillations in Nd-Doped Yttrium Aluminum, Yttrium Gallium and Gadolinium Garnets**
J. E. Geusic, H. M. Marcos, and L. G. Van Uitert (BTL)
Appl. Phys. Letters, Vol. 4, pp. 182–184, May 15, 1964 (360 watt tungsten lamp, room temp., CW output from YAlG:Nd, other pulsed data)

4.24 Study of Relaxation Processes in Nd Using Pulsed Excitation
G. E. Peterson and P. M. Bridenbaugh (BTL)
J. Opt. Soc. Am., Vol. 54, pp. 644–650, May 1964 (resonance coupling, Nd + Yb)

4.25 Radiationless Resonance Energy Transfer from UO_2^{2+} to Nd^{3+} in Coactivated Barium Crown Glass
H. W. Gandy, R. J. Ginther, and J. F. Weller (U. S. NRL)
Appl. Phys. Letters, Vol. 4, pp. 188–190, June 1, 1964 (green-UV absorption in UO_2^{++} transfers energy to Nd)

4.26 Continuous Sun-Pumped Room Temperature Glass Laser Operation
G. R. Simpson (American Optical)
Appl. Optics, Vol. 3, pp. 783–784, June 1964 (neodymium, parabolic mirror and two aplanatic refractors, 15 per cent over threshold)

4.27 Laser Oscillations at 0.918, 1.057 and 1.401 Microns in Nd^{3+}-Doped Borate Glasses
A. David Pearson, S. P. S. Porto, and W. R. Northover (BTL)
J. Appl. Phys., Vol. 35, pp. 1704–1706, June 1964 (low expansion Calibo glass, about 40 μsec lifetime)

4.28 Photolytic and Reduction Coloring at $CaWO_4$:Nd
D. C. Cronemeyer and M. W. Beaubien (Bendix)
J. Appl. Phys., Vol. 35, pp. 1779–1785, June 1964 (intense UV radiation produces brownish color, varied recovery, $CaWO_4$ refractive index 1.9 to 2.0)

4.29 Oscillatory Character of $CaWO_4$:Nd^{3+} Laser Output
E. Bernal, J. F. Ready, and D. Chen (Honeywell Research Center)
Proc. IEEE, Vol. 52, pp. 710–711, June 1964 (regular spiking, single axial mode; beats with irregular spiking)

4.30 Application of Resonance Cooperation of Rare-Earth Ions Nd^{3+} and Yb^{3+} to Lasers ($Na_{0.5}RE_{0.5}WO_4$)
G. E. Peterson and P. M. Bridenbaugh (BTL)
Appl. Phys. Letters, Vol. 4, pp. 201–202, June 15, 1964

4.31 Nonradiative Energy Exchange and Laser Oscillation in Yb^{3+}-, Nd^{3+}-Doped Borate Glass
A. David Pearson and S. P. S. Porto (BTL)
Appl. Phys. Letters, Vol. 4, pp. 202–204, June 15, 1964

4.32 Lattice Energy Transfer and Stimulated Emission from CeF_3:Nd^{3+}
J. R. O'Connor (Lincoln Lab., M.I.T.) and W. A. Hargreaves (Optovac)
Appl. Phys. Letters, Vol. 4, pp. 208–209, June 15, 1964 (41-J threshold)

4.33 Higher Oscillation Modes in Nd Glass Laser
S. Tatuoka (Japan Broadcasting Co.)
Appl. Optics, Vol. 3, pp. 986–987, August 1964 (higher order ring patterns radiated, streak camera used, see Appl. Optics, p. 1287, November 1964)

4.34 Neodymium Fluorescence in the 5 to 6 Micron Region
F. Varsanyi (BTL)
Physics Letters, Vol. 11, pp. 193–194, August 1, 1964

4.35 Energy Transfer in Silicate Glass Coactivated with Cerium and Neodymium
H. W. Gandy, R. J. Ginther, and J. F. Weller (N.R.L.)
Physics Letters, Vol. 11, pp. 213–214, August 1, 1964 (neodymium luminescence from nonradiative energy transfer process)

4.36 Single and Double Axial Mode Operation of a Nd-Optical Maser
H. Manger and H. Rothe (Tech. Hoch., Karlsruhe, Germany)
Physics Letters, Vol. 12, pp. 182–183, October 1, 1964 ($CaWO_4$: Nd, 5.9 and 11.95-mm thick etalons)

4.37 Quantitative Analysis of Single-Mode Operation of a Solid State Laser
W. Kaiser and D. Pohl (Tech. Hoch., Munich, Germany)
Physics Letters, Vol. 12, pp. 185–186, October 1, 1964 (spiking of Nd glass)

4.38 Amplification in a Fiber Laser
C. J. Koester and E. Snitzer (American Optical)
Appl. Optics, Vol. 3, pp. 1182–1186, October 1964 (47-dB gain in one meter long neodymium-glass fiber, gain vs. time delay)

4.39 Generation of Giant Pulses from a Neodymium Laser by a Reversibly Bleachable Absorber
B. H. Soffer and R. H. Hoskins (Korad)
Nature, Vol. 204, p. 276, October 17, 1964

4.40 Giant Pulse Laser Action and Pulse Width Narrowing in Neodymium-Doped Borate Glass
J. R. Sanford, J. H. Wenzel (G.E.), and G. J. Wolga (Cornell U.)
J. Appl. Phys., Vol. 35, pp. 3422–3423, November 1964 (varied reflectance, narrower pulse with higher inversion)

4.41 Energy Levels and Crystal-Field Calculations of Neodymium in Yttrium Aluminum Garnet
J. A. Koningstein and J. E. Geusic (BTL)
Phys. Rev., Vol. 136, pp. A711–A716, November 2, 1964

4.42 Cross-Pumped Cr^{3+}–Nd^{3+}: YAG Laser System
Z. J. Kiss and R. C. Duncan (RCA)
Appl. Phys. Letters, Vol. 5, pp. 200–202, November 15, 1964 (red flourescence of Cr^{3+} pumps Nd^{3+})

4.43 Dynamics of Energy Transfer from 3d to 4f Electrons in $LaAlO_3$: $Cr^{3+}Nd^{3+}$
Z. J. Kiss (RCA)
Phys. Rev. Letters, Vol. 13, pp. 654–656, November 30, 1964 (red fluorescence of Cr pumps Nd, disproves an earlier paper)

4.44 Fluorescence and Stimulated Emission from La_2O_3:Nd^{+3}
R. H. Hoskins and B. H. Soffer (Korad)
J. Appl. Phys., Vol. 36, pp. 323–324, January 1965 (77°K, 230 j threshold for 1.079μ)

4.45 Gain-Delay Characteristics of a Pulsed Neodymium-Glass Laser Oscillator-Amplifier Chain
K. F. Tittel and J. P. Chernoch (G.E.)
Proc. IEEE, Vol. 53, pp. 82–83, January 1965

4.46 Quasi-Continuous Operation of a $CaWO_4$:Nd^{3+} Maser using Long Duration Pumping Pulses
H. Manger (Technische Hochschule Karlsruhe, Germany)
Proc. IEEE, Vol. 53, pp. 83–84, January 1965 (100 m sec possible)

4.47 Laser Action in Uranyl-Sensitized Nd-Doped Glass
N. T. Melamed and C. Hirayama (Westinghouse), and P. W. French (Pittsburgh Plate Glass)
Appl. Phys. Letters, Vol. 6, pp. 43–45, February 1, 1965 (self-Q-switching, rather irregular multiple spikes)

4.48 An Injection Laser Pump for Nd^{+3} Doped Hosts
R. H. Harada and C. K. Suzuki (Autonetics)
Appl. Optics, Vol. 4, pp. 225–227, February 1965 (possible pumping at 0.87μ)

4.49 Sensitization of Nd^{3+} Luminescence by Mn^{2+} and Ce^{3+} in Glasses
Shigeo Shionoya and Eiichiro Nakazawa (U. of Tokyo, Japan)
Appl. Phys. Letters, Vol. 6, pp. 117–118, March 15, 1965 (calcium phosphate glass, 1.06μ luminescence)

4.50 Continuous Room-Temperature Nd³⁺:CaMoO₄ Laser
      R. C. Duncan (RCA)
      J. Appl. Phys., Vol. 36, pp. 874–875, Part 1, March 1965 (1 j
      pulse threshold, 1200 watts for cw)

4.51 Investigation of Stimulated Emission of CaF₂–Nd³⁺ (Type II)
      Crystals at Room Temperature
      Yu. K. Voron'ko, A. A. Kaminskiĭ, L. S. Kornienko, V. V. Osiko,
      A. M. Prokhorov, and V. T. Udovenchik (Moscow State U.
      and Lebedev Physics Inst., Academy of Sciences, U.S.S.R.)
      JETP Letters, Vol. 1, pp. 39–42, April 15, 1965 (Nd emits from
      1.037 to 1.0885 μ, longest wavelength from CaF₂ host, table of
      15 hosts)

4.52 A Repetitively Q-Switched, Continuously Pumped YAG:Nd Laser
      J. E. Geusic, M. L. Hensel and R. G. Smith (BTL)
      Appl. Phys. Letters, Vol. 6, pp. 175–177, May 1, 1965 (1000
      watts pump, 50 to 600 rps rotating mirror)

4.53 The Operational Characteristics of a CW Nd:CaWO₄ Laser in
      the Range of Dry Ice to Room Temperature
      H. R. Aldag, R. S. Horwath and C. B. Zarowin (Sperry)
      Appl. Optics, Vol. 4, pp. 559–563, May 1965 (about 1200 watts
      threshold at room temperature)

4.54 Neodymium Glass Laser with Single Pulse Duration Close to the
      Limit
      V. I. Malyshev, A. S. Markin, V. S. Petrov, I. I. Levkoev, and
      A. F. Vompe (Lebedev Physics Inst., Academy of Sciences,
      U.S.S.R.)
      JETP Letters, Vol. 1, pp. 159–160, June 15, 1965 (3000 J input,
      pentacarbocyanin analog Q-switch, 10 ns 50 MW output pulse
      using 55-cm-long cavity, 330 ns using 300-cm-long cavity)

4.55 Fluorescent Properties of Trivalent Neodymium in Lanthanum
      and Yttrium Orthoniobates and Tantalates
      G. A. Wesselink and A. Bril (Philips, The Netherlands)
      Philips Res. Repts., Vol. 20, pp. 269–277, June 1965 (lifetimes
      about 150 μsec, quantum efficiencies about 50%, favorable for
      laser application)

4.56 Optical Maser Action of Nd³⁺ in a Potassium Silicate Glass
      C. Hirayama, N. T. Melamed, and E. W. Sucov (Westing-
      house)
      Phys. and Chem. of Glasses, Vol. 6, pp. 104–107, June 1965
      (2.85% Nd₂O₃, 810 μsec lifetime, poor optical quality, about
      50J threshold)

4.57 Neodymium-Glass Laser with Pulsed Q Switching
      N. G. Basov, V. S. Zuev, and Yu. V. Senat-skiĭ (Lebedev Physics
      Inst., Academy of Sciences, U.S.S.R.)
      JETP Letters, Vol. 2, pp. 35–36, July 15, 1965 (oscillator and
      amplifier chain, 4 J output, amplifier rod damaged)

4.58 Energy Distribution in a Glass: Nd³⁺ Laser Rod
      N. F. Borrelli and M. L. Charter (Corning Glass Works)
      J. Appl. Phys., Vol. 36, pp. 2172–2174, July 1965 (gain and heat-
      ing are nonuniform across diameter)

4.59 Excitation of Auxiliary Off-Axis Laser Modes
      M. P. Vanyukov, V. I. Isaenko, L. A. Luizova, and O. A.
      Shorokhov (Vavilov State Inst. of Optics, U.S.S.R.)
      Soviet Physics JETP, Vol. 21, pp. 1–3, July 1965 (rod axis in-
      clined to Fabry-Perot cavity axis, far-field pattern)

4.60 Change of Optical Path Length in Laser Rods Within the Pump-
      ing Period
      H. Welling, C. J. Bickart, and H. G. Andresen (U.S. Army
      Electronics Command Labs.)
      IEEE J. Quantum Electronics, Vol. QE-1, pp. 223–224, August
      1965 (interferometric method, ruby and Nd glass, distortion
      during and after pump pulse)

4.61 Fluorescence Conversion Efficiency of Neodymium Glass
      L. G. DeShazer and L. G. Komai (Hughes)
      J. Opt. Soc. Am., Vol. 55, pp. 940–944, August 1965 (0.26 effi-
      ciency for 1.06 μ)

4.62 Spectral Investigation of the Stimulated Radiation of Nd³⁺ in
      CaF₂
      A. A. Kaminskiĭ, L. S. Kornienko, and A. M. Prokhorov
      (Moscow State U., U.S.S.R.)
      Soviet Physics JETP, Vol. 21, pp. 318–322, August 1965 (below
      100°K 5 new lines between 10 448 and 10 650 Å)

4.63 Laser Action in Neodymium-Doped Glass Produced Through
      Energy Transfer
      N. T. Melamed, C. Hirayama, and E. K. Davis (Westinghouse)
      Appl. Phys. Letters, Vol. 7, pp. 170–172, September 15, 1965
      (0.2 ms transfer time, pulsed operation, Mn²⁺ sensitized phos-
      phate glass)

4.64 The Polarization of Light from Nd³⁺-Glass Lasers
      Sun Lu and T. A. Rabson (Rice U.)
      Appl. Phys. Letters, Vol. 7, pp. 219–220, October 15, 1965 (beam
      more polarized near threshold, each spike generally polarized)

4.65 Giant Pulses from Neodymium Doped Calcium Tungstate by
      Gain-Switching
      R. A. Clay and D. Findlay (Royal Radar Establishment, Eng-
      land)
      Physics Letters, Vol. 19, pp. 212–213, October 15, 1965 (rotating
      prism between two laser rods, 2.5 MW peak power)

4.66 On the Spectral Properties of a Q-Switched Nd³⁺ Doped Laser
      Emission
      M. Michon, J. Ernest, R. Dumanchin, J. Hanus, and S. Raynaud
      (C.G.E. France)
      Physics Letters, Vol. 19, pp. 217–219, October 15, 1965 (200 Å
      wide for 1.1 J output from Nd-doped alkali silicate glass rod)

4.67 Influence of the 4 I¹¹/² Level Lifetime on the Effective Use of the
      Population Inversion in a Q-Spoiled Neodymium Doped Glass
      Laser
      M. Michon, J. Ernest, J. Hanus, and R. Auffret (C.G.E., France)
      Physics Letters, Vol. 19, pp. 219–220, October 15, 1965 (level 2
      lifetime is about 60 ns)

4.68 Cutoff of Ruby Laser Emission by Pulsed Irradiation
      D. M. J. Compton, R. A. Cesena, J. F. Bryant, and B. L.
      Gehman (General Dynamics)
      Proc. IEEE, Vol. 53, pp. 1668–1669, October 1965 (30 MeV
      Linac pulses of 200 mA, similar results with Nd laser)

4.69 Regular Periodical Spiking of a Neodymium-Glass Laser
      A. J. Casella (Pennsylvania State U.)
      Proc. IEEE, Vol. 53, pp. 1782–1783, November 1965 (6 percent
      doping, medium reflectivity plane mirror, 7 percent above
      threshold)

4.70 Lifetime of the Excited State ⁴F₃/₂ of the Nd³⁺ Ion in CaF₂ and
      CaWO₄
      A. A. Kaminskiĭ, L. S. Kornienko, and A. M. Prokhorov
      (Nuclear Physics Institute, Moscow State U., U.S.S.R.)
      Soviet Physics JETP, Vol. 21, pp. 844–847, November 1965
      (varied concentration and temperature, up to 1.65 ms, ap-
      paratus described)

4.71 Spectral Composition of Generation of Neodymium Glass in a
      Dispersion Resonator
      V. L. Broude, V. I. Kravchenko, N. F. Prokopyuk, and M. S.
      Soskin (Physics Institute, Ukrainian Academy of Sciences)
      JETP Letters, Vol. 2, pp. 324–326, December 1, 1965 (spectral
      shift with inclined plane mirror)

4.72 Saturation Operation and Gain Coefficient of a Neodymium-Glass
      Amplifier
      C. G. Young and J. W. Kantorski (American Optical)

Appl. Optics, Vol. 4, pp. 1675–1677, December 1965 (106 dB small-signal round-trip gain, spectral narrowing, amplified spontaneous emission, saturation onset at 13 J/cm², 7.5 percent gain cm⁻¹/Jcm⁻², 140 J output)

4.73   A Q-Switched Neodymium Glass Laser
N. G. Basov, V. S. Zuev, and Yu. V. Senat-skiĭ (Lebedev Physics Inst., Academy of Sciences, U.S.S.R.)
Soviet Physics JETP, Vol. 21, pp. 1047–1048, December 1965 (rotating prism, three amplifier rods in cascade)

4.74   A Study of the YAlG:Nd Oscillator
J. E. Geusic, H. M. Marcos, and L. G. Van Uitert (B.T.L.)
Proc. Physics of Quantum Electronics Conf., McGraw-Hill Book Co., New York, pp. 725–734, 1966

4.75   Mode Locking of a Nd³⁺-Doped Glass Laser
A. J. DeMaria, C. M. Ferrar, and G. E. Danielson, Jr. (United Aircraft Res. Labs.)
Appl. Phys. Letters, Vol. 8, pp. 22–24, January 1, 1966 (pulsed, 12.2 cm long rod, 47 Mc/s acoustic modulator within laser cavity, 0.5 ns pulses at 94 Mc/s rate)

4.76   Gd₂(MoO₄)₃: A Ferroelectric Laser Host
Hans J. Borchardt and Paul E. Bierstedt (E. I. du Pont de Nemours)
Appl. Phys. Letters, Vol. 8, pp. 50–52, January 15, 1966 (3% Nd doped, ferroelectric below 159°C, 350 J threshold at −138°C, direct modulation possible)

4.77   Frequency Control of a Nd³⁺ Glass Laser
Elias Snitzer (American Optical)
Appl. Optics, Vol. 5, pp. 121–125, January 1966 (thin plate within laser cavity, aligned and tilted, less than 0.1 Å emission linewidth)

4.78   Efficient High Energy Laser Radiation Utilizing a Coaxial Optical Pump
J. P. Lesnick and C. H. Church (Westinghouse)
IEEE J. Quantum Electronics, Vol. QE-2, pp. 16–17, January 1966 (94-cm long Nd-doped glass, 5.1% slope, 800 J output, 150 kJ max input to lamp)

4.79   Temperature-Dependent Concentration Quenching of Fluorescence by Cross Relaxation of Nd³⁺ in LaF₃
C. K. Asawa and M. Robinson (Hughes)
Phys. Rev., Vol. 141, pp. 251–258, January 1966

4.80   A Room-Temperature Continuous CaWO₄:Nd³⁺ Laser
A. A. Kaminskiĭ, L. S. Kornienko, G. V. Maksimova, V. V. Osiko, A. M. Prokhorov, and G. P. Shipulo (Moscow State U. and Lebedev Physics Institute, Academy of Sciences, U.S.S.R.)
Soviet Physics JETP, Vol. 22, pp. 22–25, January 1966 (2.6 kW threshold, water cooled, 1° beam at 30% above threshold)

4.81   Effect of Surface Finish on Temperature Distributions in Optically Pumped Nd⁺⁺⁺:Glass Laser Rods
L. J. Aplet, E. B. Jay, and W. R. Sooy (Hughes)
Appl. Phys. Letters, Vol. 8, pp. 71–73, February 1, 1966 (up to 5°C radial temp. difference, stress birefringence)

4.82   Analysis of the Optical Spectra of CaF₂:Nd³⁺ (Type 1) Crystals
Yu. K. Voron'ko, A. A. Kaminskiĭ, and V. V. Osiko (Lebedev Physics Inst., Academy of Sciences, U.S.S.R.)
Soviet Physics JETP, Vol. 22, pp. 295–300, February 1966 (concentration and temperature varied)

4.83   Nd³⁺ Glass Laser Exhibiting Regular Spikes
R. Polloni, C. A. Sacchi, and O. Svelto (Istituto di Fisica del Politecnico, Milano, Italy)
J. Appl. Phys., Vol. 37, pp. 1931–1932, March 15, 1966 (near concentric laser cavity)

4.84   Experimental Confirmation of Standing Waves in Laser Resonators
A. M. Ledger (Canadian Westinghouse, Canada)
Appl. Optics, Vol. 5, pp. 476–477, March 1966 (evaporated pattern on thin metallic film at small angle within laser cavity)

4.85   Solar-Pumped Modulated Laser
C. W. Reno (RCA)
RCA Review, Vol. 27, pp. 149–157, March 1966 (Dy and Nd lasers, transmitted television picture)

4.86   Use of a Laser Operating in the Spike Mode to Obtain a High-Temperature Plasma
M. P. Vanyukov, V. I. Isaenko, V. V. Lyubimov, V. A. Serebryakov, and O. A. Shorokhov
JETP Letters, Vol. 3, p. 205, April 15, 1966 (60 cm long, 4.5 cm diameter Nd glass rod, 1400 J output, 1.2 ms pulse, focused density of 3 GW/cm², photo of air plasma)

4.87   Inhomogeneities in Optical Crystals Resulting from Nonplanar Solid-Melt Interface during Growth
James F. Nester (Perkin-Elmer)
J. Appl. Phys., Vol. 37, pp. 2002–2004, April 1966 (CaWO₄:Nd³⁺ crystals, interferogram, radially asymmetric temperature gradient, laser beam 10 times diffraction limit)

4.88   Etude Optique du Faisceau Émis par un Laser de Grande Intensité
J. de Metz, A. Terneaud, et P. Veyrie (AEC, France)
Appl. Optics, Vol. 5, pp. 819–822, May 1966 (5% Nd glass, Q-sw oscillator, 4 amplifiers, 30 J in 25 ns)

4.89   High-Repetition Pulsed Nd³⁺ Glass Laser
Toshiro Kamogawa, Hiroaki Kotera, and Heijiro Hayami (Matsushita Electric, Japan)
Japanese J. Appl. Phys., Vol. 5, p. 449, May 1966 (40 μs pump pulse, K₂CrO₄ filter, up to 200 pps)

4.90   Self Q-Switched Nd³⁺ Glass Laser
W. Shiner, E. Snitzer, and R. Woodcock (American Optical)
Physics Letters, Vol. 21, pp. 412–413, June 1, 1966 (unfiltered pump light, color centers formed, regular spikes at about 100 kc/s)

4.91   A Sun-Pumped cw One-Watt Laser
C. G. Young (American Optical)
Appl. Optics, Vol. 5, pp. 993–997, June 1966 (Nd-doped YAG, water cooled, 61-cm diameter collector)

4.92   Theoretical Assessment of a High Power Continuous-Wave 4-Level Solid Laser
A. C. Selden (Atomic Weapons Research Estab., England)
British J. Appl. Phys., Vol. 17, pp. 729–736, June 1966 (tens of watts predicted from Nd doped glass)

4.93   Giant-Pulse Laser Activity in Neodymium-Doped Silicate Glass: The Energy Conversion Process
J. H. Wenzel (GE)
J. Appl. Phys., Vol. 37, pp. 3100–3110, July 1966 (standing waves in resonator, radial variation in gain)

4.94   A High-Gain Room-Temperature Liquid Laser: Trivalent Neodymium in Selenium Oxychloride
Adam Heller (GT & E Labs.)
Appl. Phys. Letters, Vol. 9, pp. 106–108, August 1, 1966 (solution preparation, 110 μs fluorescence decay time)

4.95   Characteristics of the Nd⁺³:SeOCl₂ Liquid Laser
Alexander Lempicki and Adam Heller (GT & E Labs.)
Appl. Phys. Letters, Vol. 9, pp. 108–110, August 1, 1966 (5 J threshold, sharp emission)

4.96   Optical Spectra of Ultrashort Optical Pulses Generated by Mode-Locked Glass:Nd Lasers

D. A. Stetser and A. J. DeMaria (United Aircraft Research Labs.)
Appl. Phys. Letters, Vol. 9, pp. 118–120, August 1, 1966 (saturable absorber, 2×10^{-13} s pulses)

4.97 Radiative Coupling Between Two Different Solid-State Lasers
H. Inaba, Y. Isawa, and N. Suda (Tohoku U., Japan)
Physics Letters, Vol. 22, pp. 293–295, August 15, 1966 (Q-sw ruby laser beam into Nd-glass laser, two-photon excitation by ruby, enhanced Nd-laser output)

4.98 A High Time Resolution Polarimeter for Laser Analysis
Sun Lu and T. A. Rabson (Rice U.)
Appl. Optics, Vol. 5, pp. 1293–1296, August 1966 (light beam polarization measured in less than 0.1 μs, spike polarization data on Nd glass laser)

4.99 Effect of Spectral Hole-Burning and Cross Relaxation on the Gain Saturation of Laser Amplifiers
Amado Y. Cabezas and Richard P. Treat (Hughes Aircraft)
J. Appl. Phys., Vol. 37, pp. 3556–3563, August 1966 (Nd laser experiments)

4.100 Erratum: Effect of Spectral Hole-Burning and Cross-Relaxation on the Gain Saturation of Laser Amplifiers
A. Y. Cabezas and R. P. Treat (Hughes Aircraft)
J. Appl. Phys., Vol. 37, p. 4598, November 1966
(J. Appl. Phys., Vol. 37, p. 3556, August 1966)

4.101 Etude de l'Emission d'un Laser au Néodyme, Déclenché par Effet Pockels
P. Wurtz
Philips Res. Repts., Vol. 21, pp. 213–245, August 1966 (fluorescence intensity, KDP Q-sw cell)

4.102 Discrimination of Axial Oscillation Modes in a Laser with External Mirrors
V. I. Malyshev and A. S. Markin (Lebedev Phys. Inst., Acad. of Sci., USSR)
Soviet Phys. JETP, Vol. 23, pp. 225–227, August 1966 (Nd-glass laser, rod position relative to mirrors, bleachable liquid, four discrete beat frequencies)

4.103 Giant Superluminescence Pulses
V. S. Zuev, V. S. Letokhov, and Yu. V. Senatskii (Lebedev Phys. Inst., Acad. of Sci., USSR)
JETP Letters, Vol. 4, pp. 125–127, September 1, 1966 (Nd glass rods 90 cm long, 40-dB single pass gain, Kerr cell shutter, 500 MW/cm² output, 9 to 12 ns pulse)

4.104 Output Spectra of Nd:YAG and Ruby Lasers and Implications for Laser Linewidth Determining Mechanisms
W. A. Specht, Jr., J. K. Neeland, and V. Evtuhov (Hughes Research)
IEEE J. Quantum Electronics, Vol. QE-2, pp. 537–541, September 1966 (inhomogeneous broadening, spatial relaxation effects

4.105 Laser Action by Enhanced Total Internal Reflection
Charles J. Koester (American Optical)
IEEE J. Quantum Electronics, Vol. QE-2, pp. 580–584, September 1966 (high index passive glass fiber core, lower index Nd-doped glass clad, optically pumped, signal in core is amplified)

4.106 The Influence of Nd^{3+} Ion Properties in a Glass Matrix on the Dynamics of a Q-Spoiled Laser
Maurice Michon (CGE, France)

IEEE J. Quantum Electronics, Vol. QE-2, pp. 612–616, September 1966 (20 Å wide spectral output, laser terminal level lifetime of 25 to 90 ns)

4.107 PTM Single-Pulse Selection from a Mode-Locked Nd^{3+}-Glass Laser Using a Bleachable Dye
A. W. Penney, Jr., and H. A. Heynau (United Aircraft)
Appl. Phys. Letters, Vol. 9, pp. 257–258, October 1, 1966 (pulse-transmission mode, 25 MW peak power, 0.2 J output, 0.8 ns duration)

4.108 Glass Lasers
E. Snitzer (American Optical)
Appl. Optics, Vol. 5, pp. 1487–1499, October 1966 (summary, Nd, Yb, and Er lasers, thermal effects, Faraday rotation glasses)

4.109 Temperature-Dependent Nd Fluorescence Parameters and Laser Thresholds
W. W. Holloway, Jr., M. Kestigian, F. F. Y. Wang, and G. F. Sullivan (Sperry Rand)
J. Opt. Soc. Am., Vol. 56, pp. 1409–1410, October 1966 (Nd-doped glass, also host crystals, 77 to 370°K)

4.110 Glass Lasers
E. Snitzer (American Optical)
Proc. IEEE, Vol. 54, pp. 1249–1261, October 1966 (summary, Nd, Yb, and Er lasers, thermal effects, Faraday rotation glasses)

4.111 The CW Pumping of YAG:Nd^{3+} by Water-Cooled Krypton Arcs
T. B. Read (AEI, England)
Appl. Phys. Letters, Vol. 9, pp. 342–344, November 1, 1966 (8104 Å from Kr, 400 W threshold, much greater output than from xenon or tungsten-iodine lamps)

4.112 Measurement of the Cross-section for Stimulated Emission in Neodymium Glass
J. G. Edwards (Nat'l Phys. Lab., England)
Nature, Vol. 212, pp. 752–753, November 12, 1966 (3.6 × 10^{-20} cm² at the peak of the fluorescence)

4.113 Mode Locking of a Q-Spoiled Nd^{3+} Doped Glass Laser by Intracavity Phase Modulation
M. Michon, J. Ernest, and R. Auffret (CGE, France)
Physics Letters, Vol. 23, pp. 457–458, November 14, 1966 (10-cm long KDP crystal, 150 MHz)

4.114 Optical Avalanche Laser
C. G. Young, J. W. Kantorski, and E. O. Dixon (American Optical)
J. Appl. Phys., Vol. 37, pp. 4319–4324, November 1966 (Nd-doped glass, five 20-cm and one 100-cm long rods, amplified spontaneous emission, 185-dB gain, spike free, 70-ns 1 GW output pulse, 1-mrad full-angle beam, unfocused air breakdown)

4.115 Unusual Crystal-Field Energy Levels and Efficient Laser Properties of YVO_4:Nd
J. R. O'Connor (M.I.T. Lincoln Lab.)
Appl. Phys. Letters, Vol. 9, pp. 407–409, December 1, 1966 (1 J threshold at 90°K)

4.116 Optical Absorption Spectra of Coupled Nd^{3+} Ions in $NdCl_3$ and $NdBr_3$
G. A. Prinz (Johns Hopkins U.)
Phys. Rev., Vol. 152, pp. 474–481, December 2, 1966

5. OTHER SOLID-STATE LASERS

5.31   Injection-Luminescence Pumping of a CaF_2:Dy^{2+} Laser
S. A. Ochs and J. I. Pankove (RCA)
Proc. IEEE, Vol. 52, pp. 713–714, June 1964 (He temp $GaAs_xP_{1-x}$ diodes for max 0.2 second pulse, 7200 Å pump, 2.36-μ output, cylindrical cluster of diodes)

5.32   Spontaneous and Stimulated Emission from Co^{2+} Ions in MgF_2 and ZnF_2
L. F. Johnson, R. E. Dietz, and H. J. Guggenheim (BTL)
Appl. Phys. Letters, Vol. 5, pp. 21–22, July 15, 1964 (cryogenic, 1.7–1.8-μ laser wavelengths)

5.33   Hot-Pressed Polycrystalline CaF_2:Dy^{2+} Laser
S. E. Hatch, W. F. Parsons, and R. J. Weagley (Eastman Kodak)
Appl. Phys. Letters, Vol. 5, pp. 153–154, October 15, 1964 (essentially theoretical density, 25 J at 77°K, 4 J at 4°K)

5.34   Energy Levels and Crystal-Field Calculations of Europium and Terbium in Yttrium Aluminum Garnet
J. A. Koningstein (BTL)
Phys. Rev., Vol. 136, pp. A717–A725, November 2, 1964

5.35   Energy Levels and Crystal-Field Calculations of Er^{3+} in Yttrium Aluminum Garnet
J. A. Koningstein and J. E. Geusic (BTL)
Phys. Rev., Vol. 136, pp. A726–A728, November 2, 1964

5.36 Yb^{3+}–Er^{3+} Glass Laser
E. Snitzer and R. Woodcock (American Optical)
Appl. Phys. Letters, Vol. 6, pp. 45–46, February 1, 1965 (room temp., 3-level Er laser, 1.54μ, energy transfer from Yb, 700j threshold, 14 ms lifetime)

5.37   Energy Transfer in Triply Activated Glasses
H. W. Gandy, R. J. Ginther, and J. F. Weller (NRL)
Appl. Phys. Letters, Vol. 6, pp. 46–49, February 1, 1965 (Ce and Nd transfers to Yb which emits at 0.976μ)

5.38   Energy Levels of Dy^{2+} in the Cubic Hosts of CaF_2, SrF_2 and BaF_2
Z. J. Kiss (RCA)
Phys. Rev., Vol. 137, pp. A1749–A1760, March 15, 1965

5.39   Zeeman Effect Studies of the $^5I_7$–$^5I_8$ Transition in CaF_2:Dy^{2+}
Z. J. Kiss, C. H. Anderson, and R. Orbach (RCA)
Phys. Rev., Vol. 137, pp. A1761–A1766, March 15, 1965

5.40   Stimulated Emission of Ho^{3+} in CaF_2 at $\lambda = 5512$ Å
Yu. K. Voron'ko, A. A. Kaminskiĭ, V. V. Osiko, and A. M. Prokhorov (Lebedev Physics Inst., Academy of Sciences, U.S.S.R.)
JETP Letters, Vol. 1, pp. 3–5, April 1, 1965 (1200 J threshold)

5.41   Plastic Deformation of Calcium Tungstate Single Crystals
A. Arbel and R. J. Stokes (Honeywell)
J. Appl. Phys., Vol. 36, pp. 1460–1468, April 1965 (deforms by slip, doped crystals used in laser applications)

5.42   Fluorescence of Tb^{3+} in CaF_2
N. Rabbiner (U.S. Army Electronics Lab.)
J. Opt. Soc. Am., Vol. 55, pp. 436–438, April 1965 (supports cubic crystal field symmetry)

5.43   Energy Transfer and CW Laser Action in Tm^{+3}:Er_2O_3
B. H. Soffer and R. H. Hoskins (Korad)
Appl. Phys. Letters, Vol. 6, pp. 200–201, May 15, 1965 (1.934μ, 2.9 ms lifetime, 77°K, 3j pulsed and 500 w CW thresholds)

5.44   Selective Excitation of Rare-Earth Ion Centers in Crystals
Yu. K. Voron'ko, A. A. Kaminskiĭ, V. V. Osiko, and A. M. Prokhorov (Lebedev Physics Inst., Academy of Sciences, U.S.S.R.)
JETP Letters, Vol. 1, pp. 120–123, May 15, 1965 (Er^{3+} in CaF_2)

5.45   Energy Transfer and Ho^{3+} Laser Action in Silicate Glass Coactivated with Yb^{3+} and Ho^{3+}
H. W. Gandy, R. J. Ginther, and J. F. Weller (U.S. Naval Res. Leb.)

Appl. Phys. Letters, Vol. 6, pp. 237–239, June 15, 1965 (laser output wavelength $> 1.9\ \mu$, 150 J threshold)

5.46   Absorption and Fluorescence of Trivalent Europium in Borate Glasses
P. K. Gallagher, C. R. Kurkjian, and P. M. Bridenbaugh (BTL)
Phys. and Chem. of Glasses, Vol. 6, pp. 95–103, June 1965 (about 2.5 millisec fluorescence lifetime at 0.614μ, about 30 Å fluorescence halfwidth at 0.579μ)

5.47   Stepwise Excitation of Fluorescence of $CaWO_4$ Activated with Er^{3+}
V. L. Bakumenko, A. N. Vlasov, E. S. Kovarskaya, G. S. Kozina, and V. N. Favorin
JETP Letters, Vol. 2, pp. 16–18, July 1, 1965

5.48   Measurement of Fluorescence Yields of Europic Ion upon Excitation to Selected Levels
William R. Dawson and John L. Kropp (TRW Space Tech. Labs.)
J. Opt. Soc. Am., Vol. 55, pp. 822–828, July 1965 (europium nitrate in alcohol, strong 6170 Å emission)

5.49   Effect of Y^{+3} on the Reduction of Sm^{+3} in CaF_2
J. R. O'Connor and R. M. Hilton (Lincoln Lab., M.I.T.)
Appl. Phys. Letters, Vol. 7, pp. 53–54, August 1, 1965

5.50   Quenching of Dy^{+2} Fluorescence by Y^{+2} in CaF_2:Dy^{+2} Lasers
J. R. O'Connor (Lincoln Lab., M.I.T.)
Appl. Phys. Letters, Vol. 7, pp. 54–55, August 1, 1965

5.51   UV Laser Emission by Crystal Excitons
E. L. Fink (General Dynamics)
Appl. Phys. Letters, Vol. 7, pp. 103–106, August 15, 1965 (1880 Å pump, 3150 Å from KBr)

5.52   Energy Transfer between Excited Er^{3+} Ions in the Alkaline Earth Fluorides
M. R. Brown (Signals Research and Development Estab., England) and W. A. Shand (U. of Aberdeen, Scotland)
Physics Letters, Vol. 18, pp. 95–96, August 15, 1965 (two 1.5 μ photons produce 0.98 μ fluorescence)

5.53   The Effects of Oxygen on the Properties of CaF_2 as a Laser Host
P. A. Forrester, G. W. Green, and D. F. Sampson (Royal Radar Estab., England)
British J. Appl. Phys., Vol. 16, pp. 1209–1210, August 1965 (experiments with Sm laser)

5.54   Coherent Oscillations from Tm^{3+}, Ho^{3+}, Yb^{3+} and Er^{3+} Ions in Yttrium Aluminum Garnet
L. F. Johnson, J. E. Geusic, and L. G. Van Uitert (BTL)
Appl. Phys. Letters, Vol. 7, pp. 127–129, September 1, 1965 (energy transfer among ions, 47 W cw and 11 J pulsed thresholds for Ho^{3+})

5.55   Energy Transfer in Y_2O_3:Tm^{3+}, Ho^{3+}
D. H. Brown (S.E.R.L., England)
Physics Letters, Vol. 18, pp. 281–282, September 1, 1965 (Tm^{3+} enhances 2-μ Ho^{3+} fluorescence)

5.56   Anomalous Power-Dependent Absorption in Er^{3+}-Doped Y_2O_3 Single Crystals
W. F. Krupke and E. R. Peressini (Aerospace)
J. Appl. Phys., Vol. 36, pp. 2970–2971, September 1965 (two-photon absorption process, irradiate by ruby laser, 5480 Å fluorescence)

5.57   The Spectrum and Temporal Characteristics of Stimulated Emission in CaF_2:Sm^{2+}
Yu. A. Anan'ev and B. M. Sedov
Soviet Physics JETP, Vol. 21, pp. 517–523, September 1965 (spikes, streak photographs)

5.58 Monopulse Generation with CaF$_2$:U^{3+} Crystals
B. A. Ermakov, A. V. Lukin, A. A. Mak, and D. S. Prilezhaev
JETP Letters, Vol. 2, pp. 239–240, October 15, 1965 (Q-switched by rotating prism, 80–90°K, 2.22 and 2.51 μ output wavelengths, 200 ns 10^{-3} J output)

5.59 Stimulated Emission from Er^{3+} Ions in CaF$_2$
Yu. K. Voron'ko, G. M. Zverev, and A. M. Prokhorov (Moscow State U., U.S.S.R.)
Soviet Physics JETP, Vol. 21, pp. 1023–1025, December 1965 (1.26 and 1.7 μ, N$_2$ temp, 1000 J threshold)

5.60 Optical Linewidth Studies of Energy-Transfer Mechanisms Between Impurity Ions
W. M. Yen, R. L. Greene, and W. C. Scott (Stanford U.) and D. L. Huber (U. of Wisconsin)
Proc. Physics of Quantum Electronics Conf., McGraw-Hill Book Co., New York, pp. 332–339, 1966

5.61 Multiple Optical Spectra of Trivalent Gadolinium in Crystals of the Fluorite Type
J. Makovsky (Hebrew U., Israel)
Proc. Physics of Quantum Electronics Conf., McGraw-Hill Book Co., New York, pp. 340–349, 1966

5.62 Radiative and Nonradiative Transitions of Rare-Earth Ions: Er^{3+} in LaF$_3$
M. J. Weber (Raytheon)
Proc. Physics of Quantum Electronics Conf., McGraw-Hill Book Co., New York, pp. 350–360, 1966

5.63 Fluorescence from Magnetic Crystals
R. E. Dietz, L. F. Johnson, and H. J. Guggenheim (B.T.L.)
Proc. Physics of Quantum Electronics Conf., McGraw-Hill Book Co., New York, pp. 361–369, 1966

5.64 Far-Infrared Solid-State Masers: A Speculative Account with Some Related Experiments
F. Varsanyi (B.T.L.)
Proc. Physics of Quantum Electronics Conf., McGraw-Hill Book Co., New York, pp. 370–375, 1966

5.65 On the Optical Spectra of Gd^{3+} in CaF$_2$
J. Makovsky (Hebrew U., Jerusalem)
Physics Letters, Vol. 19, pp. 647–649, January 1, 1966

5.66 Gd$_2$(MoO$_4$)$_3$: A Ferroelectric Laser Host
Hans J. Borchardt and Paul E. Bierstedt (E. I. du Pont de Nemours)
Appl. Phys. Letters, Vol. 8, pp. 50–52, January 15, 1966 (3% Nd doped, ferroelectric below 159°C, 350 J threshold at −138°C, direct modulation possible)

5.67 Solar-Pumped Modulated Laser
C. W. Reno (RCA)
RCA Review, Vol. 27, pp. 149–157, March 1966 (Dy and Nd lasers, transmitted television pictures)

5.68 Luminescence and Generation in CaF$_2$:Dy^{2+} Crystals Excited with a Ruby Laser
E. M. Zolotov, A. M. Prokhorov, and G. P. Shipulo (Lebedev Physics Institute, U.S.S.R.)
Soviet Phys. JETP, Vol. 22, pp. 498–500, March 1966 (2.36 μ output follows ruby spike pumping)

5.69 Analysis of the Optical Spectra of Pr^{3+}, Nd^{3+}, Eu^{3+} and Er^{3+} in Fluorite Crystals (Type 1) by the Concentration Series Method
Yu. K. Voron'ko, A. A. Kaminskii, and V. V. Osiko (Lebedev Physics Institute, U.S.S.R.)
Soviet Phys. JETP, Vol. 22, pp. 501–504, March 1966

5.70 Vanadium Charge Compensator Site in Laser CaWO$_4$
C. Kikuchi and N. Mahootian (U. of Michigan), and W. Viehmann and R. T. Farrar (Harry Diamond Labs.)
Proc. 8th Annual Electron and Laser Beam Symposium, April 6–8, 1966. Sponsored by U. of Michigan and IEEE, pp. 117–138.

5.71 Efficient, High-Power Coherent Emission from Ho^{3+} Ions in Yttrium Aluminum Garnet, Assisted by Energy Transfer
L. F. Johnson, J. E. Geusic, and L. G. Van Uitert (B.T.L.)
Appl. Phys. Letters, Vol. 8, pp. 200–202, April 15, 1966 (5% efficiency, 15 W out, 2.123 μ output)

5.72 Direct Observation of Sink-Terminated Concentration Quenching of Gd^{3+} Fluorescence in Calibo-1 Glass
A. David Pearson and G. E. Peterson (B.T.L.)
Appl. Phys. Letters, Vol. 8, pp. 210–212, April 15, 1966

5.73 Coherent Emission from Ho^{3+} Ions in Yttrium Iron Garnet
L. F. Johnson, J. P. Remeika, and J. F. Dillon, Jr. (B.T.L.)
Physics Letters, Vol. 21, pp. 37–39, April 15, 1966 (pulsed, 2.09-μ emission, magnetic modulation of amplitude and frequency)

5.74 Anomalous Fluorescent Decay of LiF:UO$_2$ under High-Intensity Excitation
O. Risgin and A. G. Becker (U. of Michigan)
Appl. Optics, Vol. 5, pp. 639–641, April 1966

5.75 Growth of Laser-Quality Rare-Earth Fluoride Single Crystals in a Dynamic Hydrogen Fluoride Atmosphere
M. Robinson and D. M. Cripe (Hughes)
J. Appl. Phys., Vol. 37, pp. 2072–2074, April 1966

5.76 Optical Absorption and Fluorescence Intensities in Several Rare-Earth-Doped Y$_2$O$_3$ and LaF$_3$ Single Crystals
William F. Krupke (Aerospace)
Phys. Rev., Vol. 145, pp. 325–337, May 6, 1966

5.77 Growth of Large Yttrium Vanadate Single Crystals for Optical Maser Studies
J. J. Rubin and L. G. Van Uitert (B.T.L.)
J. Appl. Phys., Vol. 37, pp. 2920–2921, June 1966 (oxy-hydrogen furnace)

5.78 Transfer of Excitation from the Crystal Lattice to Rare Earth Ions
M. E. Zhabotinskii, Yu. P. Rudnitskii, V. V. Tsapkin, and G. V. Éllert (Acad. of Sci., USSR)
Soviet Phys. JETP, Vol. 22, pp. 1155–1158, June 1966

5.79 Optical Er^{3+} Centers in Cubic Crystals of the Fluorite Type
Yu. K. Voron'ko, A. A. Kaminskii, and V. V. Osiko (Lebedev Phys. Inst., Acad. of Sci., USSR)
Soviet Phys. JETP, Vol. 23, pp. 10–15, July 1966

5.80 Energy Transfer and CW Laser Action in Ho^{3+}: Er$_2$O$_3$
R. H. Hoskins and B. H. Soffer, Korad)
IEEE J. Quantum Electronics, Vol. QE-2, pp. 253–255, August 1966 (thresholds of 5 J pulsed and 200 W CW at 77°K, 2.12 μ emission)

5.81 Coherent Emission from Ho^{3+} Ions in Ferrimagnetic Yttrium Iron Garnet (YIG)
J. F. Dillon, Jr., L. F. Johnson, and J. P. Remeika (BTL)
IEEE J. Quantum Electronics, Vol. QE-2, pp. 295–296, August 1966 (abstract, pulsed, near 2.09 μ)

5.82 Phonon-Terminated Optical Masers
L. F. Johnson, H. J. Guggenheim, and R. A. Thomas (BTL)
Phys. Rev., Vol. 149, pp. 179–185, September 9, 1966 (thermally tunable, Ni^{2+}, Co^{2+}, and V^{2+} ions in rutile and perovskite fluorides, about 85°K, CW threshold power of 65 W into tungsten lamp, continuous spiking, discontinuous output wavelength range from 1.12 to 2.16 μ)

5.83 CaF$_2$:Sm^{2+} Laser with Ruby Laser Excitation
V. K. Koniukhov, V. M. Marchenko, and A. M. Prokhorov

(Lebedev Phys. Inst., Acad. of Sci., USSR)
IEEE J. Quantum Electronics, Vol. QE-2, pp. 541–542, September 1966 (emission at 708, 720, and 729 nm, 20 percent efficient, axial direction pumping through mirror)

5.84 Laser Emission at 1.06 μ from Nd^{3+}-Yb^{3+} Glass
E. Snitzer (American Optical)
IEEE J. Quantum Electronics, Vol. QE-2, pp. 562–566, September 1966 (both Nd and Yb can lase at 1.06 μ)

5.85 The Nature of the Laser Transition in CdS Crystal at 90°K with Two-Photon Excitation
L. A. Kulewsky and A. M. Prokhorov (Lebedev Phys. Inst., Acad. of Sci., USSR)
IEEE J. Quantum Electronics, Vol. QE-2, pp. 584–586, September 1966 (ruby laser pump, 4946 Å output)

5.86 CaF_2Dy^2 Giant Pulse Laser with High Repetition Rate
V. V. Kostin, L. A. Kulevsky, T. M. Murina, A. M. Prokhorov, and A. A. Tikhonov (Lebedev Phys. Inst., Acad. of Sci., USSR)
IEEE J. Quantum Electronics, Vol. QE-2, pp. 611–612, September 1966 (rotating mirror, continuous discharge xenon lamp, 2.36-μ 1-MW output, 500 pps)

5.87 Energy Transfer between the Sm^{3+}, Eu^{3+}, Tb^{3+}, and Dy^{3+} Ions in Sodium Rare-Earth Tungstates
W. W. Holloway, Jr., and M. Kestigian (Sperry Rand)
J. Opt. Soc. Am., Vol. 56, pp. 1171–1174, September 1966

5.88 Cr^{3+} and Rare Earth Doped $LiNbO_3$
G. Burns, D. F. O'Kane, and R. S. Title (IBM Yorktown)
Physics Letters, Vol. 23, pp. 56–57, October 3, 1966 (also Nd doped, sharp fluorescence)

5.89 Internal Q Switching of Ho^{3+}-Stimulated Emission in Iron-Containing Glasses
H. W. Gandy, R. J. Ginther, and J. F. Weller (NRL)
Appl. Phys. Letters, Vol. 9, pp. 277–279, October 15, 1966 (2.08μ output, 77°K)

5.90 Transition Intensities Between Excited States of Rare Earth Ions in Crystals
William F. Krupke (Aerospace)
IEEE J. Quantum Electronics, Vol. QE-2, pp. 698–699, October 1966

6. He–Ne GAS LASER

6.1 Isolation of Axi-symmetrical Optical-Resonator Modes
W. W. Rigrod (BTL)
Appl. Phys. Letters, Vol. 2, pp. 51–53, February 1, 1963

6.2 Frequency Shifts in Light Diffracted by Ultrasonic Waves in Liquid Media
H. Cummins, N. Knable, L. Gampel, and Y. Yeh (Columbia U.)
Appl. Phys. Letters, Vol. 2, pp. 62–64, February 1, 1963 (Note: Erratum on p. 90)

6.3 Rotation Rate Sensing with Traveling-Wave Ring Lasers
W. M. Macek and D. T. M. Davis, Jr. (Sperry Gyroscope)
Appl. Phys. Letters, Vol. 2, pp. 67–68, February 1, 1963

6.4 Output Power of the 6328 Å Gas Maser
A. D. White, E. I. Gordon, and J. D. Rigden (BTL)
Appl. Phys. Letters, Vol. 2, pp. 91–93, March 1, 1963

6.5 Observation of New Visible Gas Laser Transitions by Removal of Dominance
Arnold L. Bloom (Spectra-Physics)
Appl. Phys. Letters, Vol. 2, pp. 101–102, March 1, 1963

6.6 Diffraction Studies with Plane-Parallel Maser Interferometer
W. W. Rigrod and A. J. Rustako, Jr. (BTL)
J. Appl. Phys., Vol. 34, pp. 967–968, Part I, April 1963

6.7 Impurity Effects in a He–Ne Laser
J. K. Powers and B. W. Harned (Philco)
Proc. IEEE, Vol. 51, pp. 605–606, April 1963

6.8 Light Scattering from Dielectric Film Laser Mirrors
Walter A. Specht, Jr. (Calif. Inst. of Tech.)
Proc. IEEE, Vol. 51, pp. 615–616, April 1963

6.9 Broadband Magnetic Field Tuning of Optical Masers
Richard L. Fork and C. K. N. Patel (BTL)
Appl. Phys. Letters, Vol. 2, pp. 180–181, May 1, 1963

6.10 Single Mode Tuning Dip in the Power Output of an He–Ne Optical Maser
R. A. McFarlane (BTL), W. R. Bennett, Jr., and W. E. Lamb, Jr. (Yale U.)
Appl. Phys. Letters, Vol. 2, pp. 189–190, May 15, 1963

6.11 The Effect of Superradiance at 3.39 μ on the Visible Transitions in the He–Ne Maser
A. D. White and J. D. Rigden (BTL)
Appl. Phys. Letters, Vol. 2, pp. 211–212, June 1, 1963 (0.59 μ laser reported)

6.12 Isotope Shift and Saturation Behavior of the 1.15-μ Transition of Ne
A. Szöke and A. Javan (MIT)
Phys. Rev. Letters, Vol. 10, pp. 521–524, June 15, 1963

6.13 Direct Modulation of a He–Ne Gas Laser
E. J. Schiel and J. J. Bolmarcich (U.S. Army)
Proc. IEEE, Vol. 51, pp. 940–941, June 1963

6.14 The Interaction of Visible and Infrared Maser Transitions in the Helium–Neon System
J. D. Rigden and A. D. White (BTL)
Proc. IEEE, Vol. 51, pp. 943–945, June 1963 (predicts 50 db/m at 3.39 μ)

6.15 Construction of a Gaseous Optical Maser Using Brewster Angle Windows
K. M. Baird, M. J. Taylor, and R. Turner (National Res. Council, Canada)
Rev. Sci. Instr., Vol. 34, p. 697, June 1963

6.16 Observation of Continuous Optical Harmonic Generation with Gas Masers
A. Ashkin, G. D. Boyd, and J. M. Dziedzic (BTL)
Phys. Rev. Letters, Vol. 11, pp. 14–17, July 1, 1963

6.17 Continuous Optical-Harmonic Generation
N. I. Adams and P. B. Schoefer (Perkin-Elmer)
Appl. Phys. Letters, Vol. 3, pp. 19–21, July 15, 1963

6.18 Some Experiments on Parallel Gas Lasers within a Common Optical Cavity
R. A. Paananen and A. Adams, Jr. (Raytheon)
Proc. IEEE, Vol. 51, pp. 1036–1037, July 1963

6.51 Neon Gas Maser Lines at 68.329 μ and 85.047 μ
      R. A. McFarlane, W. L. Faust, C. K. N. Patel, and C. G. B.
      Garrett (BTL)
           Proc. IEEE, Vol. 52, p. 318, March 1964

6.52 New He–Ne and Ne Laser Lines in the Infrared
      Vern N. Smiley (U.S. Navy Electr. Lab.)
      Appl. Phys. Letters, Vol. 4, pp. 123–124, April 1, 1964 (4 lines
      from 1.83 to 2.4 μ)

6.53 Noise Measurement in an He–Ne Laser Amplifier
      R. A. Paananen, H. Statz, D. L. Bobroff, and A. Adams, Jr.
      (Raytheon)
      Appl. Phys. Letters, Vol. 4, pp. 149–151, April 15, 1964

6.54 Erratum: Noise Measurement in an He–Ne Laser Amplifier
      R. A. Paananen, H. Statz, D. L. Bobroff, and A. Adams, Jr.
      (Raytheon)
      Appl. Phys. Letters, Vol. 4, p. 184, May 15, 1964

6.55 Interferometric Investigation of Modes in Optical Gas Masers
      Thomas G. Polanyi and William R. Watson (G. T. & E. Labs.)
      J. Opt. Soc. Am., Vol. 54, pp. 449–454, April 1964 (most c/2L
      and c/4L modes observed in 3 Gc fluorescent linewidth)

6.56 Microwave Determination of Average Electron Energy and
      Density in He–Ne Discharges
      E. F. Labuda and E. I. Gordon (BTL)
      J. Appl. Phys., Vol. 35, pp. 1647–1648, May 1964 (electron
      density about $10^{11}/cm^3$)

6.57 Verdet Constant of the "Active Medium" in a Laser Cavity
      I. Tobias (Rutgers U.) and R. A. Wallace (G.T.&E.)
      Phys. Rev., Vol. 134, pp. A549–A552, May 4, 1964 (axial mag-
      netic field, 0.633-μ He–Ne laser, 2.8-kc/s per Oe)

6.58 Analysis of Combination Tones in a Short Gas Laser
      W. J. Witteman and J. Haisma (Philips Res. Labs., Netherlands)
      Phys. Rev. Letters, Vol. 12, pp. 617–619, June 1, 1964 (hole
      burning, anomalous dispersion)

6.59 CW Optical Maser Action up to 133 μ (0.133 mm) in Neon Dis-
      charges
      C. K. N. Patel, W. L. Faust, R. A. McFarlane, and C. G. B.
      Garrett (BTL)
      Proc. IEEE, Vol. 52, p. 713, June 1964 (14 lines, 50.7 to 132.8
      microns, max 5 nW at 88.5 microns)

6.60 Anomalous Behavior of the 6402.84 Å Gas Laser
      A. D. White (BTL)
      Proc. IEEE, Vol. 52, p. 721, June 1964 (less output power with
      plasma longer than one meter)

6.61 The Amplitude Fluctuations of Optical Maser Light
      R. L. Bailey and J. H. Sanders (Oxford U., England)
      Physics Letters, Vol. 10, pp. 295–296, June 15, 1964 (upper limit
      for HeNe at 0.63 μ)

6.62 Locking of He–Ne Laser Modes Induced by Synchronous Intra-
      cavity Modulation
      L. E. Hargrove, R. L. Fork, and M. A. Pollack (BTL)
      Appl. Phys. Letters, Vol. 5, pp. 4–5, July 1, 1964 (56-Mc/s
      synchronous modulation)

6.63 Population Pulsations and Lifetimes in He–Ne Lasers
      R. L. Fork, L. E. Hargrove, and M. A. Pollack (BTL)
      Appl. Phys. Letters, Vol. 5, pp. 5–7, July 1, 1964

6.64 Measurement of He–Ne Laser Linewidth Utilizing the Doppler
      Effect
      P. J. Magill and T. Young (IBM)
      Appl. Phys. Letters, Vol. 5, pp. 13–15, July 1, 1964 (linewidth
      less than 25 c/s)

6.65 Line Strengths for Noble-Gas Maser Transitions; Calculations of
      Gain/Inversion at Various Wavelengths
      W. L. Faust and R. A. McFarlane (BTL)
      J. Appl. Phys., Vol. 35, pp. 2010–2015, July 1964 (α ranges from
      5 × 10^{-4} to 0.16 (Xe) cm^{-1})

6.66 Laser Oscillation at 1.06 Microns in He–Ne
      R. McClure, R. Pizzo, M. Schiff, and C. B. Zarowin (Sperry
      Gyroscope)
      Proc. IEEE, Vol. 52, p. 851, July 1964

6.67 Simultaneous Laser Oscillation on the Neon Doublet at 1.1523μ
      W. R. Bennett, Jr. and J. W. Knutson, Jr. (Yale U.)
      Proc. IEEE, Vol. 52, pp. 861–862, July 1964 (lines 2.26 Å (51
      Gc) apart, speculate usefulness in long distance interferometry
      and light velocity determination)

6.68 Cross-Correlation Between Discharge Current Noise and Laser
      Light Noise in He–Ne Lasers
      L. J. Prescott and A. van der Ziel (U. of Minnesota)
      Appl. Phys. Letters, Vol. 5, pp. 48–49, August 1, 1964

6.69 Dispersion Characteristics and Frequency Stabilization of an He-
      Ne Gas Laser
      W. R. Bennett, Jr. (Yale U.), S. F. Jacobs, J. T. LaTourrette,
      and P. Rabinowitz (TRG)
      Appl. Phys. Letters, Vol. 5, pp. 56–58, August 1, 1964 (10^{-8} in
      eight hours at 3.39 μ, ±60 kc deviation near peak output)

6.70 Lifetime of Helium-Neon Lasers
      R. Turner, K. M. Baird, M. J. Taylor, and C. J. van der Hoeven
      (Natl. Res. Council, Canada)
      Rev. Sci. Instr., Vol. 35, pp. 996–1001, August 1964 (cleanup
      problem, over one-thousand hours attained)

6.71 Frequency Pushing and Frequency Pulling in a He-Ne Gas Optical
      Maser
      R. A. McFarlane (BTL)
      Phys. Rev., Vol. 135, pp. A543–A550, August 3, 1964 (axial beats,
      hole repulsion)

6.72 Frequency Stabilization of Single Mode Gas Lasers
      A. D. White, E. I. Gordon, and E. F. Labuda (BTL)
      Appl. Phys. Letters, Vol. 5, pp. 97–98, September 1, 1964 (split
      laser transition profile for discriminator, magnetic field or iso-
      topes)

6.73 New Ne Laser Transitions in Near Infrared
      R. der Agobian, J. L. Otto, R. Cagnard, and R. Echard (Lab.
      Central de Telecommunications, France)
      J. Appl. Phys., Vol. 35, p. 2787, September 1964

6.74 2s-2p and 3p-2s Transitions of Neon in a Laser Ten Meters Long
      R. N. Zitter (BTL)
      J. Appl. Phys., Vol. 35, pp. 3070–3071, October 1964 (many new
      lines between 0.8865 and 1.7162 μ; also 1.0621 μ)

6.75 Detection of Spontaneous Emission Noise in He-Ne Lasers
      L. J. Prescott and A. van der Ziel (U. of Minnesota)
      Physics Letters, Vol. 12, pp. 317–319, October 15, 1964 (negligi-
      ble above 100 kc away)

6.76 Gas Lasers in Magnetic Fields
      S. A. Ahmed, R. C. Kocher, and H. J. Gerritsen (RCA)
      Proc. IEEE, Vol. 52, pp. 1356–1357, November 1964 (output
      can be enhanced by axial field)

6.77 Stability and Amplitude-Modulation Characteristic of the Dis-
      charge Current, in a DC-Pumped Helium-Neon Gas Laser Tube
      V. Met (Watkins-Johnson)
      Proc. IEEE, Vol. 52, pp. 1357–1358, November 1964

6.78 Frequency Splitting and Mode Competition in a Dual-Polarization
      He-Ne Gas Laser
      W. M. Doyle and M. B. White (Philco)
      Appl. Phys. Letters, Vol. 5, pp. 193–195, November 15, 1964
      (birefringent element in laser cavity, variable aperture effects)

6.79 FM Oscillation of the He-Ne Laser
      S. E. Harris and R. Targ (Sylvania)
      Appl. Phys. Letters, Vol. 5, pp. 202–204, November 15, 1964
      (KDP crystal inside laser cavity)

6.80 FM Laser Oscillation—Theory
S. E. Harris and O. P. McDuff (Stanford U.)
Appl. Phys. Letters, Vol. 5, pp. 205–206, November 15, 1964

6.81 Hanle Effects in the He–Ne Laser
W. Culshaw and J. Kannelaud (Lockheed)
Phys. Rev., Vol. 136, pp. A1209–A1221, November 30, 1964 (depolarization of resonance radiation in a weak magnetic field, 1.15-μ wavelength)

6.82 Moving Striations in a He–Ne Laser
A. Garscadden, P. Bletzinger (Aerospace), and E. M. Friar (WPAFB)
J. Appl. Phys., Vol. 35, pp. 3432–3433, December 1964 (striation frequency depends on current)

6.83 Microwave Electron Cyclotron Resonance Pumping of a Gas Laser
S. A. Ahmed and R. Kocher (RCA)
Proc. IEEE, Vol. 52, pp. 1737–1738, December 1964 (He–Ne, 2.45-Gc magnetron)

6.84 Cross Modulation of Gaseous Lasers by High-Intensity Light
B. Pariser and T. C. Marshall (Columbia U.)
Proc. IEEE, Vol. 52, pp. 1740–1741, December 1964 (flash lamp with filters increases or decreases output)

6.85 Gain Saturation and a Method of Specifying Output Power Characteristics of a Laser Tube
T. Takahashi, T. Shigematsu, and K. Kakizaki (Toshiba, Japan)
Proc. IEEE, Vol. 52, pp. 1741–1742, December 1964 (experiments on He–Ne Laser)

6.86 Optical Gain in Neon and Helium/Neon Pulsed Discharges
D. M. Clunie and N. H. Rock (England)
Physics Letters, Vol. 13, pp. 213–214, December 1, 1964 (multipass system, 3.6 dB/m from Ne, 30 watts saturated power from 1.15-μ He–Ne)

6.87 Further Measurements on the Noise of a D.C. Excited He–Ne Laser Oscillator
P. T. Bolwijn (State U., Netherlands)
Physics Letters, Vol. 13, pp. 311–312, December 15, 1964 (less than 100-kc width)

6.88 Inversion Mechanisms in Gas Lasers
W. R. Bennett, Jr. (Yale U.)
Appl. Optics, Suppl. 2: Chemical Lasers, pp. 3–33, 1965 (comprehensive, known gas laser transitions tabulated, gases and molecules)

6.89 Laser Mode Locking by an External Doppler Cell
L. C. Foster, M. D. Ewy, and C. B. Crumly (Zenith)
Appl. Phys. Letters, Vol. 6, pp. 6–8, January 1, 1965 (He–Ne, 57 Mc/s sonic quartz cell)

6.90 Asymmetric Visible Super-Radiant Emission from a Pulsed Neon Discharge
D. M. Clunie, R. S. A. Thorn, and K. E. Trezise (S.E.R.L., England)
Physics Letters, Vol. 14, pp. 28–29, January 1, 1965 (more 5400 Å emission from ungrounded end of discharge tube)

6.91 Frequency Spectra of He–Ne Optical Masers with External Concave Mirrors
Teiji Uchida (Nippon Elec., Japan)
Appl. Optics, Vol. 4, pp. 129–131, January 1965 (beating of axial and transverse modes)

6.92 A Helium-Neon Laser Amplifier
L. E. S. Mathias and N. H. Rock (S.E.R.L., England)
Appl. Optics, Vol. 4, pp. 133–135, January 1965 (multiple non-intersecting passes, 16 db gain in 68.4 m path length at 1.15 μ, 20 mw max output)

6.93 Resonant Energy Transfer Studies in a Helium-Neon Gas Discharge
J. T. Massey, A. G. Schulz, B. F. Hochheimer, and S. M. Cannon (Johns Hopkins U.)
J. Appl. Phys., Vol. 36, pp. 658–659, February 1965

6.94 Super-Radiant Neon Transitions
H. G. Heard (h nu systems)
Proc. IEEE, Vol. 53, p. 173, February 1965 (reclassification of transitions)

6.95 Relation between the Transient Laser Oscillations and the Plasma Quantities in a He–Ne Gaseous Laser
K. Toyoda and C. Yamanaka (Osaka U., Japan)
Japanese J. Appl. Phys., Vol. 4, pp. 226–227, March 1965 (enhanced 1.15μ output during afterglow)

6.96 Optical Resonator Effects on the Population Distribution in He–Ne Gas Lasers Determined from Side Light Measurements
A. L. Waksberg and A. I. Carswell (RCA Victor, Canada)
Appl. Phys. Letters, Vol. 6, pp. 137–138, April 1, 1965 (chopped laser at 0.63μ, compared with side emission and wavelength, varied pressure)

6.97 Locking of He–Ne Laser Modes by Intracavity Acoustic Modulation in Coupled Interferometers
M. DiDomenico, Jr., and V. Czarniewski (BTL)
Appl. Phys. Letters, Vol. 6, pp. 150–152, April 15, 1965 (triple mirror Fabry-Perot)

6.98 Characteristics of Mode-Coupled Lasers
M. H. Crowell (BTL)
IEEE J. Quantum Electronics, Vol. QE-1, pp. 12–20, April 1965 (modes in a gas laser, time-varying loss, operates like a pulse regenerative oscillator, He–Ne and argon ion lasers)

6.99 Vacuum U. V. Measurements of Helium-Neon Laser Discharge
N. Suzuki (Shimadzu Seisakusho, Japan)
Japanese J. Appl. Phys., Vol. 4, pp. 285–291, April 1965 (6328 Å laser emission reduced by 600 Å band emission from He molecule, pulsed laser more efficient with pulses shorter than 10μsec)

6.100 Mode Selection and Mode Volume Enhancement in a Gas Laser with Internal Lens
T. Li and P. W. Smith (BTL)
Proc. IEEE, Vol. 53, pp. 399–400, April 1965 ("cat's eye" configuration, HeNe laser)

6.101 Spectral Characteristics of a Gas Laser with Traveling Wave
S. N. Bagaev, V. S. Kuznetsov, Yu. V. Troitskiĭ, and B. I. Troshin (Inst. of Semiconductor Physics, Siberian Div., Academy of Sciences, U.S.S.R.)
JETP Letters, Vol. 1, pp. 114–116, May 15, 1965 (0.63 μ wavelength, triangular resonator)

6.102 Gas Mixtures and Pressures for Optimum Output Power of rf-Excited Helium-Neon Gas Lasers at 632.8 nm
K. D. Mielenz and K. Nefflen (NBS)
Appl. Optics, Vol. 4, pp. 565–567, May 1965

6.103 Characteristics of a Simple Single-Mode He–Ne Laser
K. M. Baird, D. S. Smith, G. R. Hanes, and S. Tsunekane (National Research Labs., Ottawa, Canada)
Appl. Optics, Vol. 4, pp. 569–571, May 1965 (14 cm between mirrors, 50μ watts output)

6.104 Relationship of Electron Parameters to External Electric Parameters in a Radio-Frequency-Excited Helium–Neon Discharge
J. T. Massey, A. G. Schulz, S. M. Cannon, and B. F. Hochheimer (Johns Hopkins U.)
J. Appl. Phys., Vol. 36, pp. 1790–1791, May 1965

6.105 Time Development of a Laser Signal
Bertram Pariser and Thomas C. Marshall (Columbia U.)
Appl. Phys. Letters, Vol. 6, pp. 232–234, June 15, 1965 (0.63-μ He–Ne laser output quenched by optical pumping, 100 to 300 μs to recover)

6.106 Mode Degeneracy-Dips on Output of Gas Laser
Masanobu Yamanaka, Mamoru Nakasuji, Yoshihiro Ohtsuka, and Hiroshi Yoshinaga (Osaka U., Japan)
Japanese J. Appl. Phys., Vol. 4, pp. 548–549, July 1965 (He–Ne laser, transverse mode interactions)

6.107 Calculation of Average Electron Energies in He–Ne Discharges
R. T. Young (Harry Diamond Labs.)
J. Appl. Phys., Vol. 36, pp. 2324–2325, July 1965 (between 60 000 and 120 000°K, cooler with larger diameter discharge)

6.108 Analysis of Gas Pressure and Composition in a Helium-Neon Laser Discharge
R. Turner and C. J. Van der Hoeven (National Research Council, Canada)
Rev. Sci. Instr., Vol. 36, pp. 1003–1005, July 1965 (output power, lifetime of laser)

6.109 Dispersion Characteristics of the 1.15-μ He–Ne Laser Line
H. S. Boyne, M. M. Birky, and W. G. Schweitzer, Jr. (Nat. Bur. Std., Washington)
Appl. Phys. Letters, Vol. 7, pp. 62–65, August 1, 1965 (perturbation by nearby neon transition)

6.110 Modulation of a Helium-Neon Laser by the Light of a Helium Discharge
J. F. Delpech (U. of Paris, France)
Electronics Letters, Vol. 1, pp. 168–169, August 1965

6.111 Characteristics of a Pulsed High-Pressure He–Ne Laser
S. Kobayashi, H. Okamoto, and M. Kamiyama (U. of Tokyo, Japan)
IEEE J. Quantum Electronics, Vol. QE-1, pp. 222–223, August 1965 (8 torr stops CW output, high power at beginning of pulsed laser, 1.15 μ)

6.112 Observation of a Superradiant Self-Terminating Green Laser Transition in Neon
Donald A. Leonard, Richard A. Neal, and Edward T. Gerry (Avco)
Appl. Phys. Letters, Vol. 7, p. 175, September 15, 1965 (5401 Å, pulsed, 1 kW output during 5 ns)

6.113 A Stable, Single-Frequency RF-Excited Gas Laser at 6328 Å
J. A. Collinson
Bell Sys. Tech. J., Vol. 44, pp. 1511–1519, September 1965 (1.6 mW, RF excitation, 6-inch-long cavity)

6.114 Detailed Experiments on Helium-Neon FM Lasers
E. O. Ammann, B. J. McMurtry, and M. K. Oshman (Sylvania)
IEEE J. Quantum Electronics, Vol. QE-1, pp. 263–272, September 1965

6.115 Enhanced Lasing of the High Pressure He–Ne Laser
K. Toyoda and C. Yamanaka (Osaka U., Japan)
IEEE J. Quantum Electronics, Vol. QE-1, pp. 281–283, September 1965 (higher outputs with 700–900 μs between pulses)

6.116 Infrared Laser Interferometry Utilizing Quantum-Electronic Cross Modulation
R. J. Freiberg, L. A. Weaver, and J. T. Verdeyen (U. of Illinois)
J. Appl. Phys., Vol. 36, p. 3352, October 1965 (detection of cross-modulation sidelight at visible wavelength)

6.117 On the Dependence of He–Ne Laser Discharge Current on Laser Action
G. Schiffner and F. Seifert (Technische Hochschule, Vienna, Austria)
Proc. IEEE, Vol. 53, pp. 1657–1658, October 1965 (current increases when laser beam is interrupted)

6.118 Cold Cathodes for Possible Use in 6328 Å Single Mode He–Ne Gas Lasers
U. Hochuli and P. Haldemann (U. of Maryland)
Rev. Sci. Instr., Vol. 36, pp. 1493–1494, October 1965 (aluminum cathodes, lifetimes over 3000 hours, 50 cm³ gas volume)

6.119 The Effect of Saturation on the Ellipticity of Modes in Gas Lasers
D. Polder and W. Van Haeringen (Philips, The Netherlands)
Physics Letters, Vol. 19, pp. 380–381, November 15, 1965

6.120 Single Mode Tuning Dip in the Modulated Power Output of Gas Lasers
P. T. Bolwijn (State U. Utrecht, The Netherlands)
Physics Letters, Vol. 19, pp. 384–385, November 15, 1965 (study of saturation behavior)

6.121 The Measurement of Magnetically Induced Mode Splitting in Lasers
M. L. Skolnick and T. G. Polanyi (Laser, Inc.), and I. Tobias (Rutgers)
Physics Letters, Vol. 19, pp. 386–387, November 15, 1965 (linear splitting of about 2 kc/O.)

6.122 Study of Helium-Neon Laser Amplification at 3.39 μ
G. K. Moeller and T. K. McCubbin, Jr. (Pennsylvania State U.)
Appl. Optics, Vol. 4, pp. 1412–1415, November 1965 (power gain, input signal power, discharge current, mixture ratio, total pressure)

6.123 Electron Energy Spectra in Neon, Xenon, and Helium-Neon Laser Discharges
J. Y. Wada and Hans Heil (Hughes)
IEEE J. Quantum Electronics, Vol. QE-1, pp. 327–335, November 1965

6.124 Higher Order Calculation of the Lamb Dip in the Output of an Optical Maser
Kiyoji Uehara and Koichi Shimoda (U. of Tokyo, Tokyo, Japan)
Japanese J. Appl. Phys., Vol. 4, pp. 921–927, November 1965 (shallower dip by including fifth order, Doppler limit)

6.125 Simple Improvement of Amplitude Stability in Helium-Neon Gas-Lasers
Viktor Met (Electro Optics Associates)
Proc. IEEE, Vol. 53, pp. 1780–1781, November 1965 (oscillating laser cavity mirror)

6.126 Quasi-Stationary Polarization of a Single Mode Gas Laser in a Magnetic Field
H. De Lang and G. Bouwhuis (Philips Research Labs., The Netherlands)
Physics Letters, Vol. 19, pp. 481–482, December 1, 1965 (1.15-μ He–Ne laser)

6.127 Saturation-Induced Anisotropy in a Gaseous Medium in Zero Magnetic Field
H. De Lang, G. Bouwhuis, and E. T. Ferguson (Philips Research Labs., The Netherlands)
Physics Letters, Vol. 19, pp. 482–484, December 1, 1965 (1.15-μ He–Ne gas laser with external anisotropic reflector)

6.128 Angular Beam Width of He–Ne Laser Radiation
D. C. W. Morley, L. Allen, and D. C. G. Jones (U. of Sussex, England)
Physics Letters, Vol. 19, pp. 484–486, December 1, 1965

6.129 Isotope Shift Measurement for 6328 Å He–Ne Laser Transition
R. H. Cordover, T. S. Jaseja, and A. Javan (M.I.T.)
Appl. Phys. Letters, Vol. 7, pp. 322–324, December 15, 1965 (875 Mc/s shift due to Ne²²)

6.130 Electrodeless Excitation of He–Ne Gas Lasers
D. S. Smith, K. M. Baird, and W. E. E. Berger (National Research Council, Canada)
Appl. Optics, Vol. 4, pp. 1673–1674, December 1965 (application to voice transmission)

Pratesi (Istituto di Fisica Superiore dell'Università, Firenze, Italy)
Nuovo Cimento, Vol. 43 B, pp. 150–165, May 11, 1966 (square wave modulated, 0.63 and 1.15 μ enhancements about same)

6.158 Observation of CW Intermodulation Effect in the 3.39-μ He–Ne Laser
Donald H. Close (Hughes)
Appl. Phys. Letters, Vol. 8, pp. 300–303, June 1, 1966

6.159 Approximate Determination of the Effective Cross Section for Excitation Transfer in a Two-Component Gas Mixture
L. Colombo, B. Marković, Z. Pavlović, and A. Peršin (Inst. Ruder Bošković, Jugoslavia)
J. Opt. Soc. Am., Vol. 56, pp. 890–893, July 1966 (He–Ne laser, varied pressures)

6.160 Laser Differential Spectrometry Measurement on Neon Depolarization
Th. Hänsch and P. Toschek (Heidelberg U., Germany)
Physics Letters, Vol. 22, pp. 150–151, August 1, 1966

6.161 Simple Alignment Procedure for a Helium-Neon Laser
J. F. Delpech (Institut d'Electronique Fondamentale, Faculté des Sciences, France)
Electronics Letters, Vol. 2, pp. 302–303, August 1966

6.162 The Optical Pumping of He and He–Ne Discharges
V. P. Chebotayev and V. N. Lisiczin (Inst. Phys. of Semi-Conductors, Novosibirski, USSR)
IEEE J. Quantum Electronics, Vol. QE-2, p. 295, August 1966 (abstract)

6.163 Characteristics of a Single-Frequency Michelson-Type He–Ne Gas Laser
M. DiDomenico, Jr. (BTL)
IEEE J. Quantum Electronics, Vol. QE-2, pp. 311–322, August 1966 (gain in both branches of Michelson interferometer)

6.164 Measurements of Argon Single-Frequency Laser Power and the 6328-Å Neon Isotope Shift using an Interferometer Laser
Peter Zory (Sperry Gyroscope)
J. Appl. Phys., Vol. 37, pp. 3643–3644, August 1966

6.165 Characteristics of Pulsed Laser Action in Helium-Neon and Helium-Argon Mixtures
E. I. Shtyrkov and E. V. Subbes
Optics and Spectroscopy, Vol. 21, pp. 143–144, August 1966 (over 200 Torr)

6.166 The Output Power of a Gas Laser Using Nearly Confocal Resonators
N. I. Kaliteevskii, M. M. Popov, Yu. A. Rymarchuk, T. B. Tolchinskaya, and M. P. Chaika
Optics and Spectroscopy, Vol. 21, pp. 152–154, August 1966 (varied resonator length)

6.167 Zeeman Laser Interferometer
J. A. Dahlquist, D. G. Peterson, and W. Culshaw (Lockheed)
Appl. Phys. Letters, Vol. 9, pp. 181–183, September 1, 1966 (He–Ne laser, left- and right-hand circularly polarized modes, detect target motions)

6.168 Magnetic Depolarization of a Vapor and Polarization of a Zeeman Laser
G. Durand (Harvard U.)
IEEE J. Quantum Electronics, Vol. QE-2, pp. 448–455, September 1966 (neon laser experiments)

6.169 Double Resonance Gas Laser Spectroscopy in Neon
T. O. Carroll and G. J. Wolga (Cornell U.)
IEEE J. Quantum Electronics, Vol. QE-2, pp. 456–460, September 1966

6.170 Information on Saturation and Gain of Gas Lasers from Modulation Experiments
P. T. Bolwijn (Utrecht State U., Netherlands)
IEEE J. Quantum Electronics, Vol. QE-2, pp. 670–671, September 1966 (1.15-μ He–Ne laser)

6.171 Comment on "Enhanced Lasing on the High Pressure He–Ne Laser"
P. Burlamacchi (Centro Microonde, Italy), and R. Pratesi (U. of Florence, Italy)
IEEE J. Quantum Electronics, Vol. QE-2, pp. 671–673, September 1966

6.172 Effect of Axial Modes on Doppler Experiments with Gas Lasers
P. T. Bolwijn, Th. H. Peek, and C. Th. J. Alkemade (Utrecht State U., Netherlands)
Physics Letters, Vol. 23, pp. 88–90, October 3, 1966 (moving mirror, power output modulation, He–Ne laser)

6.173 The Onset of Oscillation in a He–Ne Laser
L. Allen, D. G. C. Jones, and M. D. Sayers (U. of Sussex, England)
IEEE J. Quantum Electronics, Vol. QE-2, pp. 690–692, October 1966 (about 100 μs risetime, function of gain)

6.174 Construction of Single-Mode D.C. Operated He/Ne Lasers
J. V. Ramsay and K. Tanaka (Nat'l Stds. Lab., Australia)
Japanese J. Appl. Phys., Vol. 5, pp. 918–923, October 1966

6.175 Tunable Infrared Maser Spectrometers
Katsumi Sakurai and Koichi Shimoda (U. of Tokyo, Japan)
Japanese J. Appl. Phys., Vol. 5, pp. 938–947, October 1966 (magnetically tunable, He–Xe at 3.5μ, He–Ne at 3.39μ, tuned up to ±4000 MHz)

6.176 Isotope Shift of Neon in the 3.39 μm Maser Transition
Katsumi Sakurai, Yoshifumi Ueda, Michio Takami, and Koichi Shimoda (U. of Tokyo, Japan)
J. Phys. Soc. Japan, Vol. 21, p. 2090, October 1966 (about 63 MHz shift in laser frequency)

6.177 Dependences of Output Powers on the Discharge Current in Single-Line and Two-Line Oscillations of He–Ne Lasers
Shigeo Asami (Nat'l Research Lab. of Metrology, Japan)
Japanese J. Appl. Phys., Vol. 5, pp. 1075–1083, November 1966

6.178 Transient Behavior of He–Ne Lasers
Misao Ohi (Nat'l Research Lab. of Metrology, Japan)
Japanese J. Appl. Phys., Vol. 5, pp. 1084–1093, November 1966

6.179 Influence of Plasma-Tube-Surface Interaction on the Helium-Neon Laser Lifetime
J. V. Martinez (Xerox)
J. Appl. Phys., Vol. 37, pp. 4477–4483, November 1966 (rf-excited laser, fused silica tube decomposes, eventually prevents laser action)

6.180 Saturation and Gain of Gas Lasers from Modulation Experiments
P. T. Bolwijn (Utrecht State U., Netherlands)
J. Appl. Phys., Vol. 37, pp. 4487–4492, November 1966 (1.15-μ He–Ne laser, excitation density modulation)

6.181 Coupling of High Peak Power Pulses from He–Ne Lasers
William H. Steier (BTL)
Proc. IEEE, Vol. 54, pp. 1604–1606, November 1966 (KDP modulator within laser cavity, 80 ns 80 mW output pulse, repetition rate up to 100 kHz)

6.182 Low-Frequency Oscillations in a He–Ne Laser
N. Konjević and K. R. Hearne (U. of Liverpool, England)
Electronics Letters, Vol. 2, p. 461, December 1966 (dc laser, 30 to 100 kHz, current density fluctuations reduced by magnetic field at cathode)

7. OTHER GAS LASERS

J. Appl. Phys., Vol. 35, pp. 1348–1349, April 1964 (pure Hg laser at 1.53 to 5.85 μ)

7.29 A Proposal for Multiple Excitation of Gaseous Lasers
G. Makhov and H. G. Heard (Energy Systems)
Proc. IEEE, Vol. 52, pp. 410–411, April 1964

7.30 Laser Action in Mercury Rare Gas Mixtures
H. G. Heard, G. Makhov, and J. Peterson (Energy Systems)
Proc. IEEE, Vol. 52, p. 414, April 1964

7.31 Continuous Visible Laser Action in Single Ionized Argon, Krypton and Xenon
E. I. Godon and E. F. Labuda (BTL) and W. B. Bridges (Hughes)
Appl. Phys. Letters, Vol. 4, pp. 178–180, May 15, 1964 (4880 Å prominent argon line, water cooled tube length 10 to 25 cm, 160 mw output)

7.32 Super-radiance, Excitation Mechanisms, and Quasi-CW Oscillation in the Visible Ar^+ Laser
W. R. Bennett, Jr., J. W. Knutson, Jr., G. N. Mercer, and J. L. Detch (Yale U.)
Appl. Phys. Letters, Vol. 4, pp. 180–182, May 15, 1964 (100 watts for 20 ns, 10 w for 1 ms, 4 w for 0.8 ms at 5145 Å)

7.33 A Stimulated Emission Source at 0.34 Millimetre Wavelength
H. A. Gebbie, N. W. B. Stone, and F. D. Findlay (NPL, England)
Nature, Vol. 202, p. 685, May 16, 1964 (HCN, pulsed, 9.3-m long discharge tube)

7.34 Interpretation of CO_2 Optical Maser Experiments
C. K. N. Patel (BTL)
Phys. Rev. Letters, Vol. 12, pp. 588–590, May 25, 1964

7.35 Erratum: Interpretation of CO_2 Optical Maser Experiments
C. K. N. Patel
Phys. Rev. Letters, Vol. 12, p. 684; June 15, 1964 (Phys. Rev. Letters, Vol. 12, p. 588, 1964)

7.36 Precise Wavelength Measurement of Infrared Optical Maser Lines
Paul G. McMullin (BTL)
Appl. Optics, Vol. 3, pp. 641–642, May 1964 (between 2.4 and 10.98 microns)

7.37 Analysis of Combination Tones in a Short Gas Laser
W. J. Witteman and J. Haisma (Philips Res. Labs., Netherlands)
Phys. Rev. Letters, Vol. 12, pp. 617–619, June 1, 1964 (hole burning, anomalous dispersion)

7.38 Strong 3.27 μ Laser Oscillation in Xenon
W. T. Walter and S. M. Jarrett (TRG)
Appl. Opt., Vol. 3, pp. 789–790, June 1964 (8 lines, 2.027 to 3.508 μ, 1.1 to 15 db/m, new 3.275 μ line has 4.7 db/m)

7.39 Laser Action in Singly Ionized Krypton and Xenon
W. B. Bridges (Hughes)
Proc. IEEE, Vol. 52, pp. 843–844, July 1964 (pulsed, Kr lines between 4577 to 7993 Å, strong lines at 4619, 4762, 4765, 5208, 5682 and 6471 Å; Xe lines between 4603 and 6271, strong lines at 4603 and 5419Å)

7.40 Visible Laser Transitions in the Spectrum of Singly Ionized Iodine
G. R. Fowles and R. C. Jensen (U. of Utah)
Proc. IEEE, Vol. 52, pp. 851–852, July 1964 (simultaneous 5760 and 6127 A lines, pulsed)

7.41 Line Strengths for Noble-Gas Maser Transitions; Calculations of Gain/Inversion at Various Wavelengths
W. L. Faust and R. A. McFarlane (BTL)

J. Appl. Phys., Vol. 35, pp. 2010-2015, July 1964 (α ranges from 5×10^{-4} to 0.16 (Xe) cm^{-1})

7.42 Gas Pumping in Continuously Operated Ion Lasers
E. I. Gordon and E. F. Labuda (BTL)
Bell Sys. Tech. J., Vol. 43, pp. 1827-1829, July 1964, Pt. 2 (external gas flow connection between anode and cathode eliminates otherwise internal pressure differential and self-quenching)

7.43 Excitation Mechanisms of Population Inversion in CO and N_2 Pulsed Lasers
P. K. Cheo and H. G. Cooper (BTL)
Appl. Phys. Letters, Vol. 5, pp. 42-44, August 1, 1964 (inversion by impact until gas breakdown, 6080 A for CO and 7753A for N_2)

7.44 Dependence of the Recovery Time for the Pulsed Carbon Monoxide Laser on Gas Pressure and Tube Bore
H. G. Cooper and P. K. Cheo (BTL)
Appl. Phys. Letters, Vol. 5, pp. 44-46, August 1, 1964 (double pulse excitation)

7.45 Laser Spectroscopy of a Pulsed Mercury-Helium Discharge
A. L. Bloom, W. E. Bell, and F. O. Lopez (Spectra-Physics, Inc.)
Phys. Rev., Vol. 135, pp. A578-A579, August 3, 1964 (19 lines between 5678 to 18,128 Å from Hg)

7.46 Laser Oscillation on $X^1\Sigma^+$ Vibrational-Rotational Transitions of CO
C. K. N. Patel and R. J. Kerl (BTL)
Appl. Phys. Letters, Vol. 5, pp. 81-83, August 15, 1964 (5-5.4 μ wavelengths, 0.1 m W peak, 50-100 μs pulses)

7.47 Laser Action in Xe in Two Distinct Current Regions of AC and DC Discharges
G. E. Courville, P. J. Walsh, and J. H. Wasko (Fairleigh Dickinson U.)
J. Appl. Phys., Vol. 35, pp. 2547-2548, August 1964 (2.026 μ plus others nearby)

7.48 Laser Oscillation on Visible and Ultraviolet Transitions of Singly and Multiply Ionized Oxygen, Carbon and Nitrogen
R. A. McFarlane (BTL)
Appl. Phys. Letters, Vol. 5, pp. 91-93, September 1, 1964 (pulsed; oxygen, 3749-6721 Å; carbon 4647 and 4650 Å; nitrogen 3479-4630 Å)

7.49 Laser Transition with Predissociating Lower State in the NO Molecule
M. Huber (U. of Basel, Switzerland)
Physics Letters, Vol. 12, pp. 102-103, September 15, 1964 (output at 10,215 Å, pulsed)

7.50 New Infrared Gas Laser Transitions by Removal of Dominance
H. Brunet and P. Laures (Comp. G.E., France)
Physics Letters, Vol. 12, pp. 106-107, September 15, 1964 (3- to 5-μ lines in xenon and neon)

7.51 Mercury-Rare Gas Visible-UV Laser
H. G. Heard and J. Peterson (Energy Systems)
Proc. IEEE, Vol. 52, pp. 1049-1050, September 1964 (argon from 3545 to 12760 Å (10 lines), mercury at 5677 and 6150 Å, pulsed, over 10 watts peak at 4765, 4880, 5677, and 6150 Å)

7.52 Comments on "Mercury-Rare Gas Visible-UV Laser"
W. B. Bridges (Hughes)
J. Peterson and H. G. Heard (h nu systems)
Proc. IEEE, Vol. 53, p. 309, March 1965 (Proc. IEEE, Vol. 52, pp. 1049-1050, September 1964; correct line assignment to argon 4880 Å emission, 7 to 10 watts output)

7.53 Orange Through Blue-Green Transitions in a Pulsed-CW Xenon Gas Laser
H. G. Heard and J. Peterson (Energy Systems)
Proc. IEEE, Vol. 52, p. 1050, September 1964 (11 lines, 4921 to 5976 Å, 1 to 10 watts peak)

7.54 On the Feasibility of Flamelasers
R. Bleekrode and W. C. Nieuwpoort (Philips Res. Lab., Neth-

erlands)
Physics Letters, Vol. 12, pp. 204-205, October 1, 1964

7.55 Relaxation Rates of the Ar⁺ Laser Levels
W. R. Bennett, Jr., P. J. Kindlmann, G. N. Mercer, and J. Sunderland (Yale U.)
Appl. Phys. Letters, Vol. 5, pp. 158-160, October 15, 1964 (about 7-to-10-ns radiative lifetimes)

7.56 Visible Laser Transitions in Ionized Iodine
G. R. Fowles and R. C. Jensen (U. of Utah)
Appl. Optics, Vol. 3, pp. 1191-1192, October 1964 (pulsed, six lines between 5407 and 7033 Å)

7.57 Optical Maser Oscillation on Iso-Electronic Transitions in Ar III and Cl II
R. A. McFarlane (BTL)
Appl. Optics, Vol. 3, p. 1196, October 1964 (pulsed; argon lines at 3511, 3514, and 3638 Å; chlorine lines at 5218, 5221, and 5392 Å)

7.58 Blue Gas Laser Using Hg^{2+}
H. J. Gerritsen and P. V. Goedertier (RCA)
J. Appl. Phys., Vol. 35, pp. 3060-3061, October 1964 (4797 Å from doubly ionized mercury, pulsed, 6149 and 5677 Å also, lases in afterglow)

7.59 Super-Radiant Yellow and Orange Laser Transitions in Pure Neon
H. G. Heard and J. Peterson (Energy Systems)
Proc. IEEE, Vol. 52, p. 1258, October 1964 (5940 and 6118 Å, 1-mm bore, 40-cm long tube, 10-to-30-kV pulses, 1-watt output)

7.60 Visible Laser Transitions in Ionized Oxygen, Nitrogen and Carbon Monoxide
H. G. Heard and J. Peterson (Energy Systems)
Proc. IEEE, Vol. 52, p. 1258, October 1964 (3-mm bore, 130-cm long tube, pulsed, many lines, one-watt output)

7.61 Oxygen Laser Aging Characteristics
G. B. Jacobs (G.E.)
Proc. IEEE, Vol. 52, pp. 1259-1260, October 1964 (life data to 200 hours)

7.62 Near Infrared Lasering in Ne-Cl_2 and He-Cl_2
R. A. Paananen and F. A. Horrigan (Raytheon)
Proc. IEEE, Vol. 52, pp. 1261-1262, October 1964 (9452 Å)

7.63 Stimulated Emission in the Far Infrared from Water Vapour and Deuterium Oxide Discharges
L. E. S. Mathias and A. Crocker (SERL, England)
Physics Letters, Vol. 13, pp. 35-36, November 1, 1964 (water out to 120 μ, deuterium oxide out to 108 μ, pulsed)

7.64 Pulsed and Continuous Molecular Far Infrared Gas Lasers
W. J. Witteman and R. Bleekrode (Philips Res. Lab., Netherlands)
Physics Letters, Vol. 13, pp. 126-127, November 15, 1964 (water vapor; 28, 78, and 118 μ wavelengths; maybe OH emissions)

7.65 Selective Excitation Through Vibrational Energy Transfer and Optical Maser Action in N_2-CO_2
C. K. N. Patel (BTL)
Phys. Rev. Letters, Vol. 13, pp. 617-619, November 23, 1964 (about 10-μ radiation, over 10^{-3} watts, discharge only in N_2)

7.66 Continuous-Wave Laser Action on Vibrational-Rotational Transitions of CO_2
C. K. N. Patel (BTL)
Phys. Rev., Vol. 136, pp. A1187-A1193, November 30, 1964 (laser outputs near 10.4 and 9.4 μ, 1 inch ID, 5-m long cell, 0.2 torr, 1-mW output)

7.67 New Laser Transitions in Iodine-Inert Gas Mixtures
R. C. Jensen and G. R. Fowles (U. of Utah)
Proc. IEEE, Vol. 52, p. 1350, November 1964 (11 lines between 0.499 and 1.06 μ, 1.5-μs pulses, delayed laser outputs)

7.68 Atomic Iodine Photodissociation Laser
J. V. V. Kasper and G. C. Pimentel (U. of Calif.)

Appl. Phys. Letters, Vol. 5, pp. 231-233, December 1, 1964 (flash photolysis, 1.30-μ wavelength, 40-μs laser output, high gain of 106 dB/m, 500 watts peak)

7.69 Laser Oscillations from Nitrous Oxide at Wavelengths Around 10.9 μ
L. E. S. Mathias, A. Crocker, and M. S. Wills (SERL, England)
Physics Letters, Vol. 13, pp. 303-304, December 15, 1964 (pulsed, transitions in vibration-rotation bands)

7.70 Inversion Mechanisms in Gas Lasers
W. R. Bennett, Jr. (Yale U.)
Appl. Optics, Suppl. 2: Chemical Lasers, pp. 3-33, 1965 (comprehensive, known gas laser transitions tabulated, gases and molecules)

7.71 Measurement of Excited State Relaxation Rates
W. R. Bennett, Jr., P. J. Kindlmann, and G. N. Mercer (Yale U.)
Appl. Optics, Suppl. 2: Chemical Lasers, pp. 34-57, 1965 (excited state lifetime measurement techniques)

7.72 Collision Lasers
G. Gould (TRG)
Appl. Optics, Suppl. 2: Chemical Lasers, pp. 59-67, 1965 (manganese and lanthanum vapors, up to 1800°C, work in progress, expect 2-3μ output)

7.73 CW Laser Action in N_2O(N_2-N_2O System)
C. K. N. Patel (BTL)
Appl. Phys. Letters, Vol. 6, pp. 12-13, January 1, 1965 (vibrationally excited N_2 energy transferred to N_2O, over 20 laser lines near 10.9μ)

7.74 Superstrahlung in Gepulsten Argon-, Krypton-, und Xenon-Entladungen
D. Rosenberger (Siemens and Halske, Germany)
Physics Letters, Vol. 14, p. 32, January 1, 1965

7.75 Laser Oscillations at Wavelengths Between 21 and 32 μ from a Pulsed Discharge through Ammonia
L. E. S. Mathias, A. Crocker, and M. S. Wills (S.E.R.L., England)
Physics Letters, Vol. 14, pp. 33-34, January 1, 1965

7.76 A Suggested Mechanism for the 337 μ CN Maser
G. W. Chantry, H. A. Gebbie and J. E. Chamberlain (National Physical Lab., England)
Nature, Vol. 205, p. 377, January 23, 1965

7.77 Stimulated Emission in Krypton and Xenon Ions by Collisions with Metastable Atoms
L. Dana and P. Laurès (Comp. G. E., France)
Proc. IEEE, Vol. 53, pp. 78-79, January 1965 (0.43-0.51μ in KrII, 0.48-0.61μ in XeII)

7.78 Infrared Laser Action and Lifetimes in Argon II
F. A. Horrigan, S. H. Koozekanani, and R. A. Paananen (Raytheon)
Appl. Phys. Letters, Vol. 6, pp. 41-43, February 1, 1965 (2 watts cw at visible wavelength, 1.09 and 1.347 μ below 0.5 mw)

7.79 Laser Oscillation on a Single Hyperfine Transition in Iodine
G. R. Fowles and R. C. Jensen (U. of Utah)
Phys. Rev. Letters, Vol. 14, pp. 347-348, March 8, 1965 (6127 Å)

7.80 75-Micron Laser
Yu. N. Petrov and A. M. Prokhorov (Lebedev Physics Inst., Academy of Sciences, U.S.S.R.)
JETP Letters, Vol. 1, pp. 24-25, April 1, 1965 (75.5778 μ from xenon, 1.8-m-long quartz discharge tube)

7.81 Laser Oscillations at Submillimetre Wavelengths from Pulsed Gas Discharges in Compounds of Hydrogen, Carbon and Nitrogen
L. E. S. Mathias, A. Crocker, and M. S. Wills (S.E.R.L., England)
Electronics Letters, Vol. 1, pp. 45-46, April 1965 (between 126

and 373 μ from dimethylamine, between 181 and 205 μ from a mixture of deuterium and bromine cyanide)

7.82 Characteristics of Mode-Coupled Lasers
M. H. Crowell (BTL)
IEEE J. Quantum Electronics, Vol. QE-1, pp. 12–20, April 1965 (modes in a gas laser, time-varying loss, operates like a pulse regenerative oscillator, He–Ne and argon ion lasers)

7.83 Evidence for Radiation Trapping as a Mechanism for Quenching and Ring-Shaped Beam Formation in Ion Lasers
P. K. Cheo and H. G. Cooper (BTL)
Appl. Phys. Letters, Vol. 6, pp. 177–178, May 1, 1965 (evidence in Hg-II, A, and Xe ion lasers)

7.84 New Lines in a Pulsed Xenon Laser
J. A. Dahlquist (Lockheed)
Appl. Phys. Letters, Vol. 6, pp. 193–194, May 15, 1965 (9 lines between 4060 and 5955Å)

7.85 Visible and UV Laser Oscillation at 118 Wavelengths in Ionized Neon, Argon, Krypton, Xenon, Oxygen and Other Gases
W. B. Bridges and A. N. Chester (Hughes)
Appl. Optics, Vol. 4, pp. 573–580, May 1965 (2677 to 7993 Å range; singly, doubly and triply ionized atoms; pulsed dc discharge)

7.86 A Compact Pulsed Gas Laser for the Far Infrared
L. N. Large and H. Hill (Services Electronics Res. Lab., Baldock, England)
Appl. Optics, Vol. 4, pp. 625–626, May 1965 (bakeable cell, 0.94 meters long)

7.87 Spectroscopy of Ion Lasers
William B. Bridges and Arthur N. Chester (Hughes)
IEEE J. Quantum Electronics, Vol. QE-1, pp. 66–84, May 1965 (review, tabulation of wavelengths)

7.88 Laser Oscillation in the Mixtures of Freon and Rare Gases
Mitsuyoshi Shimazu and Yasuzi Suzaki (Hitachi, Japan)
Japanese J. Appl. Phys., Vol. 4, pp. 381–382, May 1965 (continuous, 9452 Å, probably chlorine emission)

7.89 Near-Infrared Oscillation in Pulsed Noble-Gas-Ion Lasers
D. C. Sinclair (U. S. Army Engineer R & D Labs.)
J. Opt. Soc. Am., Vol. 55, pp. 571–572, May 1965 (18 wavelengths between 10,950 and 7828 Å, 8 strong transitions)

7.90 High-Gain Laser Transition in Lead Vapor
G. R. Fowles and W. T. Silfvast (U. of Utah)
Appl. Phys. Letters, Vol. 6, pp. 236–237, June 15, 1965 (7229 Å laser output, 800 to 900°C oven temperature, quartz tube, pulsed, gain of 20 dB/m)

7.91 Investigation of the Operating Characteristics of the 3.5 μ Xenon Laser
Peter O. Clark (Hughes)
IEEE J. Quantum Electronics, Vol. QE-1, pp. 109–113, June 1965 (superradiant laser source and laser amplifier, pressure range, several discharge diameters)

7.92 Laser Action in the Ionic Spectra of Zinc and Cadmium
G. R. Fowles and W. T. Silfvast (U. of Utah)
IEEE J. Quantum Electronics, Vol. QE-1, p. 131, June 1965 (7758 and 4924 Å from Zn, 5378 and 5337 Å from Cd, pulsed, 200–400°C oven)

7.93 Spectroscopy of Mercury-Helium Discharge and 6150 Å Laser Oscillation
Norihito Suzuki (Shimadzu Seisakusho Ltd., Japan)
Japanese J. Appl. Phys., Vol. 4, pp. 452–457, June 1965 (pulsed, inversion from collision, 6150 Å from Hg II)

7.94 Ultraviolet Ion Laser Transitions between 2300 and 4000 Å
P. K. Cheo and H. G. Cooper (BTL)
J. Appl. Phys., Vol. 36, pp. 1862–1865, June 1965 (N, O, and noble gases; 55 lines, up to triply ionized atoms, pulsed)

7.95 Saturation of the Molecular Nitrogen Second Positive Laser Transition
Donald A. Leonard (Avco-Everett)
Appl. Phys. Letters, Vol. 7, pp. 4–6, July 1, 1965 (up to 200 kW output at 3371 Å, 20 ns pulses, crossed field, 10^{-3} radian beamwidth)

7.96 Pulsed-Molecular-Nitrogen Laser Theory
Edward T. Gerry (Avco-Everett)
Appl. Phys. Letters, Vol. 7, pp. 6–8, July 1, 1965 (excitation mechanism, power density)

7.97 CW High Power N_2–CO_2 Laser
C. K. N. Patel (BTL)
Appl. Phys. Letters, Vol. 7, pp. 15–17, July 1, 1965 (about 11.9 W from 2 transitions near 10.6 μ, 3 percent overall efficiency, air plus CO_2)

7.98 Effect of Foreign Gases on the CO_2 Laser: R-Branch Transitions
John A. Howe (BTL)
Appl. Phys. Letters, Vol. 7, pp. 21–22, July 1, 1965 (N_2 and air best additive gas)

7.99 Pulsed Operation of the Neutral Xenon Laser
P. O. Clark (Hughes)
Physics Letters, Vol. 17, pp. 190–192, July 15, 1965 (very different from CW operation, 1.73 to 9 μ, pulse length and current dependent)

7.100 R Branch Laser Action in N_2O
J. A. Howe (BTL)
Physics Letters, Vol. 17, pp. 252–253, July 15, 1965 (pulsed, about 10.5 μ wavelength)

7.101 Mechanism of Population Inversion at 6149 Å in the Mercury Ion Laser
D. J. Dyson (Ferranti Ltd., Scotland)
Nature, Vol. 207, pp. 361–363, July 24, 1965 (pulsed, collision energy transfer from He ion)

7.102 Laser Action in Ionized Sulfur and Phosphorus
G. R. Fowles, W. T. Silfvast, and R. C. Jensen (U. of Utah)
IEEE J. Quantum Electronics, Vol. QE-1, pp. 183–184, July 1965 (pulsed cold cathode discharges, tube inner-wall coated with element, 5320 to 5647 Å from sulfur, 6024, 6043, and 7845 Å from phosphor)

7.103 Transition Probabilities for Some Ar II Laser Lines
H. Statz, F. A. Horrigan, and S. H. Koozekanani (Raytheon), C. L. Tang (Cornell U.), and G. F. Koster (M.I.T.)
J. Appl. Phys., Vol. 36, pp. 2278–2286, July 1965 (predict 10 W cm^{-3} in 2-mm-diameter tube, cooling problem)

7.104 Increasing Continuous Laser-Action on CO_2 Rotational Vibrational Transitions through Selective Depopulation of the Lower Laser Level by Means of Water Vapour
W. J. Witteman (Philips, Netherlands)
Physics Letters, Vol. 18, pp. 125–127, August 15, 1965

7.105 A Metallic Plasma Tube for Ion Lasers
J. Dane Rigden (Perkin-Elmer)
IEEE J. Quantum Electronics, Vol. QE-1, p. 221, August 1965 (14 aluminum disks each with 3-mm central hole for plasma, disks electrically isolated by O-rings, over one watt output from argon laser)

7.106 Electron-Beam Pumping of Noble-Gas Ion Lasers
J. M. Hammer and C. P. Wen (RCA)
Appl. Phys. Letters, Vol. 7, pp. 159–161, September 15, 1965 (electron tube transverse to laser beam, both inside Fabry-Perot cavity, modulation)

7.107 Continuous-Duty Argon Ion Lasers
       E. F. Labuda, E. I. Gordon, and R. C. Miller (BTL)
       IEEE J. Quantum Electronics, Vol. QE-1, pp. 273–279, September 1965 (axial magnetic field, high discharge current, metal-wall vessel, 1 W out)

7.108 Proposed Gas Maser Pumping Scheme for the Far Infrared
       Willard H. Wells (Jet Propulsion Lab., California Inst. of Tech.)
       J. Appl. Phys., Vol. 36, pp. 2838–2843, September 1965 (rotational state interaction between HCl and HF, HF emission at 81.25 μ)

7.109 Laser Lines in Hg I
       K. Bockasten, M. Garavaglia, B. A. Lengyel, and T. Lundholm (Uppsala U., Sweden)
       J. Opt. Soc. Am., Vol. 55, pp. 1051–1053, September 1965 (region of 1–2.5 μ)

7.110 Transition Probabilities in the Ar I Spectrum
       R. H. Garstang and Janet Van Blerkom (U. of Colorado)
       J. Opt. Soc. Am., Vol. 55, pp. 1054–1057, September 1965

7.111 Investigation of a Neon-Hydrogen Laser at Large Discharge Currents
       V. P. Chebotaev and L. S. Vasilenko (Academy of Sciences, U.S.S.R.)
       Soviet Physics JETP, Vol. 21, pp. 515–516, September 1965 (hollow cathode, 0.94 to 1.4 μ wavelength)

7.112 On the Possibility of Stimulated Emission in the Far Ultraviolet
       P. A. Bazhulin, I. N. Knyazev, and G. G. Petrash (Academy of Sciences, U.S.S.R.)
       Soviet Physics JETP, Vol 21, pp. 649–650, September 1965 (considers H_2 molecule)

7.113 Ring Discharge Excitation of Gas Ion Lasers
       W. E. Bell (Spectra-Physics)
       Appl. Phys. Letters, Vol. 7, pp. 190–191, October 1, 1965 (pulsed, gas discharge is secondary of transformer, 1 Mc/s generator; Kr^+, Hg^+, and S^+ laser)

7.114 UV and Visible Laser Oscillations in Fluorine, Phosphorus, and Chlorine
       P. K. Cheo and H. G. Cooper (BTL)
       Appl. Phys. Letters, Vol. 7, pp. 202–204, October 1, 1965 (pulsed, 3347–6671 Å from P, 2759–6371 Å from F, 2632–4768 Å from Cl)

7.115 Stimulated Emission in the System CO/CO_2
       R. A. McFarlane and J. A. Howe (BTL)
       Physics Letters, Vol. 19, pp. 208–210, October 15, 1965 (1784 to 1953 cm^{-1} from CO, 1031 to 1083 cm^{-1} from CO_2, maximum output at CO_2 flow rate of 100 cm^3 min^{-1})

7.116 Laser Gain Measurements in a Xenon-Krypton Discharge
       H. Brunet (C.G.E., France)
       Appl. Optics, Vol. 4, p. 1354, October 1965 (41 dB/m from Xe at 5.57 μ, Kr enhances 3.68 μ output from Xe)

7.117 Bistable Traveling-Wave Oscillations of Ion Ring Laser
       W. W. Rigrod and T. J. Bridges (BTL)
       IEEE J. Quantum Electronics, Vol. QE-1, pp. 298–303, October 1965 (4879.9 Å from argon, spontaneous and controlled one-way oscillation)

7.118 Output Spectra of the Argon Ion Laser
       T. J. Bridges and W. W. Rigrod (BTL)
       IEEE J. Quantum Electronics, Vol. QE-1, pp. 303–308, October 1965 (pure near threshold, random oscillation sequence much above threshold, also 3 distinct maxima envelope)

7.119 Laser Oscillations in Silicon Tetrachloride Vapor
       Mitsuyoshi Shimizu and Yasuzi Suzaki (Hitachi Central Research Lab., Japan)
       Japanese J. Appl. Phys., Vol. 4, p. 819, October 1965 (1.2 and 1.58 μ from Si atom, 1.98 and 2.02 μ from Cl, electrodeless high frequency discharge)

7.120 Comments on the Mechanism of the 337-Micron CN Laser
       H. P. Broida (U. of California, Santa Barbara), K. M. Evenson (NBS, Boulder), and T. T. Kikuchi (GM Defense Research Labs.)
       J. Appl. Phys., Vol. 36, p. 3355, October 1965 (suggests other transitions from 169 to 896 μ)

7.121 High-Power Brewster Window Laser at 10.6 Microns
       T. J. Bridges and C. K. N. Patel (BTL)
       Appl. Phys. Letters, Vol. 7, pp. 244–245, November 1, 1965 (CO_2 and N_2, stationary gas, 2–5 W, 2–5 percent, 5-mm-thick KCl window, operating variables)

7.122 CW Laser on Vibrational-Rotational Transitions of CO
       C. K. N. Patel (BTL)
       Appl. Phys. Letters, Vol. 7, pp. 246–247, November 1, 1965 (143 lines between 5.0 and 6.2 μ, continuous-flow N_2-CO system)

7.123 Laser Oscillation in the Visible Spectrum of Singly Ionized Pure Bromine Vapor
       William M. Keeffe and Walter J. Graham (U.S. Naval Research Lab.)
       Appl. Phys. Letters, Vol. 7, pp. 263–264, November 15, 1965 (pulsed, five blue-green wavelengths, strongest at 5182 and 5332 Å)

7.124 CW Laser Oscillation in an N_2-CS_2 System
       C. K. N. Patel (BTL)
       Appl. Phys. Letters, Vol. 7, pp. 273–274, November 15, 1965 (10 wavelengths between 11.48 and 11.55 μ, 10 mW output)

7.125 High-Power Laser Action in CO_2-He Mixtures
       G. Moeller and J. Dane Rigden (Perkin-Elmer)
       Appl. Phys. Letters, Vol. 7, pp. 274–276, November 15, 1965 (18 watts, 4 percent efficiency)

7.126 Electron Energy Spectra in Neon, Xenon, and Helium-Neon Laser Discharges
       J. Y. Wada and Hans Heil (Hughes)
       IEEE J. Quantum Electronics, Vol. QE-1, pp. 327–335, November 1965

7.127 Gain and Bandwidth Narrowing in a Regenerative He–Xe Laser Amplifier
       Vern N. Smiley, Adolph L. Lewis, and David K. Forbes (U.S. Navy Electronics Lab., San Diego)
       J. Opt. Soc. Am., Vol. 55, pp. 1552–1553, November 1965 (2.03 μ wavelength, 27 dB gain, magnetostrictively tuned amplifier)

7.128 CW High-Power CO_2-N_2-He Laser
       C. K. N. Patel, P. K. Tien, and J. H. McFee (BTL)
       Appl. Phys. Letters, Vol. 7, pp. 290–292, December 1, 1965 (106 watt CW, 6.2 percent efficiency; 183 watt peak, 7.4 percent efficiency)

7.129 Infrared Laser Oscillation in HBr and HI Gas Discharges
       S. M. Jarrett, J. Nunez, and G. Gould (TRG)
       Appl. Phys. Letters, Vol. 7, pp. 294–296, December 1, 1965 (CW output; 2.28, 2.35, and 2.83 μ from Br; 2.75, 3.23, and 3.43 μ from I)

7.130 Pulsed Laser Transitions in Manganese Vapor
       M. Piltch, W. T. Walter, N. Solimene, and G. Gould (TRG), and W. R. Bennett, Jr. (Yale U.)
       Appl. Phys. Letters, Vol. 7, pp. 309–310, December 1, 1965 (0.53 to 14 μ, 37 dB/m gain at 5341 Å, 300 W peak in 20 ns pulse, 1300°C)

7.131   Erratum: Pulsed Laser Transitions in Manganese Vapor
        M. Piltch, W. T. Walter, N. Solimene, and G. Gould (TRG),
        and W. R. Bennett, Jr., (Yale U.)
        Appl. Phys. Letters, Vol. 9, p. 253, September 15, 1966 (Appl.
        Phys. Letters, Vol. 7, p. 309, 1965)

7.132   A Study of Modes in a Br_2-Ar Laser
        K. K. Chow and T. B. Ramachandran (Microwave Associates)
        Appl. Optics, Vol. 4, pp. 1670–1671, December 1965 (doublets of
        beats)

7.133   Visible Laser Transitions in Ionized Selenium, Arsenic, and Bro-
        mine
        W. E. Bell, A. L. Bloom, and J. P. Goldsborough (Spectra-
        Physics)
        IEEE J. Quantum Electronics, Vol. QE-1, p. 400, December 1965
        (pulsed, RF ring discharge, 5097 and 5229 Å from SeII, 5498
        to 6171 Å from AsII, 5185 and 5334 Å from BrII)

7.134   Laser Emission in Ionized Mercury: Isotope Shift, Linewidth,
        and Precise Wavelength
        R. L. Byer, W. E. Bell, E. Hodges, and A. L. Bloom (Spectra-
        Physics)
        J. Opt. Soc. Am., Vol. 55, pp. 1598–1602, December 1965 (6151 Å
        emission, hollow cathode laser, pulsed)

7.135   Far-Infrared Masers and Their Applications to Spectroscopy
        C. G. B. Garrett (B.T.L.)
        Proc. Physics of Quantum Electronics Conf., McGraw-Hill Book
        Co., New York, pp. 557–566, 1966

7.136   Vibration Energy Transfer—An Efficient Means of Selective Ex-
        citation in Molecules
        C. K. N. Patel (B.T.L.)
        Proc. Physics of Quantum Electronics Conf., McGraw-Hill Book
        Co., New York, pp. 643–654, 1966

7.137   Stimulated-Emission Spectroscopy of Some Diatomic Molecules
        R. A. McFarlane (B.T.L.)
        Proc. Physics of Quantum Electronics Conf., McGraw-Hill Book
        Co., New York, pp. 655–663, 1966

7.138   Excitation Mechanisms of the Argon-Ion Laser
        E. I. Gordon, E. F. Labuda, Richard C. Miller, and C. E. Webb
        (B.T.L.)
        Proc. Physics of Quantum Electronics Conf., McGraw-Hill Book
        Co., New York, pp. 664–673, 1966

7.139   Transition Probabilities, Lifetimes and Related Considerations in
        Ionized Argon Lasers
        H. Statz, F. A. Horrigan, and S. H. Koozekanani (Raytheon),
        C. L. Tang (Cornell U.), and G. F. Koster (M.I.T.)
        Proc. Physics of Quantum Electronics Conf., McGraw-Hill Book
        Co., New York, pp. 674–687, 1966

7.140   Emission-Line Widths of Ion Lasers
        A. L. Bloom, R. L. Byer, and W. E. Bell (Spectra-Physics)
        Proc. Physics of Quantum Electronics Conf., McGraw-Hill Book
        Co., New York, pp. 688–689, 1966

7.141   Ion Laser Oscillations in Sulfur
        H. G. Cooper and P. K. Cheo (B.T.L.)
        Proc. Physics of Quantum Electronics Conf., McGraw-Hill Book
        Co., New York, pp. 690–697, 1966

7.142   Pressure Dependence of the Iodine Photodissociation Laser Peak
        Output
        M. A. Pollack (B.T.L.)
        Appl. Phys. Letters, Vol. 8, pp. 36–38, January 15, 1966 (xenon
        flash lamp, four alkyl iodides, maximum power with 20 to 130
        torr, 1.31 μ output)

7.143   Rotational Structure of Ultraviolet Generation of Molecular Ni-
        trogen

V. M. Kaslin and G. G. Petrash (Lebedev Physics Institute,
Academy of Sciences, U.S.S.R.)
JETP Letters, Vol. 3, pp. 55–58, January 15, 1966 (nitrogen
laser, pulsed, band emissions between 3371 and 3577Å)

7.144   Stimulated Emission up to 0.538 mm Wavelength from Cyanic
        Componds
        H. Steffen, J. Steffen, J.-F. Moser and F. K. Kneubühl (Tech-
        nischen Hochschule Zurich, Switzerland)
        Physics Letters, Vol. 20, pp. 20–21, January 15, 1966 (also 0.310
        and 0.336 mm)

7.145   Pulsed-Laser Action in Atomic Copper Vapor
        William T. Walter, Martin Piltch, Nicholas Solimene, and Gor-
        don Gould (TRG)
        Bull. Am. Phys. Soc., Vol. 11, p. 113–GF14, January 1966 (5105
        and 5782 Å, 58 dB/m gain, 1.2 kW in 20 ns pulse, 1500°C heated
        tube, 10^{-3} efficiency)

7.146   Q-Switched Molecular Gas Lasers with High Peak Power
        M. A. Kovacs, G. W. Flynn, C. K. Rhodes, and A. Javan
        (M.I.T.)
        Bull. Am. Phys. Soc., Vol. 11, p. 128–HG6, January 1966 (CO_2-
        H_2-He laser, 1.1 mJ per pulse, also 3W CW)

7.147   Laser Action in Sulfur using Hydrogen Sulfide
        Ramon U. Martinelli and H. J. Gerritsen (RCA)
        J. Appl. Phys., Vol. 37, pp. 444–445, January 1966 (continuous,
        1.0455 μ)

7.148   Vibrational-Rotational Laser Action in Carbon Monoxide
        C. K. N. Patel (B.T.L.)
        Phys. Rev., Vol. 141, pp. 71–83, January 1966 (pulsed, many
        lines in 5.0–5.4 μ region, excitation mechanisms)

7.149   Stimulated Emission from Molecular Hydrogen and Deuterium
        in the Near Infrared
        P. A. Bazhulin, I. N. Knyazev, and G. G. Petrash (Lebedev
        Physics Institute, Academy of Sciences, U.S.S.R.)
        Soviet Physics JETP, Vol. 22, pp. 11–16, January 1966 (pulsed,
        0.83 to 1.3 μ for H_2, 0.83 and 0.95 μ for D_2, 0.92 μ for HD)

7.150   Q Switching of Molecular Laser Transitions
        M. A. Kovacs, G. W. Flynn, and A. Javan (M.I.T.)
        Appl. Phys. Letters, Vol. 8, pp. 62–63, February 1, 1966 (CO_2
        and N_2O)

7.151   Observation of Laser Action in the R-Branch of CO_2 and N_2O
        Vibrational Spectra
        G. Moeller and J. Dane Rigden (Perkin-Elmer)
        Appl. Phys. Letters, Vol. 8, pp. 69–70, February 1, 1966

7.152   The Identification of Some Si and Cl Laser Lines Observed by
        Cheo and Cooper
        H. P. Palenius (U. of Lund, Lund, Sweden)
        Appl. Phys. Letters, Vol. 8, pp. 82–83, February 15, 1966

7.153   On the Emission Spectrum of the Negative Glow Plasma of a
        Hollow-Cathode Discharge in Magnetic Field
        Constantin Popovici and Maria Somesan (Physics Institute of
        the Romanian Academy, Bucharest, Romania)
        Appl. Phys. Letters, Vol. 8, pp. 103–105, March 1, 1966 (He
        and Mo)

7.154   Determination of Vibration-Rotational Line Strengths and Widths
        in CO_2 Using a CO_2-N_2 Laser
        T. K. McCubbin, Jr., Ronald Darone, and James Sorrell (Penn-
        sylvania State U.) ·
        Appl. Phys. Letters, Vol. 8, pp. 118–119, March 1, 1966

7.155   High Peak Power Pulsed 10 μ CO_2 Laser
        C. Frapard, M. Roulot and X. Ziegler (Centre de Recherches
        de la Compagnie Générale d'Electricité, France)

Physics Letters, Vol. 20, pp. 384–385, March 1, 1966 (up to 825 W, laser pulse lengths up to 500 μs)

7.156 RF Induction Excitation of CW Visible Laser Transitions in Ionized Gases
J. P. Goldsborough, E. B. Hodges, and W. E. Bell (Spectra-Physics)
Appl. Phys. Letters, Vol. 8, pp. 137–139, March 15, 1966

7.157 Laser Oscillation in Atomic Cl in HCl and HI Gas Discharges
S. M. Jarrett, J. Nuñez, and G. Gould (TRG)
Appl. Phys. Letters, Vol. 8, pp. 150–151, March 15, 1966

7.158 The Lifetimes of Ionized Argon States
J. Bakos, J. Szigeti and L. Varga (Central Res. Inst. of Physics, Hungary)
Physics Letters, Vol. 20, pp. 503–504, March 15, 1966 (8.4 to 10.3 ns)

7.159 Isotope Shifts and the Role of Fermi Resonance in the CO_2 Infrared Maser
Irwin Wieder and Gregor B. McCurdy (Interphase Corp.-West)
Phys. Rev. Letters, Vol. 16, pp. 565–567, March 28, 1966

7.160 A New Xenon Laser Oscillation at 5401 Å
R. H. Neusel (TRW)
IEEE J. Quantum Electronics, Vol. QE-2, p. 70, March 1966 (pulsed)

7.161 Nitrogen Laser Action in a Supersonic Flow
J. Wilson (Avco)
Appl. Phys. Letters, Vol. 8, pp. 159–161, April 1, 1966

7.162 Observation of New Br II Laser Transitions
W. M. Keeffe and W. J. Graham (U. S. Naval Res. Lab.)
Physics Letters, Vol. 20, p. 643, April 1, 1966 (pulsed, 6117 and 6168 Å)

7.163 Spontaneous-Emission Line Shape of Ion Laser Transitions
W. R. Bennett, Jr., E. A. Ballik, and G. N. Mercer (Yale U.)
Phys. Rev. Letters, Vol. 16, pp. 603–605, April 4, 1966 (argon)

7.164 Temperatures, Lorentzian Widths, and Drift Velocities in the Argon-Ion Laser
E. A. Ballik, W. R. Bennett, Jr., and G. N. Mercer (Yale U.)
Appl. Phys. Letters, Vol. 8, pp. 214–216, April 15, 1966

7.165 Experimental and Theoretical Investigations of the Spontaneous Infrared Radiation Fluctuations of a High-Pressure, High-Current Argon Discharge
W. Gubler and M. J. O. Strutt (Eidgenössische Technische Hochschule, Zürich, Switzerland)
Appl. Optics, Vol. 5, pp. 659–665, April 1966

7.166 New CW Laser Transitions in Argon, Krypton, and Xenon
W. B. Bridges and A. S. Halstead (Hughes)
IEEE J. Quantum Electronics, Vol. QE-2, p. 84, April 1966

7.167 Inversion Mechanisms, Population Densities and Coupling-Out of a High-Power Molecular Laser
W. J. Witteman
Philips Res. Repts., Vol. 21, pp. 73–84, April 1966 (CO_2 + N_2 + H_2O, 20 W output with 10% reflectivity from tilted Ge plates)

7.168 Submillimeter Gas Laser
G. T. Flesher and W. M. Muller (General Motors)
Proc. IEEE, Vol. 54, pp. 543–546, April 1966 (water vapor and CN)

7.169 Water Vapor Gas Laser Operating at 118-Microns Wavelength
D. P. Akitt, W. Q. Jeffers, and P. D. Coleman (U. of Illinois)
Proc. IEEE, Vol. 54, pp. 547–551, April 1966

7.170 Continuous Wave Submillimeter Oscillation in H_2O, D_2O, and CH_3CN

Walter W. Müller and Gail T. Flesher (GM Def. Res. Labs.)
Appl. Phys. Letters, Vol. 8, pp. 217–218, May 1, 1966 (0.79106, 0.11865 and 0.22034 mm from H_2O; 0.0843, 0.1077 and 0.1716 mm from D_2O; 0.3113 and 0.3367 mm from CH_3CN)

7.171 Erratum: Continuous-Wave Submillimeter Oscillation in H_2O, D_2O, and CH_3CN
Walter M. Müller and Gail T. Flesher (GM Defense Research Labs.)
Appl. Phys. Letters, Vol. 9, p. 218, September 1, 1966 (Appl. Phys. Letters, Vol. 8, p. 217, 1966)

7.172 Cyclotron Resonance Excitation of Gas-Ion Laser Transitions
John P. Goldsborough (Spectra-Physics)
Appl. Phys. Letters, Vol. 8, pp. 218–219, May 1, 1966 (laser output depends on transverse magnetic field, argon, less than 1 torr, 1 kw 2.45 Gc/s magnetron excitation)

7.173 Measurement of 10.6-μ CO_2 Laser Transition Probability and Optical Broadening Cross Sections
Edward T. Gerry and Donald A. Leonard (Avco Everett)
Appl. Phys. Letters, Vol. 8, pp. 227–229, May 1, 1966

7.174 A New Krypton Laser Oscillation at 5016.4 Å
R. H. Neusel (TRW Systems)
IEEE J. Quantum Electronics, Vol. QE-2, p. 106, May 1966

7.175 On the Possibility of Anomalous Diffusion Effects in Continuous-Duty Ion Lasers in Magnetic Fields
Edward J. Powers (U. of Texas)
Proc. IEEE, Vol. 54, pp. 804–805, May 1966

7.176 Comments on a New Laser Emission at 0.774 mm Wavelength from ICN
H. Steffen, J. Steffen, J. F. Moser, and F. K. Kneubühl (Eidgenossischen Technischen Hochschule, Zurich, Switzerland)
Physics Letters, Vol. 21, pp. 425–426, June 1, 1966 (pulsed)

7.177 OCS Molecular Laser
Thomas F. Deutsch (Raytheon)
Appl. Phys. Letters, Vol. 8, pp. 334–335, June 15, 1966 (with gas additives, pulsed, wavelengths near 8.39 and 8.25 μ, 29 wavelengths between 5.1 and 5.7 μ)

7.178 Laser Action in Singly Ionized Ge, Sn, Pb, In, Cd and Zn
W. T. Silfvast, G. R. Fowles, and B. D. Hopkins (U. of Utah)
Appl. Phys. Letters, Vol. 8, pp. 318–319, June 15, 1966 (pulsed, wavelengths between 4415 and 7611 Å)

7.179 The Influence of Hydrogen on the Output of a N_2-CO_2 Laser
D. Rosenberger (Siemens and Halske, Germany)
Physics Letters, Vol. 21, pp. 520–521, June 15, 1966 (enhances power output)

7.180 New Laser Transitions in Antimony and Tellurium
W. E. Bell, A. L. Bloom and J. P. Goldsborough (Spectra-Physics)
IEEE J. Quantum Electronics, Vol. QE-2, p. 154, June 1966 (Sb at 6130 Å, Te at 5576, 5708 and 6350 Å)

7.181 Progress in Ionized-Argon Lasers
Roy A. Paananen (Raytheon)
IEEE Spectrum, Vol. 3, No. 6, pp. 88–99, June 1966

7.182 High-Power Infrared Laser with Adjustable Coupling-Out
W. J. Witteman and G. v. d. Goot (Philips, Netherlands)
J. Appl. Phys., Vol. 37, p. 2919, June 1966 (CO_2 laser, 20 W output, 12% efficiency, tilted Ge plate for reflective coupling)

7.183 Hyperfine Spectrum of Xenon in the 3.5-μm Maser Transition
Katsumi Sakurai and Koichi Shimoda (U. of Tokyo, Japan)
J. Phys. Soc. Japan, Vol. 21, p. 1214, June 1966 (magnetically tuned, strong asymmetry)

7.184 Laser Action on Several Hyperfine Transitions in Mn I
W. T. Silfvast and G. R. Fowles (U. of Utah)
J. Opt. Soc. Am., Vol. 56, pp. 832–833, June 1966 (intense 5341 Å transition, other visible and infrared lines)

7.185 Emission Spectra during Glow-Discharge Decomposition of Ammonia
D. C. Carbaugh and J. M. Marchello (U. of Maryland)
J. Opt. Soc. Am., Vol. 56, pp. 836–837, June 1966 (potential laser capabilities)

7.186 Continuously-Operated Ultraviolet Lasers
Roy Paananen (Raytheon)
Appl. Phys. Letters, Vol. 9, pp. 34–35, July 1, 1966 (Ne at 3324 Å and 3378 Å, Kr at 3507 Å, Ar at 3511 Å, 26-inch long ceramic tube)

7.187 Effects of Cascade in the Excitation of the Ar II Laser
R. I. Rudko and C. L. Tang (Cornell U.)
Appl. Phys. Letters, Vol. 9, pp. 41–44, July 1, 1966 (not insignificant)

7.188 High Pressure, High Magnetic Field Effects in Continuous Argon Lasers
I. Gorog and F. W. Spong (RCA)
Appl. Phys. Letters, Vol. 9, pp. 61–63, July 1, 1966 (4880 Å, radiation trapping, up to 1800 gauss, up to 1.1 Torr)

7.189 On the Excitation of the 4.38-eV Level of Lead in a Hollow Cathode Discharge in Magnetic Field
Maria Someşan and Constantin Popovici (Acad. of Sci. Romania)
Appl. Phys. Letters, Vol. 9, pp. 65–67, July 1, 1966

7.190 Sub-millimetre Maser Amplification and Continuous Wave Emission
H. A. Gebbie, N. W. B. Stone, W. Slough, and J. E. Chamberlain (Nat'l Phys. Lab., England), and W. A. Sheraton (G. & E. Bradley, England)
Nature, Vol. 211, p. 62, July 2, 1966 (CN discharge, 0.337 mm)

7.191 Beam Laser for the Infrared Band
N. G. Basov, A. I. Oraevskii, and V. A. Shcheglov (Lebedev Phys. Inst., Acad. of Sci., USSR)
JETP Letters, Vol. 4, pp. 41–42, July 15, 1966 (3 to 20 μ band proposed, thermal pumping, propose N_2O and HCN)

7.192 Anomale Wechselwirkung Von Gekoppelten Schwingungen Im Kontinuierlichen Argon-Ionen-Laser
D. Rosenberg (Siemens and Halske, Germany)
Physics Letters, Vol. 22, pp. 54–55, July 15, 1966 (polarity and relaxation times anomalies)

7.193 Laser Amplifier Noise at 3.5 Microns in Helium-Xenon
J. W. Klüver (BTL)
J. Appl. Phys., Vol. 37, pp. 2987–2999, July 1966 (unsaturated gain of 17 dB, saturated gain of 1 dB, cone angle of noise, good noise properties of amplifier)

7.194 Laser Transition at 651.6 nm in Ionized Iodine
C. S. Willet and O. S. Heavens (U. of York, England)
Optica Acta, Vol. 13, pp. 271–273, July 1966 (pulsed, helium and iodine mixture)

7.195 High Temperature Alumina Discharge Tube for Pulsed Metal Vapor Lasers
M. Piltch and G. Gould (TRG)
Rev. Sci. Instr., Vol. 37, pp. 925–927, July 1966 (up to 1400°C)

7.196 Vibrational Relaxation Measurements in CO_2 using an Induced-Fluorescence Technique
L. O. Hocker, M. A. Kovacs, C. K. Rhodes, G. W. Flynn, and A. Javan (M.I.T.)

Phys. Rev. Letters, Vol. 17, pp. 233–235, August 1, 1966 (10.6 μ Q-switching experiments)

7.197 New Near Infrared Laser Lines in Argon I
K. Bockasten, T. Lundholm, and O. Andrade (U. of Uppsala, Sweden)
Physics Letters Vol. 22, pp. 145–146, August 1, 1966 (pulsed, 3 lines near 12 μ, probably superradiant)

7.198 Competition, Hysteresis and Reactive Q-Switching in CO_2 Lasers at 10.6 Microns
T. J. Bridges (BTL)
Appl. Phys. Letters, Vol. 9, pp. 174–176, August 15, 1966 (high repetition rate (30 to 60 kHz) average power equal to CW power)

7.199 Precision Spectroscopy of New Infrared Emission Systems of Molecular Nitrogen
R. A. McFarlane (BTL)
IEEE J. Quantum Electronics, Vol. QE-2, pp. 229–232, August 1966 (four groups of lines between 5.3 and 8.1 μ)

7.200 Generation of New Infrared Maser Frequencies by Isotopic Substitution
G. B. McCurdy and I. Wieder (Interphase Corporation-West)
IEEE J. Quantum Electronics, Vol. QE-2, pp. 294–295, August 1966 (abstract, CO_2 experiments)

7.201 A High Power Pulsed Nitrogen Laser
J. D. Shipman and A. C. Kolb (NRL)
IEEE J. Quantum Electronics, Vol. QE-2, p. 298, August 1966 (abstract, 700 kW, 3 ns pulse)

7.202 Gas Pumping in Repetitively Pulsed Ion Lasers
R. H. Neusel (TRW/Systems)
IEEE J. Quantum Electronics, Vol. QE-2, pp. 331–333, August 1966 (axial pressure differential near anode, by-pass connection desirable)

7.203 A New Pulsed Ion Laser Transition in Nitrogen at 3995 Å
R. B. Allen, R. B. Starnes, and A. A. Dougal (U. of Texas)
IEEE J. Quantum Electronics, Vol. QE-2, p. 334, August 1966

7.204 New Laser Oscillations in Krypton and Xenon
R. H. Neusel (TRW/Systems)
IEEE J. Quantum Electronics, Vol. QE-2, p. 334, August 1966 (4615 Å from krypton, 4723 and 4748 Å from xenon, pulsed)

7.205 A Pulsed, Coaxial Transmission Line Gas Laser
M. Geller, D. E. Altman, and T. A. DeTemple (USN Electronics Lab.)
J. Appl. Phys., Vol. 37, pp. 3639–3640, August 1966 (low inductance, N_2 laser, 10 ns output pulses)

7.206 Measurements of Argon Single-Frequency Laser Power and the 6328-Å Neon Shift using an Interferometer Laser
Peter Zory (Sperry Gyroscope)
J. Appl. Phys., Vol. 37, pp. 3643–3644, August 1966

7.207 Some Results of a Study of the Pulsed Argon Laser
R. K. Leonov, E. D. Protsenko, and Yu. M. Sapunov
Optics and Spectroscopy, Vol. 21, pp. 141–142, August 1966

7.208 Characteristics of Pulsed Laser Action in Helium-Neon and Helium-Argon Mixtures
E. I. Shtyrkov and E. V. Subbes
Optics and Spectroscopy, Vol. 21, pp. 143–144, August 1966 (over 200 Torr)

7.209 Pulsed Stimulated Emission in a Hydrogen-Atom Beam Laser
G. M. Strakhovskiĭ and A. V. Uspenskiĭ (Lebedev Phys. Inst., Acad. of Sci., USSR)

Soviet Phys. JETP, Vol. 23, pp. 247–249, August 1966 (analytical, two relaxation times)

7.210 Laser Action in Optically-Pumped CN
M. A. Pollack (BTL)
Appl. Phys. Letters, Vol. 9, pp. 230–232, September 15, 1966 (5.2 μ output, flash photolysis)

7.211 New Visible CW Laser Lines in Singly-Ionized Chlorine
C. B. Zarowin (IBM Watson)
Appl. Phys. Letters, Vol. 9, pp. 241–242, September 15, 1966 (4132, 4740, and 5103 Å, inductively-pumped plasma, 13.5-MHz 4-kW RF power supply)

7.212 New Generation Lines of a Pulsed Iodine-Vapor Laser
V. M. Koval'chuk and G. G. Petrash (Lebedev Phys. Inst., Acad. of Sci., USSR)
JETP Letters, Vol. 4, pp. 144–146, September 15, 1966 (4533, 4674, 4934, and 10 417 Å; 0.33 mJ output per pulse, 2.2 kW peak power)

7.213 Electron Temperature in the Electric Discharge Used for the Argon Ion Laser
V. F. Kitaeva, Yu. I. Osipov, and N. N. Sobolev (Lebedev Phys. Inst., Acad. of Sci., USSR)
JETP Letters, Vol. 4, pp. 146–148, September 15, 1966 (between 3.3×10^4 to 9.9×10^4 °K)

7.214 Recent Developments in CO_2 Lasers
J. D. Rigden and G. Moeller (Perkin-Elmer)
IEEE J. Quantum Electronics, Vol. QE-2, pp. 365–368, September 1966

7.215 Pulsed and Steady-State Infrared Emission Studies of CO_2 Laser Systems
M. J. Weber and T. F. Deutsch (Raytheon)
IEEE J. Quantum Electronics, Vol. QE-2, pp. 369–375, September 1966 (dc discharge, wavelength range between 2.5 and 15 μ)

7.216 Rate Determining Processes for the Production of Radiation in High Power Molecular Lasers
W. J. Witteman (Philips, Netherlands)
IEEE J. Quantum Electronics, Vol. QE-2, pp. 375–378, September 1966 (CO_2 laser, sealed off system)

7.217 Progress and Applications of Q-Switching Techniques Using Molecular Gas Lasers
G. W. Flynn, L. O. Hocker, A. Javan, M. A. Kovacs, and C. K. Rhodes (M.I.T.)
IEEE J. Quantum Electronics, Vol. QE-2, pp. 378–381, September 1966 (CO_2–N_2–He, rotating mirror, 10 kW peak from 3-W CW laser, 5×10^{-4} s lifetime)

7.218 High Resolution Spectroscopy by Zeeman-Tuned Infrared Laser
Henri Brunet (CGE, France)
IEEE J. Quantum Electronics, Vol. QE-2, pp. 382–384, September 1966 (up to 1800 gauss, 0.1 cm^{-1} shift, xenon and neon lasers)

7.219 Generation of New Infrared Maser Frequencies by Isotopic Substitution
G. B. McCurdy and I. Wieder (Carver Corp.)
IEEE J. Quantum Electronics, Vol. QE-2, pp. 385–387, September 1966 (10.6 μ output from $C^{12}O_2^{16}$ shifts to 9.4 μ using $C^{12}O_2^{18}$)

7.220 The Theory of Doppler and Impact Broadening of Spectral Lines and Pressure Effects on the Power Output of a Gas Laser
S. Rautian and I. Sobel'man (Lebedev Phys. Inst., Acad. of Sci., USSR)
IEEE J. Quantum Electronics, Vol. QE-2, pp. 446–448, September 1966

7.221 Efficient Pulsed Gas Discharge Lasers
W. T. Walter, N. Solimene, M. Piltch, and G. Gould (TRG)
IEEE J. Quantum Electronics, Vol. QE-2, pp. 474–479, September 1966 (cyclic excitation and relaxation, atomic vapor discharges, 5105 and 5782 Å from copper, 8542 and 8662 Å from calcium, gains of 58 dB/m)

7.222 Spectroscopic Studies of Gas Discharges Used for Argon Ion Lasers
V. F. Kitaeva, Yu. I. Osipov, and N. N. Sobolev (Lebedev Phys. Inst., Acad. of Sci., USSR)
IEEE J. Quantum Electronics, Vol. QE-2, pp. 635–637, September 1966

7.223 Laser Lines in Atomic and Molecular Hydrogen
K. Bockasten, T. Lundholm, and O. Andrade (U. of Uppsala, Sweden)
J. Opt. Soc. Am., Vol. 56, pp. 1260–1261, September 1966 (eight lines between 8349 and 18 751 Å)

7.224 Self-Locking of Modes in the Argon Ion Laser
O. L. Gaddy and E. M. Schaefer (U. of Illinois)
Appl. Phys. Letters, Vol. 9, pp. 281–282, October 15, 1966 (subnanosecond pulses every 4.2 ns period)

7.225 NO Molecular Laser
Thomas F. Deutsch (Raytheon)
Appl. Phys. Letters, Vol. 9, pp. 295–297, October 15, 1966 (vibrational-rotational transitions, dissociation of NOCl in pulsed discharge, emission between 5.84 and 6.43 μ)

7.226 A Mechanism Ensuring Level Population Inversion in CO_2 Lasers
N. N. Sobolev and V. V. Sokovikov (Lebedev Phys. Inst., Acad. of Sci., USSR)
JETP Letters, Vol. 4, pp. 204–207, October 15, 1966 (explains high efficiency and high power output, involves N_2, CO_2, CO, and He)

7.227 Mechanism of the Submillimeter Laser Emissions from the CN-Radical
H. Steffen, P. Schwaller, J.-F. Moser, and F. K. Kneubühl (Swiss Fed. Inst. of Tech.)
Physics Letters, Vol. 23, pp. 313–314, October 31, 1966 (emission at 0.676 mm, laser mechanism proposed)

7.228 Characteristics of a Pulsed ArII Ion Laser Using the External Spark Gap
S. Kobayashi, T. Izawa, K. Kawamura, and M. Kamiyama (U. of Tokyo, Japan)
IEEE J. Quantum Electronics, Vol. QE-2, pp. 699–700, October 1966 (spark gap-capacitor discharge, 100 pps, varied pressure)

7.229 Threshold Properties of Continuous Duty Rare Gas Ion Laser Transitions
E. F. Labuda and A. M. Johnson (BTL)
IEEE J. Quantum Electronics, Vol. QE-2, pp. 700–701, October 1966 (argon, krypton, xenon, neon, 4370 to 8716 Å)

7.230 New Laser Wavelengths in Krypton
T. H. E. Cottrell, D. C. Sinclair, and J. M. Forsyth (U. of Rochester)
IEEE J. Quantum Electronics, Vol. QE-2, p. 703, October 1966 (pulsed, 6072 and 6417 Å)

7.231 Tunable Infrared Maser Spectrometers
Katsumi Sakurai and Koichi Shimoda (U. of Tokyo, Japan)
Japanese J. Appl. Phys., Vol. 5, pp. 938–947, October 1966 (magnetically tunable, He–Xe at 3.5 μ, He–Ne at 3.39 μ, tuned up to ±4000 MHz)

7.232 Transition Probabilities between Laser States in Carbon Dioxide
H. Statz (Raytheon), C. L. Tang (Cornell U.), and G. F. Koster (M.I.T.)
J. Appl. Phys., Vol. 37, pp. 4278–4284, October 1966

7.233 Discharge Dynamics in the Large Diameter Magnetically Confined Ion Gas Laser
S. A. Ahmed and T. J. Faith, Jr. (RCA)
Proc. IEEE, Vol. 54, pp. 1470–1471, October 1966 (argon, pressure differential within tube, pressure effects)

7.234 Direct Electron Excitation Cross Sections Pertinent to the Argon Ion Laser
W. R. Bennett, Jr., G. N. Mercer, P. J. Kindlmann, B. Wexler, and H. Hyman (Yale U.)
Phys. Rev. Letters, Vol. 17, pp. 987–991, November 7, 1966

7.235 Notes on the Submillimeter Laser Emission from Cyanic Compounds
W. Prettl and L. Genzel (U. of Freiburg, Germany)
Physics Letters, Vol. 23, pp. 443–444, November 14, 1966 (other lines between 119 and 334 μ)

7.236 An Investigation of the Effect of Gas Additives on the Electron Temperature and Density in a CO_2 Laser Discharge
P. O. Clark and M. R. Smith (Hughes Research)
Appl. Phys. Letters, Vol. 9, pp. 367–369, November 15, 1966 (3 eV electron temperature, 3×10^9 e cm^{-3}, added He, radial profiles)

7.237 Pulsed Operation of CO_2-N_2-He Lasers
P. O. Clark and M. R. Smith (Hughes Research)
Appl. Phys. Letters, Vol. 9, pp. 369–372, November 15, 1966 (almost 10^2 increase in power compared to CW operation, 50 to 200 μs pulses)

7.238 Evaluation of Absorption and Gain from Spontaneous Emission Profiles
E. A. Ballik and C. J. Elliott (Yale U.)
Appl. Optics, Vol. 5, pp. 1858–1861, November 1966 (argon discharge, 4401 Å transition)

7.239 Intra-Cavity Color Selection in Ion Lasers
C. P. Wen, J. M. Hammer, I. Gorog, F. W. Spong, and J. A. van Raalte (RCA)
IEEE J. Quantum Electronics, Vol. QE-2, pp. 711–713, November 1966 (electrooptic crystal in laser cavity; red, green, and blue from Kr$^+$ laser)

7.240 Heterodyne Detection and Linewidth Measurement with High Power CO_2 Lasers
R. A. Brandewie, W. T. Haswell, III, and R. H. Harada (Autonetics)
IEEE J. Quantum Electronics, Vol. QE-2 pp. 756–757, November 1966 (Cu-doped Ge mixer, near 4°K, linewidth less than 710 Hz)

7.241 New Laser Oscillations in Xenon and Krypton
R. H. Neusel (TRW/Systems)
IEEE J. Quantum Electronics, Vol. QE-2, p. 758, November 1966 (4049, 6238, and 6176 Å from Xe, 6037, 6310, 6312, and 6602 Å from Kr)

7.242 Relative Intensities of the $5s^2P_{3/2} \rightarrow 4p^2D_{5/2}$ and the $5s^4P_{5/2} \rightarrow 4p^4D_{7/2}$ Transitions in the Ar II Laser
H. Marantz, R. I. Rudko, and C. L. Tang (Cornell U.)
Appl. Phys. Letters, Vol. 9, pp. 409–411, December 1, 1966

7.243 CO_2 Laser Self-Modulation Characteristics
G. B. Jacobs (GE)
Appl. Optics, Vol. 5, pp. 1960–1961, December 1966

7.244 Polarization Characteristics of an Ionized-Gas Laser in a Magnetic Field
Douglas C. Sinclair (U. of Rochester)
J. Opt. Soc. Am., Vol. 56, pp. 1727–1731, December 1966 (argon, output power increases with field, weak Faraday rotation)

7.245 Interpretation of Oscillation Lines in Ar-Br$_2$ Laser
L. N. Tunitsky and E. M. Cherkasov (Lebedev Phys. Inst., Acad. of Sci., USSR)
J. Opt. Soc. Am., Vol. 56, pp. 1783–1784, December 1966 (four wavelengths near 8446Å due to oxygen impurity)

8. ORGANIC AND CHEMICAL LASERS

8.1 Optical Maser Action in an Eu^{+3}-containing Organic Matrix
N. E. Wolff and R. J. Pressley (RCA)
Appl. Phys. Letters, Vol. 2, pp. 152–154, April 15, 1963

8.2 Stimulated Processes in Organic Compounds
Alexander Lempicki and Harold Samelson (G. T. & E. Labs.)
Appl. Phys. Letters, Vol. 2, pp. 159–161, April 15, 1963

8.3 Triplet Excitons and Delayed Fluorescence in Anthracene Crystals
R. G. Kepler, J. C. Caris, P. Avakian and E. Abramson (E. I. du Pont)
Phys. Rev. Letters, Vol. 10, pp. 400–402, May 1, 1963

8.4 Fluorescence Lifetime of the Europium Dibenzoylmethides
Max Metlay (G.E.)
J. Chem. Phys., Vol. 39, pp. 491–492, July 15, 1963

8.5 Stimulated Emission in Rare-Earth Chelate (europium benzoylacetonate) in a Capillary Tube
Erhard J. Schimitschek (U.S. Navy Electr. Lab.)
Appl. Phys. Letters, Vol. 3, pp. 117–118, October 1, 1963

8.6 Triplet Energy Transfer and Triplet–Triplet Interaction in Aromatic Crystals

Joshua Jortner, Sang-il Choi, Joseph L. Katz, and Stuart A. Rice (U. of Chicago)
Phys. Rev. Letters, Vol. 11, pp. 323–326, October 1, 1963

8.7 Study of Anthracene Fluorescence Excited by the Ruby Giant-Pulse Laser
J. L. Hall (J.I.L.A.), D. A. Jennings, and R. M. McClintock (N.B.S.)
Phys. Rev. Letters, Vol. 11, pp. 364–366, October 15, 1963

8.8 How Rare Earth Chelate Lasers Work
Max Metlay (G.E.)
Electronics, Vol. 36, No. 46, pp. 67–69, November 15, 1963

8.9 Franck-Condon Pumping of Vibrational Energy Levels
M. W. Muller, A. Sher, R. Solomon, and D. G. Low (Varian)
Proc. Third Int. Conf. Quantum Electronics, Columbia U. Press, N. Y., Vol. 1, pp. 685–690, 1964

8.10 Rare Earth Chelates and the Molecular Approach to Lasers
H. Lyons (Lyons Research Assoc.) and M. L. Bhaumik (Electro-Optical Sys.)
Proc. Third Int. Conf. Quantum Electronics, Columbia U. Press, N. Y., Vol. 1, pp. 699–708, 1964

8.11 Investigations of a Europium Chelate Solution as a Potential Liquid Optical Maser
R. C. Ohlmann, E. P. Riedel, R. G. Charles, and J. M. Feldman (Westinghouse)
Proc. Third Int. Conf. Quantum Electronics, Columbia U. Press, N. Y., Vol. 1, pp. 779-785, 1964

8.12 Refractive Gradient Effects in Proposed Liquid Lasers
H. Winston and R. A. Gudmundsen (Quantum Tech. Labs.)
Appl. Optics, Vol. 3, pp. 143-146, January 1964

8.13 Stimulated Emission in Aromatic Organic Compounds
D. L. Stockman, W. R. Mallory, and K. F. Tittel (G.E.)
Proc. IEEE, Vol. 52, pp. 318-319, March 1964

8.14 Double-Photon Excitation of Fluorescence in Anthracene
M. Iannuzzi and E. Polacco (Oxford U., England)
Phys. Rev. Letters, Vol. 13, pp. 371-372, September 21, 1964 (analytical)

8.15 Laser Action in Europium Dibenzoylmethide
E. J. Schimitschek and R. B. Nehrich, Jr, (U.S.N. Electronics Lab.)
J. Appl. Phys., Vol. 35, pp. 2786-2787, September 1964 (spiking and beam pattern)

8.16 Laser Action in Europium Chelates Prepared with Ammonia
R. B. Nehrich, E. J. Schimitschek, and J. A. Trias (U. S. NEL)
Physics Letters, Vol. 12, pp. 198-199, October 1, 1964 (185-J threshold)

8.17 Room-Temperature Operation of a Europium Chelate Liquid Laser
H. Samelson, A. Lempicki, C. Brecher, and V. Brophy (G. T. and E. Labs.)
Appl. Phys. Letters, Vol. 5, pp. 173-174, November 1, 1964 (tetrakis form of chelate, 1700-J threshold)

8.18 Laser Action in Rare Earth Chelates
A. Lempicki, H. Samelson, and C. Brecher (GT & E)
Appl. Optics, Suppl. 2: Chemical Lasers, pp. 205-213, 1965 (review)

8.19 Mechanism of Energy Transfer in Some Rare-Earth Chelates
M. L. Bhaumik and M. A. El-Sayed (Electro-Optical Systems)
Appl. Optics, Suppl. 2: Chemical Lasers, pp. 214-215, 1965

8.20 Chemical Lasers I
W. C. Nieuwpoort and R. Bleekrode (Philips Res. Labs., Eindhoven, The Netherlands)
Z. angewandte Mathematik und Physik, Vol. 16, No. 1, pp. 101-106, January 25, 1965

8.21 Chemical Lasers II
R. Bleekrode and W. C. Nieuwpoort (Philips Res. Labs., Eindhoven, The Netherlands)
Z. angewandte Mathematik und Physik, Vol. 16, No. 1, pp. 107-110, January 25, 1965 (oxy-acetylene flame investigation)

8.22 Stimulated Emission in an Europium Chelate Solution at Room Temperature
E. J. Schimitschek, J. A. Trias, and R. B. Nehrich, Jr. (U. S. Navy Elec. Lab.)
J. Appl. Phys., Vol. 36, pp. 867-868, Part 1, March 1965 (thresholds of 150j at −40°C and 800j at 30°C, 6118 Å, concentration effect)

8.23 HCl Chemical Laser
J. V. V. Kasper and G. C. Pimentel (U. of Calif.)
Phys. Rev. Letters, Vol. 14, pp. 352-354, March 8, 1965 (about 3.8μ, vibrational excitation in chemical reaction, pulsed, 200j threshold)

8.24 Effect of Organic Cations on the Laser Threshold of Solutions of Europium Tetrakis Benzoyltrifluoroacetonate
E. P. Riedel and R. G. Charles (Westinghouse)
J. Appl. Phys., Vol. 36, pp. 3954-3955, December 1965 (laser action from 14 organic salts of Eu (BTF)₄⁻)

8.25 Deuterium Isotope Effect on the Performance of Europium Chelate Lasers
Daniel L. Ross, Joseph Blanc, and Robert J. Pressley (RCA)
Appl. Phys. Letters, Vol. 8, pp. 101-102, February 15, 1966

8.26 Chemical Lasers
George C. Pimentel (U. of California)
Scientific American, Vol. 204, No. 4, pp. 32-39, April 1966

8.27 Laser Oscillation in Chemically Formed CO
M. A. Pollack (B.T.L.)
Appl. Phys. Letters, Vol. 8, pp. 237-238, May 1, 1966 (flash photolysis, $CO_2 + O_2$, 2000 J input, 31 lines between 1761 and 1961 cm⁻¹)

8.28 Erratum: Laser Oscillation in Chemically Formed CO
M. A. Pollack (BTL)
Appl. Phys. Letters, Vol. 9, p. 74, July 1, 1966 (Appl. Phys. Letters, Vol. 8, p. 237, 1966)

8.29 Chemical Pumping through Thermal Decomposition of Dimethyl Peroxide
J. R. Henderson and M. Muramoto (Douglas Aircraft)
Appl. Optics, Vol. 5, pp. 831-834, May 1966 (unsuccessful laser experiment)

8.30 Gas Laser Excited in the Process of Photodissociation
T. L. Andreeva, V. A. Dudkin, V. I. Malyshev, G. V. Mikhaĭlov, V. N. Sorokin, and L. A. Novikova (Lebedev Physics Institute, Academy of Sciences, U.S.S.R.)
Soviet Physics JETP, Vol. 22, pp. 969-970, May 1966 (atomic iodine emission from CH₃I and CF₃I, 40 μs pump pulse, 220 J threshold)

8.31 High-Energy Atomic Iodine Photodissociation Laser
A. J. DeMaria and C. J. Ultee (United Aircraft)
Appl. Phys. Letters, Vol. 9, pp. 67-69, July 1, 1966 (CF₃I laser, 1.15 μ, 65 J output, 10⁵ W for 1.5 μs)

8.32 Erratum: High-Energy Atomic Iodine Photodissociation Laser
A. J. DeMaria and C. J. Ultee (United Aircraft)
Appl. Phys. Letters, Vol. 9, p. 218, September 1, 1966 (Appl. Phys. Letters, vol. 9, p. 67, 1966)

8.33 Molecular Laser Action in Nitric Oxide by Photodissociation of NOCl
M. A. Pollack (BTL)
Appl. Phys. Letters, Vol. 9, pp. 94-96, July 15, 1966 (15 emission lines between 5.59 and 6.30 μ, 4 μs output pulse)

8.34 An Optical Resonator for a Laser with a Liquid Active Substance
V. P. Bykov
Soviet Phys. JETP, Vol. 23, pp. 94-96, July 1966 (design considerations, europium chelates, self focusing)

8.35 Recirculating Liquid Laser
Erhard J. Schimitschek, Richard B. Nehrich and John A. Trias USN Electronics Lab.)
Appl. Phys. Letters, Vol. 9, pp. 103-104, August 1, 1966 (europium chelate, 10°C, 13.5 cm long cell, 140 J threshold)

8.36 Fluorescence Quantum Efficiencies of Octa-coordinated Europium Homogeneous and Mixed Chelates in Organic Solvents
N. Filipescu and G. W. Mushrush (George Wash. U.), and C. R. Hurt and N. McAvoy (Goddard)
Nature, Vol. 211, pp. 960-961, August 27, 1966

8.37 End-Pumped Stimulated Emission from a Thiacarbocyanine Dye
P. P. Sorokin, W. H. Culver, E. C. Hammond, and J. R. Lankard (IBM)
IBM J. Res. & Dev., Vol. 10, p. 401, September 1966 (Q-sw ruby pump 0.63 MW output at 0.816 μ, 14 percent efficiency)

9. GaAs AND GaAs₁₋ₓPₓ INJECTION LUMINESCENT DEVICES

9.58 Absorption Data of Laser-Type GaAs at 300° and 77°K
      W. J. Turner and W. E. Reese (IBM)
      J. Appl. Phys., Vol. 35, pp. 350–352, February 1964

9.59 Breakdown and Hysteresis in Forward Biased p–n GaAs
      Luminescent Junctions
      R.C.C. Leite and A. Yariv (BTL)
      Proc. IEEE, Vol. 52, pp. 191–192, February 1964

9.60 Quantum Efficiency of a New GaAs Spontaneous Infrared
      Source
      W. N. Carr and G. E. Pittman (Texas Instruments)
      Proc. IEEE, Vol. 52, pp. 204–205, February 1964

9.61 Modification of the Near-Field Pattern of a GaAs Laser by a
      Magnetic Field
      A. B. Fowler and E. J. Walker (IBM)
      J. Appl. Phys., Vol. 35, pp. 1241–1242, April 1964

9.62 Harmonic Generation in Injection Lasers
      A. W. Smith, M. I. Nathan, J. A. Armstrong, A. E. Michel,
      and K. Weiser (IBM)
      J. Appl. Phys., Vol. 35, pp. 733–734, Part I, March 1964

9.63 Delay between Current Pulse and Light Emission of a Gallium
      Arsenide Injection Laser
      K. Konnerth and C. Lanza (IBM)
      Appl. Phys. Letters, Vol. 4, pp. 120–121, April 1, 1964 (rise
      time less than 0.2 ns infers modulation of about 5 Gc)

9.64 Shaped Electroluminescent GaAs Diodes
      A. R. Franklin and R. Newman (Sperry Rand)
      J. Appl. Phys., Vol. 35, pp. 1153–1155, April 1964

9.65 Refractive Index of GaAs
      D. T. F. Marple (G.E.)
      J. Appl. Phys., Vol. 35, pp. 1241–1242, April 1964

9.66 Light Emitting Characteristics of a GaAs Diode Having Nega-
      tive Resistance
      Tatsuo Yamamoto (Shizuoka U., Japan)
      Proc. IEEE, Vol. 52, p. 409, April 1964

9.67 GaAs p–si–n Negative Resistance Infrared Emitting Diode at
      Liquid N$_2$ and Room Temperatures
      S. W. Ing, Jr., H. A. Jensen, and B. Stern (G.E.)
      Appl. Phys. Letters, Vol. 4, pp. 162–164, May 1, 1964

9.68 Pressure Effect on Resistivity of Ga(As$_{1-x}$P$_x$)
      G. E. Fenner (G.E.)
      Phys. Rev., Vol. 134, pp. A1113–A1118, May 18, 1964

9.69 Effect of Impurity Distribution on Simultaneous Laser Action in
      GaAs at 0.84 and 0.88μ
      H. Nelson and G. C. Dousmanis (RCA)
      Appl. Phys. Letters, Vol. 4, pp. 192–194, June 1, 1964 (possible
      Nd pump, possible internal mixing)

9.70 Interferometric Measurement of Linewidth and Noise in GaAs
      Lasers
      J. A. Armstrong and A. W. Smith (IBM)
      Appl. Phys. Letters, Vol. 4, pp. 196–198, June 1, 1964 (linewidth
      less than 50 Mc/s, 1-mW output, 15°K, low noise)

9.71 Radiation Transfer by a Light Pipe Between Media with High
      Indices of Refraction
      M. A. Gilleo (Amelco Semiconductor)
      Appl. Optics, Vol. 3, pp. 765–767, June 1964 (GaAs experiment,
      As$_2$S$_3$ light pipes)

9.72 Characteristics of a Continuous High Power GaAs Junction Laser
      W. E. Engeler and M. Garfinkel (G. E.)
      J. Appl. Phys., Vol. 35, pp. 1734–1741, June 1964 (20°K, 3.1-watt
      output, 46 per cent efficiency, radiation pattern)

9.73 Carrier Transport Across Electroluminescent p-n Junctions in GaAs
      J. I. Pankove (RCA)

J. Appl. Phys., Vol. 35, pp. 1890–1892, June 1964 (relationship
between junction voltage and emission spectrum, 30-meV vari-
ance)

9.74 Output Power from GaAs Lasers at Room Temperature
      C. C. Gallagher, P. C. Tandy (AFCRL), B. S. Goldstein, and
      J. D. Welch (Lincoln Lab., MIT)
      Proc. IEEE, Vol. 52, pp. 717–719, June 1964 (14-watts peak,
      10-ns pulse)

9.75 Triangular Injection Lasers
      J. C. Marinace, A. E. Michel, and M. I. Nathan (IBM)
      Proc. IEEE, Vol. 52, pp. 722–723, June 1964 (equilateral triangle)

9.76 Absorption Edge Measurements in Compensated GaAs
      G. Lucovsky (Philco)
      Appl. Phys. Letters, Vol. 5, pp. 37–39, July 15, 1964 (edge shift
      with compensation)

9.77 Cooperative Effect in GaAs Lasers
      A. B. Fowler (IBM)
      J. Appl. Phys., Vol. 35, pp. 2275–2276, July 1964 (short and
      long units as pair)

9.78 Light-Emitting, Formed-Point-Contact Gallium Arsenide and Gal-
      lium Arsenide-Phosphide Diodes
      L. U. Kibler, C. A. Burrus, and R. F. Trambarulo (BTL)
      Proc. IEEE, Vol. 52, pp. 850–851, July 1964 (very low capaci-
      tance, shortest wavelength about 0.7μ)

9.79 Modification of the Threshold Current and Near-Field Emission
      Pattern of a GaAs Laser by an Adsorbed Dielectric Layer
      E. J. Walker and A. E. Michel (IBM)
      J. Appl. Phys., Vol. 35, pp. 2285–2289, August 1964 (threshold
      current fluctuates with time)

9.80 New Junction Laser Resonant Structures
      M. Garfinkel, W. E. Engeler, and D. J. Locke (G. E.)
      J. Appl. Phys., Vol. 35, pp. 2321–2323, August 1964 (unusual
      radiation patterns)

9.81 Electroluminescent Gallium Arsenide Diodes with Negative Re-
      sistance
      K. Weiser and R. S. Levitt (IBM)
      J. Appl. Phys., Vol. 35, pp. 2431–2438, August 1964

9.82 Use of Electron Probes in the Study of Recombination Radiation
      D. B. Wittry and D. F. Kyser (USC)
      J. Appl. Phys., Vol. 35, pp. 2439–2442, August 1964 (wide varia-
      tion in radiation among different specimens due to surface de-
      formation and to nonuniform distribution of impurities)

9.83 GaAs Laser Linewidth Measurements by Heterodyne Detection
      J. W. Crowe and R. M. Craig, Jr. (IBM)
      Appl. Phys. Letters, Vol. 5, pp. 72–74, August 15, 1964 (7.2-cm
      long cavity with external mirror, 1.9-Gc beat, 50 Mc full width)

9.84 High-Order Transverse Modes in GaAs Lasers
      K. Weiser and F. Stern (IBM)
      Appl. Phys. Letters, Vol. 5, pp. 115–116, September 15, 1964
      (pulsed, two beams 30° apart)

9.85 Common Occurrence of Artifacts or "Ghost" Peaks in Semicon-
      ductor Injection and Electroluminescence Spectra
      W. N. Carr and J. R. Biard (Texas Instruments)
      J. Appl. Phys., Vol. 35, pp. 2776–2777, September 1964

9.86 Optical Generation Spectrum for the Electron Thermal-Injection
      Mechanism in GaAs Diodes
      W. N. Carr and J. R. Biard (Texas Instruments)
      J. Appl. Phys., Vol. 35, pp. 2777–2779, September 1964

9.87 Electron-Beam Pumped GaAs Laser
      C. E. Hurwitz and R. J. Keyes (MIT)
      Appl. Phys. Letters, Vol. 5, pp. 139–141, October 1, 1964 (liquid
      helium temperature, 50-kV beam, 0.2-μs pulse, 1000 pps)

9.88 Recombination Scheme and Intrinsic Gap Variation in GaAs$_{1-x}$P$_x$
      Semiconductors from Electron Beam and p-n Diode Excitation
      D. A. Cusano, G. E. Fenner, and R. O. Carlson (G. E.)
      Appl. Phys. Letters, Vol. 5, pp. 144–146, October 1, 1964

9.89 Injection Lasers Far Above Threshold
F. N. Hooge and H. Kalter (Philips Res. Lab., Netherlands)
Physics Letters, Vol. 12, pp. 191–192, October 1, 1964 (modes disappear in GaAs laser)

9.90 The Role of Diffusion Current in the Electroluminescence of GaAs Diodes
M. F. Millea and L. W. Aukerman (Aerospace)
Appl. Phys. Letters, Vol. 5, pp. 168–169, October 15, 1964

9.91 CW Operation of GaAs Injection Lasers
M. F. Lamorte, R. B. Liebert, and T. Gonda (RCA)
Proc. IEEE, Vol. 52, pp. 1257–1258, October 1964 (calculations show CW operation not possible above 200°K with present threshold current density)

9.92 Temperature Dependence of Threshold Current in GaAs Lasers
G. C. Dousmanis, H. Nelson, and D. L. Staebler (RCA)
Appl. Phys. Letters, Vol. 5, pp. 174–176, November 1, 1964 (T³ dependence at higher temperatures, almost constant current below 10°K)

9.93 Fluorescence of GaAs Under Intense Electron Excitation
R. V. Babcock (Westinghouse)
J. Appl. Phys., Vol. 35, pp. 3354–3357, November 1964 (77° to 300°K, varying emission wavelength)

9.94 Some Observations on Triangular GaAs Lasers
I. Ladany (U. S. Naval Res. Lab.)
Proc. IEEE, Vol. 52, pp. 1353–1354, November 1964

9.95 Room-Temperature GaAs Laser Voice-Communication System
D. Karlsons, C. W. Reno, and W. J. Hannan (RCA)
Proc. IEEE, Vol. 52, pp. 1354–1355, November 1964 (30 A., 30-ns pulses, 1.5 watts out, 20 kc average rep rate)

9.96 High-Efficiency Injection Laser at Room Temperature
H. Nelson, J. I. Pankove, F. Hawrylo, G. C. Dousmanis, and C. Reno (RCA)
Proc. IEEE, Vol. 52, pp. 1360–1361, November 1964 (epitaxially grown diode, 100 μs, 120-A pulses, 60 watts out, 30 pps)

9.97 Internal Frequency Modulation of GaAs Junction Lasers by Changing the Index of Refraction Through Electron Injection
G. E. Fenner (G. E.)
Appl. Phys. Letters, Vol. 5, 198–199, November 15, 1964 (pulsed dual diode, different currents, continuous mode shift up to 1 Å or 45 Gc)

9.98 High Power CW Operation of GaAs Injection Lasers at 77°K
J. C. Marinace (IBM)
IBM J. Res. and Dev., Vol. 8, pp. 543–544, November 1964 (large heat-sink mount, one watt at 77°K, 0.45 watt at 90°K)

9.99 Light Emission from Reverse Biased GaAs and InP p-n Junctions
A. E. Michel, M. I. Nathan, and J. C. Marinace (IBM)
J. Appl. Phys., Vol. 35, pp. 3543–3547, December 1964 (150 Å emission width)

9.100 Angular Distribution of Radiation from GaAs Injection Lasers
M. M. Antonoff (Sperry Gyroscope)
J. Appl. Phys., Vol. 35, pp. 3623–3624, December 1964 (analysis supports dielectric-slab model)

9.101 Evidence of Stimulated Emission in Ruby-Laser-Pumped GaAs
J. J. Schlickman, M. E. Fitzgerald, and R. H. Kingston (MIT)
Proc. IEEE, Vol. 52, pp. 1739–1740, December 1964 (threshold ruby power of 10⁴ W/cm²)

9.102 A Solid-State Room-Temperature Operated GaAs Laser Transmitter
G. F. Dalrymple, B. S. Goldstein, and T. M. Quist (MIT)
Proc. IEEE, Vol. 52, pp. 1742–1743, December 1964 (transistor switch, 40-ns drive pulse, 100 A, 4 watts out, 20 ns out)

9.103 Intensity Fluctuations in a GaAs Laser
J. A. Armstrong and A. W. Smith (IBM)
Phys. Rev. Letters, Vol. 14, pp. 68–70, January 18, 1965 (single mode cw laser, 10° K, noise bandwidth of 1300 ± 200 Mc/s)

9.104 Erratum: Intensity Fluctuations in a GaAs Laser
J. A. Armstrong and A. W. Smith (IBM)
Phys. Rev. Letters, Vol. 14, p. 208, February 8, 1965 (Phys. Rev. Letters, Vol. 14, p. 68; January 18, 1965)

9.105 Interactions between Closely Coupled GaAs Injection Lasers
C. E. Kelly (IBM)
IEEE Trans. on Electron Devices, Vol. ED-12, pp. 1–4, January 1965 (pulsed, 2° and 77° K, lowers threshold, produces quenching, single block)

9.106 Modification of the Threshold Current of GaAs Laser by a Reflective Coating on One End
Y. Nannichi (Nippon Electric, Japan)
Japanese J. Appl. Phys., Vol. 4, pp. 53–55, January 1965 (reduces with increasing reflectivity)

9.107 Inherent Properties of a Tunnel-Injection Laser
G. Wade, C. A. Wheeler, and R. G. Hunsperger (Cornell U.)
Proc. IEEE, Vol. 53, pp. 98–99, January 1965

9.108 Pressure Dependence of the Emission from Ga(As$_{1-x}$P$_x$) Electroluminescent Diodes
G. E. Fenner (G.E.)
Phys. Rev., Vol. 137, pp. A1000–A1007, February 1, 1965 (increased pressure shortens wavelengths)

9.109 Vapor-Liquid-Solid Growth of Gallium Phosphide
N. Holonyak, Jr., C. M. Wolfe and J. S. Moore (U. of Illinois)
Appl. Phys. Letters, Vol. 6, pp. 64–65, February 15, 1965 (whiskers)

9.110 Injection Luminescence in GaAs Transistors
M. H. Norwood, H. Strack and W. G. Hutchinson (Texas Instruments)
Appl. Phys. Letters, Vol. 6, pp. 71–73, February 15, 1965 (77 to 300°K, recombination radiation)

9.111 Temperature Dependence of the Multimode Behavior of GaAs Lasers
J. M. Lavine and A. A. Iannini (Raytheon)
J. Appl. Phys., Vol. 36, pp. 402–405, February 1965 (fewer modes at higher temperature)

9.112 Improving the External Efficiency of Electroluminescent Diodes
S. V. Galginaitis (G.E.)
J. Appl. Phys., Vol. 36, pp. 460–461, February 1965 (noncoherent, TIR paraboloid structure)

9.113 GaAs Injection Laser with Novel Mode Control and Switching Properties
M. I. Nathan, J. C. Marinace, R. F. Rutz, A. E. Michel, and G. J. Lasher (IBM)
J. Appl. Phys., Vol. 36, pp. 473–480, February 1965 (control by current distribution, dual units, bistable)

9.114 Thermal Conductivity of GaAs and GaAs$_{1-x}$P$_x$ Laser Semiconductors
R. O. Carlson, G. A. Slack, and S. J. Silverman (G.E.)
J. Appl. Phys., Vol. 36, pp. 505–507, February 1965 (3–300°K)

9.115 Tunneling-Assisted Radiative Recombination in GaAs-Diffused Junctions
J. E. Ripper and R. C. C. Leite (BTL)
Proc. IEEE, Vol. 53, p. 160, February 1965 (analytical)

9.116 Improved Performance of GaAs$_{1-x}$P$_x$ Laser Diodes
J. J. Tietjen and S. A. Ochs (RCA)
Proc. IEEE, Vol. 53, pp. 180–181, February 1965 (77°K, 1 μs pulses, 2-5 × 10³ amp/cm² threshold)

9.117 Laser Emission from n-Type GaAs Excited by Fast Electrons
D. A. Cusano and J. D. Kingsley (G.E.)
Appl. Phys. Letters, Vol. 6, pp. 91–93, March 1, 1965 (20 and 77°K, 16–30 kV)

9.118 Injection Mechanisms in GaAs Diffused Electroluminescent Junctions
R. C. C. Leite, J. C. Sarace, D. H. Olson, B. G. Cohen, and J. M. Whelan (BTL), and A. Yariv (California Inst. of Tech.)
Phys. Rev., Vol. 137, pp. A1583–A1590, March 1, 1965 (two of three different mechanisms tested)

9.119 Saturation of the Optical Absorption in GaAs
       A. E. Michel and M. I. Nathan (IBM)
       Appl. Phys. Letters, Vol. 6, pp. 101–102, March 15, 1965 (transmission slope increases with intensity)

9.120 Radiative Recombination in GaP p-n and Tunnel Junctions
       R. A. Logan, M. Gershenzon, F. A. Trumbore, and H. G. White (BTL)
       Appl. Phys. Letters, Vol. 6, pp. 113–115, March 15, 1965 (efficient junctions)

9.121 Effects of Roughening Cleaved Surfaces on the Characteristics of GaAs Injection Laser Diodes
       J. Nishizawa (Tohoku U., Japan) and I. Sasaki and Kaoru Takahashi (Semiconductor Res. Inst., Japan)
       Appl. Phys. Letters, Vol. 6, pp. 115–116, March 15, 1965 (increased power, 60 Å shorter wavelength)

9.122 On the Parameters Which Affect the C. W. Output of GaAs Lasers
       M. Ciftan and P. P. Debye (Raytheon)
       Appl. Phys. Letters, Vol. 6, pp. 120–121, March 15, 1965 (1.7 cm long diode, helium temp., 7.2 to 12 watts cw output, mode beat freq linewidth less than 150 kc)

9.123 Photon-Assisted Tunneling in GaAs Diodes
       T. N. Morgan and M. I. Nathan (IBM)
       Bull. Am. Phys. Soc., Series II, Vol. 10, p. 389–LC11, March 1965 (electroluminescent diodes)

9.124 Electroluminescence and Lasing Action in GaAs$_x$P$_{1-x}$
       M. Pilkuhn and H. Rupprecht (IBM)
       J. Appl. Phys., Vol. 36, pp. 684–688, Part 1, March 1965 ($x >$ 0.55, shortest wavelength at 77° K was 6380 Å)

9.125 Self-Induced Oscillations in the Stimulated Light Emission from GaAs Injection Lasers
       R. S. Levitt and M. H. Pilkuhn (IBM)
       J. Appl. Phys., Vol. 36, p. 859, Part 1, March 1965 (cyclic laser quenching due to heating, 1–25 Mc rate at 77°K)

9.126 Electric Field Effect on the Refractive Index in GaAs
       B. O. Seraphin and N. Bottka (Michelson Lab.)
       Appl. Phys. Letters, Vol. 6, pp. 134–136, April 1, 1965 (predict index change up to -1.8% at 2×10^5 v/cm)

9.127 Intrinsic and Extrinsic Photomixing in Semi-Insulating GaAs
       R. J. Strain and C. C. Tooke (Standard Telecommuncation Labs., England)
       Appl. Phys. Letters, Vol. 6, pp. 157–158, April 15, 1965 (mixing axial modes of either 0.63 or 1.15 μ gas lasers)

9.128 Theory of the Effect of Strain on GaAs Electroluminescent Diodes
       P. R. Emtage (Westinghouse)
       J. Appl. Phys., Vol. 36, pp. 1408–1411, April 1965

9.129 On the Visibility of 8400 Å Light from GaAs Laser Diodes
       Y. Nannichi (Nippon Electric, Japan)
       Japanese J. Appl. Phys., Vol. 4, p. 308, April 1965 (eye excited by intense infrared)

9.130 The Emission Process through Donor States in GaAs Laser Diodes
       O. Ohtsuki, T. Kotani, Y. Iwai, and I. Tsurumi (Kobe Kogyo, Japan)
       Japanese J. Appl. Phys., Vol. 4, pp. 314–315, April 1965 (two kinds of radiative transition processes)

9.131 Junction Heating in GaAs Injection Lasers
       K. Konnerth (IBM)
       Proc. IEEE, Vol. 53, pp. 397–398, April 1965 (wavelength increases about 100 Å, junction temperature increases up to 60°K within pulse)

9.132 Stimulated Cathode-Luminescence in n-Type GaAs at 77°K
       P. D. Coleman and G. E. Bennett (U. of Illinois)
       Proc. IEEE, Vol. 53, pp. 419–420, April 1965 (pulsed)

9.133 Transient Thermal Effects in Gallium Arsenide Injection Lasers
       C. H. Gooch (S.E.R.L., England)
       Physics Letters, Vol. 16, pp. 5–6, May 1, 1965 (55 Å shift in 18 μsec, about 20° temp rise, Joule and contact resistance heating)

9.134 Intensity Fluctuations and Correlations in a GaAs Laser
       A. W. Smith and J. A. Armstrong (IBM)
       Physics Letters, Vol. 16, pp. 38–39, May 1, 1965 (lasing and non-lasing modes)

9.135 Generation in GaAs under Two-Photon Optical Excitation of Neodymium-Glass Laser Emission
       N. G. Basov, A. Z. Grasyuk, I. G. Zubarev, and V. A. Katulin (Lebedev Physics Inst., Academy of Sciences, U.S.S.R.)
       JETP Letters, Vol. 1, pp. 118–120, May 15, 1965 (Q-sw Nd glass laser pump)

9.136 Interband Magneto-Optical Absorption in Gallium Arsenide
       M. V. Hobden (Royal Radar Estab., England)
       Physics Letters, Vol. 16, pp. 107–108, May 15, 1965 (absorption edge shift, up to 120 k gauss, 20°K)

9.137 Threshold Dependency on Reabsorption Loss in Injection Lasers
       M. F. Lamorte, H. Junker, and T. Gonda (RCA)
       IEEE J. Quantum Electronics, Vol. QE-1, pp. 103–104, May 1965 (between 77 to 300°K in GaAs degenerate junctions)

9.138 Radiative Recombination between Deep-Donor-Acceptor Pairs in GaP
       M. Gershenzon, F. A. Trumbore, R. M. Mikulyak, and M. Kowalchik (BTL)
       J. Appl. Phys., Vol. 36, pp. 1528–1537, May 1965 (many bands just below band gap)

9.139 Spectral Distribution of Room Temperature GaAs-Junction Luminescence as a Function of Base Thickness
       G. Lucovsky and A. J. Varga (Philco)
       Proc. IEEE, Vol. 53, pp. 491–492, May 1965 (smaller photon energy from thicker base)

9.140 Das Verhalten von GaAs-Laserdioden bei hohen Strahlungsleistungen
       H. J. Henkel, E. Klein, and H. Kuckuck (Siemens-Schuckertwerke AG, Germany)
       Solid-State Electronics, Vol. 8, pp. 475–478, May 1965 (500 amp 100 ns pulses, radiated 80 watts)

9.141 Magnetic Field Effect on the Current Distribution in p-n Junctions
       M. Garfinkel and W. E. Engeler (G.E.)
       J. Appl. Phys., Vol. 36, pp. 1877–1882, June 1965 (redistributes current, measurements on GaAs laser diode)

9.142 Location of the Source of Recombination Radiation in Ga(As$_{1-x}$P$_x$) p-n Junctions by Electron Bombardment
       C. M. Wolfe, M. D. Sirkis, C. J. Nuese, N. Holonyak, Jr., and O. L. Gaddy (U. of Illinois), and O. T. Purl and W. E. Kunz (Watkins-Johnson)
       J. Appl. Phys., Vol. 36, p. 2087, June 1965 (in junction region slightly into p-type side)

9.143 Luminescence and Recombination through Defects in p-n Junctions
       T. N. Morgan (IBM)
       Phys. Rev., Vol. 139, pp. A294–A299, July 5, 1965 (deep centers at high temperatures, copper-doped GaAs)

9.144 Effects of Optical Interaction of Two Diode Lasers
       P. G. Eliseev, A. A. Novikov, and V. B. Fedorov (Inst. of Precision Mechanics and Computation Techniques, Academy of Sciences, U.S.S.R.)
       JETP Letters, Vol. 2, pp. 36–39, July 15, 1965 (short and long GaAs diodes)

9.145 Influence of Pulse Duration on the Spectra of GaAs Lasers above Threshold
F. N. Hooge, H. Kalter, and A. M. H. Hoppenbrouwers (Philips, The Netherlands)
Physics Letters, Vol. 17, pp. 254–255, July 15, 1965 (2.5 μs and 30 ns, longer wavelength with longer pulses)

9.146 Thermal Problems of the Injection Laser
R. W. Keyes
IBM J. Res. and Dev., Vol. 9, pp. 303–314, July 1965 (pulsed and continuous, maximum of stimulated power in a pulse)

9.147 Properties of GaAs Diodes with P–P°–N Structures
K. Weiser
IBM J. Res. and Dev., Vol. 9, pp. 315–326, July 1965 (electrical and electroluminescent properties, fabrication, few ns switching time)

9.148 Time-Resolved Spectral Output of Pulsed GaAs Lasers
T. Gonda, H. Junker, and M. F. Lamorte (RCA)
IEEE J. Quantum Electronics, Vol. QE-1, pp. 159–163, July 1965 (longer wavelength during pulse, 100 Å shift)

9.149 Optical Gain and Losses of Epitaxial and Diffused GaAs Injection Lasers
M. H. Pilkuhn, H. Rupprecht, and J. Woodall (IBM)
IEEE J. Quantum Electronics, Vol. QE-1, p. 184, July 1965 (lower threshold from epitaxial diodes)

9.150 Silicon-Controlled-Rectifier Long-Pulse Driver for Injection Lasers
M. C. Teich, D. A. Berkley, and G. J. Wolga (Cornell U.)
Rev. Sci. Instr., Vol. 36, pp. 973–974, July 1965 (longer than 30 μs, up to 20 amps)

9.151 RF Transmission Line for Low Temperature Applications
Ronald A. Laing (Tulane U.)
Rev. Sci. Instr., Vol. 36, p. 1045, July 1965

9.152 Ultrafast Infrared Emitting Diodes
P. Leclerc (Compagnie Générale de Télégraphie Sans Fil, France) and C. Zajde (Ecole Normale Supérieure, Laboratoire de l'Accélérateur Linéaire, France)
Rev. Sci. Instr., Vol. 36, pp. 1056–1057, July 1965 (matched mount, subnanosecond risetimes)

9.153 Electroluminescence Near Band Gap in Gallium Phosphide Containing Shallow Donor and Acceptor Levels
L. M. Foster and M. Pilkuhn (IBM)
Appl. Phys. Letters, Vol. 7, pp. 65–67, August 1, 1965

9.154 Photomixing in a GaAs$_x$P$_{1-x}$–GaAs Heterodiode
T. B. Ramachandran, K. K. Chow, W. J. Moroney, and P. Olendzensky (Microwave Associates)
J. Appl. Phys., Vol. 36, pp. 2594–2595, August 1965 (bromine-argon laser, beats in 13–22 Gc/s region)

9.155 Identification of Laser Transitions in Electron-Beam-Pumped GaAs
D. A. Cusano (GE)
Appl. Phys. Letters, Vol. 7, pp. 151–152, September 15, 1965 (varied majority dopant concentration)

9.156 Junction Heating of GaAs Injection Lasers during Continuous Operation
M. H. Pilkuhn and H. S. Rupprecht
IBM J. Res. and Dev., Vol. 9, pp. 400–404, September-November 1965 (ΔT, quantum efficiencies, pulsed and CW)

9.157 Mode Confinement and Gain in Junction Lasers
W. W. Anderson (Stanford U.)
IEEE J. Quantum Electronics, Vol. QE-1, pp. 228–236, September 1965 (three-laser media, gain-phase relations)

9.158 Quantum Efficiency and Radiative Lifetime in p-Type Gallium Arsenide
Jüri Vilms and William E. Spicer (Stanford Electronics Lab.)
J. Appl. Phys., Vol. 36, pp. 2815–2821, September 1965 (28 percent and 3×10^{-10} seconds at 80°K, 8 percent and 2×10^{-8} seconds at 300°K)

9.159 Laser-Action Threshold in Electron-Beam Excited Gallium Arsenide
Claude A. Klein (Raytheon)
Appl. Phys. Letters, Vol. 7, pp. 200–202, October 1, 1965 (suggests lower threshold at room temperature using heavily compensated n-type units)

9.160 Intensity Fluctuations in GaAs Laser Emission
J. A. Armstrong and Archibald W. Smith (IBM)
Phys. Rev., Vol. 140, pp. A155–A164, October 4, 1965 (below and above threshold, noise consistent with van der Pol's oscillator model)

9.161 Evidence for Avalanche Injection Laser in p-Type GaAs
K. Weiser and J. F. Woods (IBM)
Appl. Phys. Letters, Vol. 7, pp. 225–228, October 15, 1965

9.162 Characteristics of a GaAs Spontaneous Infrared Source with 40 Percent Efficiency
William N. Carr (Texas Instruments)
IEEE Trans. on Electron Devices, Vol. ED-12, pp. 531–535, October 1965

9.163 Transient Change of Emission in Pulsed GaAs Laser
Manabu Saji and Yoshio Inuishi (Osaka U., Japan)
Japanese J. Appl. Phys., Vol. 4, p. 820, October 1965 (about 10 Å longer wavelength during 2 μs pulse, mechanisms)

9.164 Radiation Damage and Annealing of GaAs Laser Diode
Manabu Saji and Yoshio Inuishi (Osaka U., Japan)
Japanese J. Appl. Phys., Vol. 4, pp. 830–831, October 1965 (^{60}Co radiation, increased threshold)

9.165 GaAs Optically Coupled Transistor with a Lasing Emitter
R. F. Rutz, M. I. Nathan, A. E. Michel, and J. C. Marinace (IBM)
Proc. IEEE, Vol. 53, p. 1664, October 1965 (high collection efficiency when lasing, 4.2°K, pulsed, also 100–1200 Mc/s oscillations)

9.166 Threshold Dependency on Reabsorption Loss in Injection Lasers
Frank Stern (IBM Zurich Research Lab., Switzerland)
IEEE J. Quantum Electronics, Vol. QE-1, p. 358, November 1965 (comment on paper by M. F. Lamorte, et al, IEEE J. Quantum Electronics, QE-1, 103–104, May 1965)

9.167 Internal Quantum Efficiency of GaAs Electroluminescent Diodes
Dale E. Hill (Monsanto)
J. Appl. Phys., Vol. 36, pp. 3405–3409, November 1965 (greater than 50 percent)

9.168 Injection Mechanisms in GaAs Diffused Electroluminescent Junctions
R. C. C. Leite, J. C. Sarace, D. H. Olson, B. G. Cohen, J. M. Whelan, and A. Yariv
Phys. Rev., Vol. 140, p. AB4, December 27, 1965 (Erratum, Phys. Rev., Vol. 137, pp. A1583–A1590, March 1, 1965)

9.169 Single-Filamentary Junction Laser
Akira Kawaji, Hiroo Yonezu and Yoshihiro Yasuoka (Nippon Electric Co., Japan)
Japanese J. Appl. Phys., Vol. 4, pp. 1024–1025, December 1965 (GaAs, two portions with small overlap)

9.170 Growth and Dislocation Structure of Single-Crystal Ga(As$_{1-x}$P$_x$)
       C. M. Wolfe, C. J. Nuese, and N. Holonyak, Jr. (U. of Illinois)
       J. Appl. Phys., Vol. 36, pp. 3790–3801, December 1965 (preparation techniques, 3000 A/cm^2 threshold current laser diodes with x = 1/3)

9.171 Uniaxial Pressure Wavelength Changes in GaAs Lasers in CW Operation
       R. H. Durrett, E. D. Jacobs, J. Winocur, and W. L. Zingery (Autonetics)
       Proc. IEEE, Vol. 53, pp. 2121–2122, December 1965 (about 2.5 Å shorter per 100 bars)

9.172 Mechanism for Radiative Recombination in GaAs P-N Junctions
       G. Lucovsky (Philco)
       Proc. Physics of Quantum Electronics Conf., McGraw-Hill Book Co., New York, pp. 467–477, 1966

9.173 Excitation Dependence of Photoluminescence in N- and P-Type Compensated GaAs
       Marshall I. Nathan and T. N. Morgan (IBM)
       Proc. Physics of Quantum Electronics Conf., McGraw-Hill Book Co., New York, pp. 478–486, 1966

9.174 Intensity Fluctuations and Correlations in a GaAs Laser
       J. A. Armstrong and Archibald W. Smith (IBM)
       Proc. Physics of Quantum Electronics Conf., McGraw-Hill Book Co., New York, pp. 701–705, 1966

9.175 Continuous Operation is Near for Uncooled Diode Lasers
       Michael F. Lamorte (RCA)
       Electronics, Vol. 39, No. 1, pp. 95–99, January 10, 1966 (basic principles, temperature, power)

9.176 Phenomena Influencing the Temperature Behavior of Stimulated Emission in GaAs p-n Junctions
       M. F. LaMorte, T. Gonda, and H. Junker (RCA)
       IEEE J. Quantum Electronics, Vol. QE-2, pp. 9–15, January 1966

9.177 Light Emission Associated with Growth Defects from Reverse-Biased GaP p-n Junctions
       M. Gershenzon and A. Ashkin (B.T.L.)
       J. Appl. Phys., Vol. 37, pp. 246–248, January 1966

9.178 Effect of Donor Impurities on the Direct-Indirect Transition in Ga(As$_{1-x}$P$_x$)
       N. Holonyak, Jr., C. J. Nuese, M. D. Sirkis, and G. E. Stillman (U. of Illinois)
       Appl. Phys. Letters, Vol. 8, pp. 83–85, February 15, 1966

9.179 The Effects of Silicon and Selenium Doping on Gallium Arsenide Laser Characteristics
       C. D. Dobson (Std. Telecomm. Labs., England)
       British J. Appl. Phys., Vol. 17, pp. 187–190, February 1966 (no significant difference)

9.180 Width of the Spontaneous Emission Region in Degenerate GaAs p-n Junctions
       H. C. Casey, Jr., R. J. Archer, R. H. Kaiser, and J. C. Sarace (B.T.L.)
       J. Appl. Phys., Vol. 37, pp. 893–898, February 1966

9.181 Room Temperature Super-Radiance Radiation in n-Type GaAs by Continuous Electron-Beam Excitation
       H. C. Casey, Jr., and R. H. Kaiser (B.T.L.)
       Appl. Phys. Letters, Vol. 8, pp. 113–115, March 1, 1966

9.182 The 1.0- and 1.28-eV Emission from GaAs Diodes
       M. F. Millea and L. W. Aukerman (Aerospace)
       J. Appl. Phys., Vol. 37, pp. 1788–1792, March 15, 1966

9.183 Optical Inhomogeneities in Gallium Arsenide
       M. E. Drougard (IBM)
       J. Appl. Phys., Vol. 37, pp. 1858–1866, March 15, 1966

9.184 Infrared Radiation from Bulk GaAs
       K. K. N. Chang, S. G. Liu, and H. J. Prager (RCA)
       Appl. Phys. Letters, Vol. 8, pp. 196–198, April 15, 1966 (high carrier concentration, nanosecond pulses, 1500 V/cm threshold, 9000 Å output)

9.185 Effect of Higher Absorption in Non-Lasing GaAs Diodes at 300°K
       T. Gonda, M. F. LaMorte, P. Nyul, and H. Junker (RCA)
       IEEE J. Quantum Electronics, Vol. QE-2, pp. 74–76, April 1966 (reduces efficiency)

9.186 Optimization of the Gallium Arsenide Injection Laser for Maximum CW Power Output
       J. Vilms, L. Wandinger, and K. L. Klohn (U. S. Army Electronics Command)
       IEEE J. Quantum Electronics, Vol. QE-2, pp. 80–83, April 1966

9.187 Double Quenching on a Selective Diffused Junction Laser
       Akira Kawaji, Hiroo Yonezu, and Yoshihiro Yasuoka (Nippon Electric, Japan)
       Japanese J. Appl. Phys., Vol. 5, pp. 340–341, April 1966 (perpendicular beams)

9.188 Dislocations and Precipitates in GaAs Injection Lasers
       M. S. Abrahams and C. J. Buiocchi (RCA)
       J. Appl. Phys., Vol. 37, pp. 1973–1977, April 1966

9.189 Radiative Pair Recombination and Surface Recombination in GaP Photoluminescence
       M. Gershenzon and R. M. Mikulyak (B.T.L.)
       Appl. Phys. Letters, Vol. 8, pp. 245–247, May 15, 1966

9.190 Optimum Design for a Room-Temperature, Pulse-Operated GaAs Injection Laser
       A. Akselrad (RCA)
       Appl. Phys. Letters, Vol. 8, pp. 250–252, May 15, 1966 (70 ns pulses, 9 W output)

9.191 Intensity Noise in Multimode GaAs Laser Emission
       A. W. Smith and J. A. Armstrong
       IBM J. Res. & Dev., Vol. 10, pp. 225–232, May 1966

9.192 Large Wavelength Changes with Cavity Q in Injection Lasers
       G. C. Dousmanis and D. L. Staebler (RCA)
       J. Appl. Phys., Vol. 37, pp. 2278–2280, May 1966 (up to ±0.6%)

9.193 Gallium Arsenide Laser Operating at Room Temperature
       N. G. Basov, Yu. P. Zakharov, T. F. Nikitina, Yu. M. Popov, G. M. Strakhovskii, V. M. Tatarenkov, and A. N. Khvoshchev (Lebedev Phys. Inst., Acad. of Sci., USSR)
       JETP Letters, Vol. 3, pp. 289–291, June 1, 1966 (pulsed, longer wavelength with more current)

9.194 High-Energy Emission in GaAs Electroluminescent Diodes
       M. I. Nathan, T. N. Morgan, G. Burns, and A. E. Michel (IBM)
       Phys. Rev., Vol. 146, pp. 570–574, June 10, 1966

9.195 Orientation Effect in GaAs Injection Lasers
       M. S. Abrahams and J. I. Pankove (RCA)
       J. Appl. Phys., Vol. 37, pp. 2596–2597, June 1966 (emission pattern depends on orientation)

9.196 Crystal Mosaic Structures and the Lasing Properties of GaAs Laser Diodes
       D. A. Shaw, K. A. Hughes, N. F. B. Neve, D. V. Sulway, and P. R. Thornton (University College of North Wales) and C. Gooch (S.E.R.L., England)
       Solid-State Electronics, Vol. 9, pp. 664–665, June 1966 (may result in filamentary action)

9.197 The Effect of Impurity Concentration on the Maximum CW Power From Gallium Arsenide Lasers at 77°K
       K. M. Hergenrother and J. M. Feldman (Northeastern U.)

Appl. Phys. Letters, Vol. 9, pp. 70–71, July 1, 1966 (optimum doping for given heat sink temperature)

9.198 Infrared and Microwave Radiations Associated with a Current-Controlled Instability in GaAs
S. G. Liu (RCA)
Appl. Phys. Letters, Vol. 9, pp. 79–81, July 15, 1966 (100 ns pulses, 0.9 μ and up to 44 GHz, similar results with GaAs$_{1-x}$P$_x$)

9.199 Avalanche Transistors to Generate Jitter-Free Nanosecond Current Pulses for Driving GaAs Laser Diodes at Low Temperatures
Y. U. Hussain (The City University, England)
Electronics Letters, Vol. 2, pp. 268–269, July 1966 (2.5 ns pulse, over 10 A peak)

9.200 Variation of the Gain Factor of GaAs Lasers with Photon and Current Densities
Yasuo Nannichi (Nippon Electric Co., Japan)
J. Appl. Phys., Vol. 37, pp. 3009–3012, July 1966 (either reflective or antireflective films on GaAs)

9.201 Performance of Room Temperature GaAs Lasers at High Pulse Repetition Rates (50 kc/s)
G. C. Dousmanis and H. E. Gross (RCA)
Proc. IEEE, Vol. 54, pp. 998–999, July 1966 (80 to 200 ns pulses, 12 W peak pulse power)

9.202 Effect of Band Tails on Stimulated Emission of Light in Semiconductors
Frank Stern (IBM, Switzerland)
Phys. Rev., Vol. 148, pp. 186–194, August 5, 1966 (temperature dependence of tails, linear gain with excitation)

9.203 Measurements of Gain Perpendicular to the Junction in Gallium Arsenide Laser Structures
T. S. Moss, G. J. Burrell, and A. Hetherington (Royal Aircraft Estab., England)
IEEE J. Quantum Electronics, Vol. QE-2, pp. 279–282, August 1966

9.204 Semiconductor Laser Amplifier
J. W. Crowe and W. E. Ahearn (IBM Watson)
IEEE J. Quantum Electronics, Vol. QE-2, pp. 283–289, August 1966 (GaAs, 3λ/4 SiO coatings)

9.205 Internal Self-Damage of Gallium Arsenide Lasers
D. P. Cooper, C. H. Gooch, and R. J. Sherwell (Services Electron. Research Lab., England)
IEEE J. Quantum Electronics, Vol. QE-2, pp. 329–330, August 1966 (due to radiation emitted by laser)

9.206 Spontaneous and Stimulated Emission from GaAs Diodes with Three-Layer Structures
M. Pilkuhn and H. Rupprecht (IBM Watson)
J. Appl. Phys., Vol. 37, pp. 3621–3628, August 1966

9.207 Efficient Electroluminescence From GaAs Diodes at 300°K
H. Rupprecht, J. M. Woodall, K. Konnerth, and D. G. Pettit (IBM Watson)
Appl. Phys. Letters, Vol. 9, pp. 221–223, September 15, 1966 (50 μ wide active region, up to 6 percent efficiency)

9.208 Continuous Coherent Radiation of Epitaxial Diodes of GaAs at 77°K
L. M. Kogan, L. D. Libov, D. N. Nasledov, T. F. Nikitina, I. N. Oraevskii, G. M. Strakhovskii, O. A. Sungurova, and B. V. Tsarenkov (Lebedev Phys. Inst., Acad. of Sci., USSR)
JETP Letters, Vol. 4, pp. 143–144, September 15, 1966 (70 mW output at 1.5 A)

9.209 Electrical and Electroluminescent Properties of Gallium Phosphide Diffused p-n Junctions
M. Gershenzon, R. A. Logan, and D. F. Nelson (BTL)
Phys. Rev., Vol. 149, pp. 580–597, September 16, 1966 (also preparation)

9.210 Radiative Recombination in Annealed Electron-Irradiated GaAs
George W. Arnold (Sandia)
Phys. Rev., Vol. 149, pp. 679–680, September 16, 1966

9.211 Linewidth Measurements of CW Gallium Arsenide Lasers at 77°K
W. E. Ahearn and J. W. Crowe (IBM Watson)
IEEE J. Quantum Electronics, Vol. QE-2, pp. 597–602, September 1966 (150 kHz or less)

9.212 Optical Gain and Losses of Solution-Grown-Diffused GaAs Injection Lasers
Wataru Susaki (Mitsubishi Electric, Japan)
Japanese J. Appl. Phys., Vol. 5, pp. 845–846, September 1966

9.213 Surface Aspects of the Thermal Degradation of GaAs p-n Junction Lasers and Tunnel Diodes
H. Kessler and N. N. Winogradoff (NBS, Washington, D. C.)
IEEE Trans. on Electron Devices, Vol. ED-13, pp. 688–691, October 1966

9.214 Semiconductor Laser Arrays and Their Characteristics
L. Wandinger and K. L. Klohn (US Army Electronics Command)
Proc. IEEE, Vol. 54, pp. 1491–1492, October 1966 (2 or 3 laser units in GaAs)

9.215 State of the Art in GaP Electroluminescent Junctions
M. Gershenzon
Bell Sys. Tech. J., Vol. 45, pp. 1599–1609, November 1966 (Zn-diffused diode, considers both red and green emissions)

9.216 Self Modulation of Emission from an Injection Semiconductor Laser
V. D. Kurnosov, V. I. Magalyas, A. A. Pleshkov, L. A. Rivlin, V. G. Trukhan, and V. V. Tsvetkov
JETP Letters, Vol. 4, pp. 303–305, December 1, 1966 (GaAs experiments, spike output, up to 20 GHz rate)

9.217 Mixing of Visible and Near-Resonance Infrared Light in GaP
W. L. Faust and Charles H. Henry (BTL)
Phys. Rev. Letters, Vol. 17, pp. 1265–1268, December 19, 1966 (sum and difference frequencies, infrared laser frequencies near lattice resonance, reststrahl)

9.218 Laser Diodes by Solution Regrowth on the (111) Plane
I. Ladany and W. A. Schmidt (NRL)
J. Appl. Phys., Vol. 37, pp. 4999–5000, December 1966 GaAs)

9.219 SiO Evaporation on a GaAs Electroluminescent Diode
T. Yamamoto and K. Kawamura (Shizuoka U., Japan)
Proc. IEEE, Vol. 54, pp. 1967–1968, December 1966 (radiation intensity varies with SiO evaporation)

9.220 Gallium Arsenide Laser with Plane Resonator
N. G. Basov, O. V. Bogdankevich, V. A. Goncharov, B. M. Lavrushin, and V. Yu. Sudzilovskii (Lebedev Phys. Inst. Acad. of Sci., USSR)
Soviet Phys. Doklady, Vol. 11, pp. 522–524, December 1966 (exciting electron beam and laser beam perpendicular to junction plane, "emitting" mirror laser)

10. OTHER INJECTION LUMINESCENT DEVICES AND ANALYSES

10.1 Injection Electroluminescence in CdS by Tunneling Films
R. C. Jaklevic, D. K. Donald, J. Lambe, and W. C. Vassell (Ford)
Appl. Phys. Letters, Vol. 2, pp. 7–9, January 1, 1963

10.2 Dielectric-Waveguide Mode of Light Propagation in p–n Junctions
Amnon Yariv and R. C. C. Leite (BTL)
Appl. Phys. Letters, Vol. 2, pp. 55–57, February 1, 1963

10.3 Observation of the Dielectric-Waveguide Mode of Light Propagation in p–n Junctions
W. L. Bond, B. G. Cohen, R. C. C. Leite, and A. Yariv (BTL)
Appl. Phys. Letters, Vol. 2, pp. 57–59, February 1, 1963

10.4 Direct Radiative Recombination in Ge Mesa Transistors
S. Wang, S. Yee, and K. Nosaka (U. of California)
Appl. Phys. Letters, Vol. 2, pp. 149–150, April 15, 1963

10.5 Doping of Semiconductors for Injection Lasers
Robert W. Keyes (IBM)
Proc. IEEE, Vol. 51, p. 602, April 1963

10.6 Experimental Observation of Heat Liberation in p–n Junctions
M. A. Melehy and E. A. Jarmoc (U. of Connecticut)
Proc. IEEE, Vol. 51, p. 614, April 1963

10.7 Maser Action in InAs Diodes
I. Melngailis (MIT)
Appl. Phys. Letters, Vol. 2, pp. 176–178, May 1, 1963 (3.14 microns)

10.8 Stimulated Light Emission from Indium Phosphide
K. Weiser and R. S. Levitt (IBM)
Appl. Phys. Letters, Vol. 2, pp. 178–179, May 1, 1963 (0.9 micron)

A Proposal for a DC Pumped Rare-Earth Laser
R. L. Bell (Varian)
J. Appl. Phys., Vol. 34, pp. 1563–1564, May 1963

10.10 Magnetically Tunable CW InAs Diode Maser
I. Melngailis and R. H. Rediker (MIT)
Appl. Phys. Letters, Vol. 2, pp. 202–204, June 1, 1963

10.11 Injection Electroluminescence in Gallium Antimonide
A. R. Calawa (MIT)
J. Appl. Phys., Vol. 34, pp. 1660–1662, June 1963

10.12 The Frequency Response of Optotransistors
H. N. Yu (IBM)
Proc. IEEE, Vol. 51, pp. 945–946, June 1963

10.13 Maximum Pulse Repetition Rate for an Injection Laser
V. Uzunoglu and M. H. White (Westinghouse)
Proc. IEEE, Vol. 51, p. 960, June 1963

10.14 Polarization in Junction Luminescence
H. F. Lockwood (G. T. & E. Labs.)
J. Appl. Phys., Vol. 34, pp. 2110–2111, July 1963

10.15 Reflection and Guiding of Light at p–n Junctions
A. Ashkin and M. Gershenzon (BTL)
J. Appl. Phys., Vol. 34, pp. 2116–2119, July 1963

10.16 On Mode Confinement in p–n Junctions
R. C. C. Leite and A. Yariv (BTL)
Proc. IEEE, Vol. 51, pp. 1035–1036, July 1963

10.17 Some Properties of InP Lasers
G. Burns, R. S. Levitt, M. I. Nathan, and K. Weiser (IBM)
Proc. IEEE, Vol. 51, pp. 1148–1149, August 1963

10.18 Semiconductor Diode Masers of (In_xGa_{1-x}) As
I. Melngailis, A. J. Strauss, and R. H. Rediker (MIT)
Proc. IEEE, Vol. 51, pp. 1154–1155, August 1963 (1.77 and 2.07 microns)

10.19 Photoinduced Recombination Radiation in InP Diodes
W. J. Turner and G. D. Pettit (IBM)
Appl. Phys. Letters, Vol. 3, pp. 102–104, September 15, 1963
(Note: Erratum in November 1, 1963, issue, p. 171)

10.20 Injection Electroluminescence in Silicon
M. A. Melehy and E. A. Jarmoc (U. of Connecticut)
Proc. IEEE, Vol. 51, p. 1365, October 1963

10.21 Silicon Carbide Diode Laser
L. B. Griffiths, A. I. Mlavsky, G. Rupprecht, A. J. Rosenberg, P. H. Smakula, and M. A. Wright (Tyco)
Proc. IEEE, Vol. 51, pp. 1374–1376, October 1963

10.22 Infrared InSb Diode in High Magnetic Fields
R. J. Phelan, A. R. Calawa, R. H. Rediker, R. J. Keyes, and B. Lax (MIT)
Appl. Phys. Letters, Vol. 3, pp. 143–145, November 1, 1963
(5.2 microns)

10.23 Super Radiant Narrowing in Fluorescence Radiation of Inverted Population
Amnon Yariv and R. C. C. Leite (BTL)
J. Appl. Phys., Vol. 34, pp. 3410–3411, November 1963

10.24 Microscopy of Internal Crystal Imperfections in Si p–n Junction Diodes by Use of Electron Beams
J. J. Lander, H. Schreiber, Jr., T. M. Buck, and J. R. Mathews (BTL)
Appl. Phys. Letters, Vol. 3, pp. 206–207, December 1, 1963

10.25 A Proposed Class of Hetero-junction Injection Lasers
Herbert Kroemer (Varian)
Proc. IEEE, Vol. 51, pp. 1782–1783, December 1963

10.26 Masers a Semi-Conducteurs
P. Aigrain (Lab. de Physique, Ecole Normale Supérieure, Paris, France)
Proc. Third Int. Conf. Quantum Electronics, Columbia U. Press, N. Y., Vol. 2, pp. 1761–1767, 1964

10.27 Inverted Populations in Semi-Conductors
N. G. Basov (Lebedev Physical Institute, Academy of Sciences, Moscow, U.S.S.R.)
Proc. Third Int. Conf. Quantum Electronics, Columbia U. Press, N. Y., Vol. 2, pp. 1769–1785, 1964

10.28 Possibilités de Laser à Semi-Conducteur
M. G. A. Bernard and G. Duraffourg (Centre Natl. d'Etudes des Telécommunications, Issy-les-Moulineaux, France)
Proc. Third Int. Conf. Quantum Electronics, Columbia U. Press, N. Y., Vol. 2, pp. 1849–1861, 1964

10.29 Conditions for Coherent Emission and Super Radiant Narrowing in p-n Injection Lasers
A. Yariv and R. C. C. Leite (BTL)
Proc. Third Int. Conf. Quantum Electronics, Columbia U. Press, N. Y., Vol. 2, pp. 1873–1878, 1964

10.30 On the Propagation of Electromagnetic Radiation in p-n Junctions
A. Yariv and R. C. C. Leite (BTL)
Proc. Third Int. Conf. Quantum Electronics, Columbia U. Press, N. Y., Vol. 2, pp. 1879–1882, 1964

10.31 Spontaneous and Stimulated Infra-red Emission from Indium Phosphide Arsenide Diodes
F. B. Alexander, V. R. Bird, D. R. Carpenter, G. W. Manley, P. S. McDermott, J. R. Peloke, H. F. Quinn, R. J. Riley, and

L. R. Yetter (IBM)
Appl. Phys. Letters, Vol. 4, pp. 13–15, January 1, 1964

10.32 Silicon Carbide Diode "Laser"
R. N. Hall (G.E.)
Proc. IEEE, Vol. 52, p. 91, January 1964

10.33 Magnetic Tuning of CW InSb Diode Laser
R. J. Phelan, Jr., and R. H. Rediker (MIT Lincoln Lab.)
Proc. IEEE, Vol. 52, pp. 91–92, January 1964

10.34 Efficiency of a p–n Diode Laser
A. C. Scott (U. of Wisconsin)
Proc. IEEE, Vol. 52, pp. 325–326, March 1964

10.35 Efficient Electroluminescence from p–n Junctions in CdTe at 77°K
G. Mandel and F. F. Morehead (IBM)
Appl. Phys. Letters, Vol. 4, pp. 143–145, April 15, 1964

10.36 Optical Absorption, Electroluminescence, and the Band Gap of BP
R. J. Archer, R. Y. Koyama, E. E. Loebner, and R. C. Lucas (Hewlett-Packard)
Phys. Rev. Letters, Vol. 12, pp. 538–540, May 11, 1964

10.37 Erratum: Optical Absorption, Electroluminescence, and the Band Gap of BP
R. J. Archer, R. Y. Koyama, E. E. Loebner, and R. C. Lucas
Phys. Rev. Letters, Vol. 12, p. 684; June 15, 1964 (Phys. Rev. Letters, Vol. 12, p. 538, 1964)

10.38 Stimulated Emission of 100 μ Radiation from Bi-Sb p-n Junctions
R. Sehr (Korad)
Proc. IEEE, Vol. 52, pp. 725–726, June 1964 (proposal)

10.39 The Mechanism of Band-Gap Laser Action in InSb Diodes
R. L. Bell and K. T. Rogers (Varian)
Appl. Phys. Letters, Vol. 5, pp. 9–11, July 1, 1964

10.40 P-N Junction Lasers
G. Burns and M. I. Nathan (IBM)
Proc. IEEE, Vol. 52, pp. 770–794, July 1964 (extensive review, over 162 references)

10.41 Efficient, Visible Electroluminescence from p-n Junctions in Zn_x-$Cd_{1-x}Te$
F. F. Morehead and G. Mandel (IBM)
Appl. Phys. Letters, Vol. 5, pp. 53–54, August 1, 1964 (4 per cent efficiency, 7060A, 77°K)

10.42 Injection Electroluminescence in p-type ZnTe
M. G. Miksic, G. Mandel, F. F. Morehead, A. A. Onton, and E. S. Schlig (IBM)
Physics Letters, Vol. 11, pp. 202–203, August 1, 1964 (emission about 5340 Å, about 100-Å linewidth, 77°K)

10.43 PbTe Diode Laser
J. F. Butler, A. R. Calawa, R. J. Phelan, Jr., T. C. Harman, A. J. Strauss, and R. H. Rediker (MIT)
Appl. Phys. Letters, Vol. 5, pp. 75–77, August 15, 1964 (6.5-μ output, 2-μs pulse, 12°K)

10.44 Luminescence and Coherent Emission in a Large-Volume Injection Plasma in InSb
I. Melngailis, R. J. Phelan, and R. H. Rediker (Lincoln Lab., MIT)
Appl. Phys. Letters, Vol. 5, pp. 99–100, September 1, 1964 (50-micron diameter coherent spots, pulsed, 10°K, 5-micron wavelength, 16-kG field)

10.45 Spectral Output of Semiconductor Lasers
H. Statz, C. L. Tang, and J. M. Lavine (Raytheon)
J. Appl. Phys., Vol. 35, pp. 2581–2585, September 1964 (standing waves in resonator, multimode oscillations)

10.46 Injection Electroluminescence in ZnSe Metal-Semiconductor Diodes

A. G. Fischer (RCA)
Physics Letters, Vol. 12, pp. 313–314, October 15, 1964 (output near 0.6μ)

10.47 PbSe Diode Laser
J. F. Butler, A. R. Calawa, R. J. Phelan, Jr., A. J. Strauss, and R. H. Rediker (MIT)
Solid State Communications, Vol. 2, pp. 303–304, October 1964 (8.5 μ, 12° K, 4 μs pulses, 3.9 amp threshold)

10.48 Significance of Band Structure in Determining Radiative Recombination and Laser Action in the Lead Salt Semiconductors
F. A. Junga, K. F. Cuff, J. S. Blakemore, and E. R. Washwell (Lockheed)
Physics Letters, Vol. 13, pp. 103–105, November 15, 1964 (lead salt semiconductors superior to III–V compounds)

10.49 Thermal Limitations on the Energy of a Single Injection Laser Light Pulse
G. J. Lasher and W. V. Smith (IBM)
IBM J. Res. and Dev., Vol. 8, pp. 532–536, November 1964 (analytical)

10.50 Planar Dielectric-Waveguide Modes
A. H. Luther (U. S. Navy Ordnance Lab.)
Proc. IEEE, Vol. 52, p. 1386, November 1964 (applicable to p-n junctions)

10.51 High-Current-Density Injection Electroluminescence in Cadmium Sulfide
D. D. O'Sullivan and E. C. Malarkey (Westinghouse)
Appl. Phys. Letters, Vol. 6, pp. 5–6, January 1, 1965 (peak emission at 4900 Å, 77° K, 1 μsec 20,000 A/cm² pulses)

10.52 Photoluminescence of Defect-Exciton Complexes in II–VI Compounds
R. E. Halsted and M. Aven (G.E.)
Phys. Rev. Letters, Vol. 14, pp. 64–65, January 18, 1965 (4° K)

10.53 Injection Lasers
F. N. Hooge (Philips Res. Labs., Eindhoven, The Netherlands)
Z. angewandte Mathematik und Physik, Vol. 16, No. 1, pp. 89–97, January 25, 1965

10.54 Theoretical Dependence of Lead Salt Laser Radiation with Applied Stress
G. W. Pratt, Jr., and J. E. Ripper (M.I.T.)
Bull. Am. Phys. Soc., Series II, Vol. 10, No. 1, p. 84–GE2, January 1965 (8.5 to 12μ shift in PbTe at 5000 atm, 4.2°K)

10.55 Experimental Dependence of PbSe Diode Radiation on Stress
A. R. Calawa, J. F. Butler, and R. H. Rediker (M.I.T.)
Bull. Am. Phys. Soc., Series II, Vol. 10, No. 1, p. 84–GE3, January 1965

10.56 Photoeffects in Lead-Telluride p-n Junctions
Robert A. Laff (IBM)
Bull. Am. Phys. Soc., Series II, Vol. 10, No. 1, p. 84–GE4, January 1965

10.57 Inherent Properties of a Tunnel-Injection Laser
G. Wade, C. A. Wheeler, and R. G. Hunsperger (Cornell U.)
Proc. IEEE, Vol. 53, pp. 98–99, January 1965

10.58 Longitudinal Injection-Plasma Laser of InSb
I. Melngailis (M.I.T.)
Appl. Phys. Letters, Vol. 6, pp. 59–60, February 1, 1965 (10°K, 45 amp 50 ns pulses, longitudinal magnetic field)

10.59 Optically Pumped Semiconductor Laser
R. J. Phelan, Jr. and R. H. Rediker (M.I.T.)
Appl. Phys. Letters, Vol. 6, pp. 70–71, February 15, 1965 (GaAs laser pump, InSb emits at 5.3 μ, 4.2°K, 50 ns pulses)

10.60 Energy Bands in PbTe
J. B. Conklin, Jr., L. E. Johnson, and G. W. Pratt, Jr. (M.I.T.)
Phys. Rev., Vol. 137, pp. A1282–A1294, February 15, 1965

10.61 Recombination Emission from Silicon Transistors
J. D. Van Wyk (U. of Pretoria, South Africa)
Proc. IEEE, Vol. 53, pp. 307–308, March 1965 (visible radiation)

10.62 Single Mode Differential Efficiency for Circular and Rectangular Laser Diodes
A. C. Scott (U. of Wisconsin)
Proc. IEEE, Vol. 53, pp. 315–316, March 1965 (omni-directional circular, bi-directional rectangular)

10.63 Properties of the PbSe Diode Laser
J. F. Butler, A. R. Calawa, and R. H. Rediker (M.I.T. Lincoln Lab.)
IEEE J. Quantum Electronics, Vol. QE-1, pp. 4–7, April 1965 (fabrication techniques, 12°K, pulsed 8.5 μ wavelength, measurements, 17 Mc/s shift per gauss)

10.64 Dependence of Threshold Currents on the Impurity Concentration in Laser Diodes
Y. Nannichi (Nippon Electric Co., Japan)
J. Appl. Phys., Vol. 36, pp. 1499–1500, April 1965 (current decreases with concentration)

10.65 Peculiar Spectral Distribution of the Light Emission from ZnTe During d.c. Pulse Excitation
A. C. Aten, J. H. Haanstra and G. Diemer (Philips, Netherlands)
Philips Res. Repts., Vol. 20, pp. 125–135, April 1965 (inhomogeneous region of point contact)

10.66 Electron Beam Pumped Lasers of PbS, PbSe, and PbTe
C. E. Hurwitz, A. R. Calawa, and R. H. Rediker (Lincoln Lab., M.I.T.)
IEEE J. Quantum Electronics, Vol. QE-1, pp. 102–103, May 1965 (liquid He temp)

10.67 Optically Pumped InAs Laser
Ivan Melngailis (Lincoln Lab., M.I.T.)
IEEE J. Quantum Electronics, Vol. QE-1, pp. 104–105, May 1965 (one watt pump threshold from GaAs junction laser, 10°K)

10.68 Theory of a Pressure-Tuned Lead Salt Laser
G. W. Pratt, Jr., and J. E. Ripper (M.I.T.)
J. Appl. Phys., Vol. 36, pp. 1525–1528, May 1965 (up to 10^4 atmospheres, 8.5 to 10.5μ from PbSe diode, sonic waves to frequency modulate)

10.69 Magnetic Properties of InAs Diode Electroluminescence
F. L. Galeener, I. Melngailis, G. B. Wright, and R. H. Rediker (M.I.T.)
J. Appl. Phys., Vol. 36, pp. 1574–1579, May 1965 (up to 109 kG, increases photon energy, decreases threshold current)

10.70 Refractive Index of InP
G. D. Pettit and W. J. Turner (IBM)
J. Appl. Phys., Vol. 36, p. 2081, June 1965 (index ranges from 3.1 to 3.4, increases with temperature and photon energy)

10.71 The Luminescence Diode Acting as a Heat Pump
P. Gerthsen and E. Kauer (Philips, Germany)
Physics Letters, Vol. 17, pp. 255–256, July 15, 1965

10.72 Efficient Visible Injection Electroluminescence from p–n Junctions in ZnSe$_x$Te$_{1-x}$
M. Aven (GE)
Appl. Phys. Letters, Vol. 7, pp. 146–148, September 15, 1965 (18 percent at 70°K, orange spectral output at 1.98 eV)

10.73 Solution of the Equation for Wave Propagation in Layered Slabs with Complex Dielectric Constants
J. W. Cooley and F. Stern
IBM J. Res. and Dev., Vol. 9, pp. 405–411, September–November 1965

10.74 Pressure-Tuned PbSe Diode Laser
J. M. Besson, J. F. Butler, A. R. Calawa, W. Paul, and R. H. Rediker (Harvard U. and M.I.T. Lincoln Lab.)
Appl. Phys. Letters, Vol. 7, pp. 206–208, October 15, 1965 (emission wavelength changed from 7.5 to 11 μ with pressure up to 7000 bars)

10.75 Recombination Radiation Stimulated in Silicon by Long-Wave Infrared Radiation
Sh. M. Kogan, T. M. Lifshitz, and V. I. Sidorov (Inst. for Radio Engineering and Electronics, Academy of Sciences, U.S.S.R.)
JETP Letters, Vol. 2, pp. 230–232, October 15, 1965 (8 to 20 μ incident, about 1-μ photons emitted)

10.76 Electron-Beam Excitation of Semiconductor Lasers
C. Benoit à la Guillaume and J. M. Debever (Laboratoire de Physique de l'Ecole Normale Supérieure, Paris, France)
Proc. Physics of Quantum Electronics Conf., McGraw-Hill Book Co., New York, pp. 397–410, 1966

10.77 The Excitation Mechanism in Electron-Beam Pumped Lasers
Claude A. Klein (Raytheon)
Proc. Physics of Quantum Electronics Conf., McGraw-Hill Book Co., New York, pp. 424–434, 1966

10.78 Laser Emission by Optical Pumping of Semiconductors
R. J. Phelan, Jr. (Lincoln Lab., M.I.T.)
Proc. Physics of Quantum Electronics Conf., McGraw-Hill Book Co., New York, pp. 435–441, 1966

10.79 Laser Action in Gallium Antimonide Diodes
R. Eymard, G. Duraffourg, C. Chipaux, and M. Bernard (Centre National d'Etudes des Télécommunications, Issy-les-Moulineaux, France)
Proc. Physics of Quantum Electronics Conf., McGraw-Hill Book Co., New York, pp. 450–457, 1966

10.80 Magnetoemission Studies of PbS, PbTe, and PbSe Diode Lasers
J. F. Butler and A. R. Calawa (Lincoln Lab., M.I.T.)
Proc. Physics of Quantum Electronics Conf., McGraw-Hill Book Co., New York, pp. 458–466, 1966

10.81 Electroluminescence of Rare-Earth Ions in Cadmium Fluoride
J. Lambe, D. K. Donald, W. C. Vassell, and T. Cole (Ford Motor Co.)
Appl. Phys. Letters, Vol. 8, pp. 16–18, January 1, 1966 (Sm, Eu and Tb; electrolytic solution for one electrode, emission from contact layer)

10.82 Growth and Properties of β-SiC Single Crystals
W. E. Nelson, F. A. Halden, and A. Rosengreen (Stanford Research Institute)
J. Appl. Phys., Vol. 37, pp. 333–336, January 1966 (injection electroluminescence between 2 and 2.3 eV, room temp., 10^{-4} efficiency)

10.83 Effective Mass in InAs and InSb from the Landau Shift of Peak Emission in Laser Diodes
A. N. Chakravarti (University College of Technology, Calcutta, India)
J. Appl. Phys., Vol. 37, p. 446, January 1966

10.84 Directional Radiation from Incoherent Electroluminescent Diodes
J. J. Schlickman, H. T. Minden, and M. M. Antonoff (Sperry Rand Research Center)
J. Appl. Phys., Vol. 37, pp. 451–452, January 1966 (far field patterns of GaAs and InGaAs diodes)

10.85 Properties of InAs Lasers
I. Melngailis and R. H. Rediker (Lincoln Lab., M.I.T.)
J. Appl. Phys., Vol. 37, pp. 899–911, February 1966

10.86 Electron-Beam Pumped Lasers of CdSe and CdS
C. E. Hurwitz (Lincoln Lab., M.I.T.)
Appl. Phys. Letters, Vol. 8, pp. 121–124, March 1, 1966

10.87 Efficient Injection Electroluminescence in ZnTe by Avalanche Breakdown

B. L. Crowder, F. F. Morehead, and P. R. Wagner (IBM)
Appl. Phys. Letters, Vol. 8, pp. 148–149, March 15, 1966

10.88 Formation of Built-in Light-emitting Junctions in Solution-grown
GaP Containing Shallow Donors and Acceptors
L. M. Foster, T. S. Plaskett, and J. E. Scardefield
IBM J. Res. & Dev., Vol. 10, pp. 114–121, March 1966

10.89 Green Luminescence from Solution-grown Junctions in GaP Containing Shallow Donors and Acceptors
M. H. Pilkuhn and L. M. Foster
IBM J. Res. & Dev., Vol. 10, pp. 122–129, March 1966

10.90 Spontaneous and Coherent Photoluminescence in $Cd_xHg_{1-x}Te$
I. Melngailis and A. J. Strauss (Lincoln Laboratory, M.I.T.)
Appl. Phys. Letters, Vol. 8, pp. 179–180, April 1, 1966 (optically excited by pulsed GaAs laser output, coherent emission at 3.8 and 4.1 μ)

10.91 Spontaneous and Stimulated Emission in Indium Phosphide Diodes
Wataru Susaki (Mitsubishi Electric Corp., Japan)
Japanese J. Appl. Phys., Vol. 5, p. 334, April 1966 (about 0.9 μ)

10.92 Determination of Coupling Efficiency in Light Emitting Diodes
Sverre T. Eng (Hughes)
Proc. IEEE, Vol. 54, pp. 688–690, April 1966

10.93 Efficient Visible Lasers of CdS_xSe_{1-x} by Electron-Beam Excitation
C. E. Hurwitz (M.I.T., Lincoln Laboratory)
Appl. Phys. Letters, Vol. 8, pp. 243–245, May 15, 1966 (11% efficient, 4900 to 6900 Å)

10.94 The Behavior of a Semiconductor in a Strong Resonant Radiation Field
V. F. Chel'tsov
Soviet Physics JETP, Vol. 22, pp. 1024–1025, May 1966

10.95 Band Structure and Laser Action in $Pb_xSn_{1-x}Te$
J. O. Dimmock, I. Melngailis, and A. J. Strauss (Lincoln Lab., M.I.T.)
Phys. Rev. Letters, Vol. 16, pp. 1193–1196, June 27, 1966 (near 15 μ, 12°K, pulsed at 3 kc/s)

10.96 Observation of Light Emission from Semiconducting CdS in the Current Oscillatory Mode at 77°K
Sinclair S. Yee (Lawrence Rad. Lab., U. of California)
Appl. Phys. Letters, Vol. 9, pp. 10–13, July 1, 1966 (about 5125 Å, electroluminescence)

10.97 Ultraviolet ZnO Laser Pumped by an Electron Beam
F. H. Nicoll (RCA)
Appl. Phys. Letters, Vol. 9, pp. 13–15, July 1, 1966 (pulsed threshold at 15 kV and 3 A/cm², about 3800 Å)

10.98 Efficient Ultraviolet Laser Emission in Electron-Beam-Excited ZnS
C. E. Hurwitz (M.I.T. Lincoln Lab.)
Appl. Phys. Letters, Vol. 9, pp. 116–118, August 1, 1966 (pulsed, 4.2 and 77°K, 3245 to 3300Å, 1.7 W with 6.5 percent power efficiency)

10.99 Laser Operation of CdSe Pumped with a Ga(AsP) Laser Diode
N. Holonyak, Jr., M. D. Sirkis, G. E. Stillman, and M. R. Johnson (U. of Illinois)
Proc. IEEE, Vol. 54, pp. 1068–1069, August 1966 (6667 Å pump, 6917 Å emitted, 77°K, pulsed)

10.100 Incoherent Source Optical Pumping of Visible and Infrared Semiconductor Lasers
R. J. Phelan (M.I.T. Lincoln Lab.)
Proc. IEEE, Vol. 54, pp. 1119–1120, August 1966 (xenon flashlamp pump, InSb and CdS_xSe_{1-x} lasers)

10.101 Recombination Kinetics and Electroluminescence from Deep Levels in the Carrier Diffusion Region of a p-n Junction
D. F. Nelson (BTL)
Phys. Rev., Vol. 149, pp. 574–579, September 16, 1966

10.102 Semiconductor Lasers and Fast Detectors in the Infrared (3 to 15 Microns)
M. Rodot, C. Verie, and Y. Marfaing (CNRS, France), and J. Besson and H. Lebloch (SAT, France)
IEEE J. Quantum Electronics, Vol. QE-2, pp. 586–593, September 1966

10.103 Semiconductor Lasers with Radiating Mirrors
N. G. Basov, O. V. Bogdankevich, and A. Z. Grasyuk (Lebedev Phys. Inst., Acad. of Sci., USSR)
IEEE J. Quantum Electronics, Vol. QE-2, pp. 594–597, September 1966 (semiconductor layer on one mirror of Fabry-Perot cavity, electron and optical beam pumping on and off axis)

10.104 The Theory of Semiconductor Lasers with Consideration of Saturation Effects
O. N. Krokhin (Lebedev Phys. Inst., Acad. of Sci., USSR)
IEEE J. Quantum Electronics, Vol. QE-2, pp. 605–607, September 1966

10.105 Theory of the Interband Magnetooptical Semiconductor Laser
B. Sacks and B. Lax (M.I.T. Nat'l Magnet Lab.)
IEEE J. Quantum Electronics, Vol. QE-2, pp. 607–610, September 1966 (injection and optical pumping)

10.106 Volume Excitation of an Ultrathin Single-Mode CdSe Laser
G. E. Stillman, M. D. Sirkis, J. A. Rossi, M. R. Johnson, and N. Holonyak, Jr. (U. of Illinois)
Appl. Phys. Letters, Vol. 9, pp. 268–269, October 1, 1966 (2 to 4 μ thick, pulsed $GaAs_{1-x}P_x$ optical pump, face pumping, 6940 and 6893 Å emission, 77°K)

10.107 Electroluminescence and Semiconductor Lasers
Henry F. Ivey (Westinghouse)
IEEE J. Quantum Electronics, Vol. QE-2, pp. 713–726, November 1966 (review, principles, all semiconductor lasers, 189 references)

10.108 The Role of Light Absorption by Free Carriers in a Semiconductor Laser
V. S. Mashkevich and V. L. Vinetskiĭ (Phys. Inst., Ukrainian SSR)
Soviet Phys. JETP, Vol. 23, pp. 935–938, November 1966 (may prevent laser action)

11. LASER AMPLIFIER

11.1 Theory of Pulse Propagation in a Laser Amplifier
Lee M. Frantz and John S. Nodvik (Space Tech. Lab.)
J. Appl. Phys., Vol. 34, pp. 2346–2349, August 1963

11.2 13-Inch Ruby Oscillator-Amplifier Chain
H. G. Heard (Energy Systems)
Proc. IEEE, Vol. 51, p. 1664, November 1963

11.3 Equivalent Circuits for Quantum Noise in Linear Amplifiers
H. A. Haus (MIT) and J. A. Mullen (Raytheon)
Proc. Third Int. Conf. Quantum Electronics, Columbia U. Press,
N. Y., Vol. 1, pp. 71–93, 1964

11.4 Criteria for Optical Maser Amplifiers and Oscillators
J. Jacobs, D. Holmes, L. Hatkin, and F. A. Brand (U. S. Army,
Fort Monmouth)
Proc. Third Int. Conf. Quantum Electronics, Columbia U. Press,
N. Y., Vol. 2, pp. 1071–1077, 1964

11.5 High Gain, High Power Pulsed Ruby Optical Amplifier
J. E. Geusic and H. E. D. Scovil (BTL)
Proc. Third Int. Conf. Quantum Electronics, Columbia U. Press,
N. Y., Vol. 2, pp. 1211–1220, 1964

11.6 Pulse Sharpening and Gain Saturation in Traveling-Wave
Masers
E. O. Schulz-DuBois
Bell Sys. Tech. J., Vol. 43, pp. 625–658, March 1964 (optical
masers also discussed)

11.7 The Effects of Saturation and Regeneration in Ruby Laser Amplifiers
J. I. Davis and W. R. Sooy (Hughes)
Appl. Optics, Vol. 3, pp. 715–718, June 1964 (thorough analytical
treatment, 5.53 J/cm² input to saturate gain)

11.8 Pulse Propagation in a Laser Amplifier
J. P. Wittke (RCA)
J. Appl. Phys., Vol. 35, pp. 1668–1672, June 1964 (theoretical,
traveling-wave, input and output pulses dissimilar, velocity
anomaly)

11.9 Gain and Reflection Characteristics of Active Interference Filter
Laser Amplifier
K. Tomiyasu (G. E.)
Proc. IEEE, Vol. 52, pp. 856–857, July 1964 (analytical, gain
enhancement, matched input possible)

11.10 Anregung von Eigenschwingungen in Passiven Optischen Resona-
toren
H. Boersch, H. Eichler, and G. Herziger (Technischen U., Berlin)
Physics Letters, Vol. 11, pp. 291–292, August 15, 1964

11.11 Umformung von Wellenformen in Passiven Laser-Resonatoren
H. Eichler and G. Herziger (Tech. U., Berlin)
Physics Letters, Vol. 12, p. 193, October 1, 1964

11.12 Amplification in a Fiber Laser
C. J. Koester and E. Snitzer (American Optical)
Appl. Optics, Vol. 3, pp. 1182–1186, October 1964 (47-dB gain
in one-meter long neodymium-glass fiber, gain vs. time delay)

11.13 Gas Laser Pre-Amplifier Performance
W. B. Bridges and G. S. Picus (Hughes)
Appl. Optics, Vol. 3, pp. 1189–1190, October 1964 (16-dB im-
provement in sensitivity at 3.5 micron using xenon laser)

11.14 High-Intensity Propagation through Absorptive or Amplifying
Media
W. J. Condell (Lab. for Physical Sciences)
J. Opt. Soc. Am., Vol. 54, pp. 1166–1167, September 1964

11.15 Optical Gain in Neon and Helium/Neon Pulsed Discharges
D. M. Clunie and N. H. Rock (England)
Physics Letters, Vol. 13, pp. 213–214, December 1, 1964 (multipass system, 3.6 dB/m from Ne, 30-watts saturated power from
1.15-μ He-Ne)

11.16 Transmission Line Formulation for Optical Maser Amplification
H. Jacobs, D. A. Holmes, L. Hatkin, and F. A. Brand (USAEL
and Carnegie Tech.)
J. Opt. Soc. Am., Vol. 54, pp. 1416–1424, December 1964 (three
and five-layer cases)

11.17 A Helium-Neon Laser Amplifier
L. E. S. Mathias and N. H. Rock (S.E.R.L., England)

Appl. Optics, Vol. 4, pp. 133–135, January 1965 (multiple non-
intersecting passes, 16 db gain in 68.4 m path length at 1.15μ,
20 mw max output)

11.18 Gain-Delay Characteristics of a Pulsed Neodymium-Glass Laser
Oscillator-Amplifier Chain
K. F. Tittel and J. P. Chernoch (G.E.)
Proc. IEEE, Vol. 53, pp. 82–83, January 1965

11.19 Amplification in a Thick Ruby Lens
E. R. Lanczi (Mitre)
Appl. Optics, Vol. 4, p. 255, February 1965 (gain of 2, also
image amplified)

11.20 Laser Amplifiers
E. L. Steele and W. C. Davis (North American)
J. Appl. Phys., Vol. 36, pp. 348–351, February 1965 (Q-sw ruby
osc, gain saturates at 5 to 7 in 15-cm long ruby amp)

11.21 Saturation Effects in Solid-State Laser Amplifiers
K. N. Seeber (Raytheon)
IEEE Trans. on Electron Devices, Vol. ED-12, pp. 63–66,
February 1965 (calculations for Nd amplifiers)

11.22 Stability of Traveling Waves in Lasers
J. A. White (NBS)
Phys. Rev., Vol. 137, pp. A1651–A1654, March 15, 1965

11.23 Infrared Laser Preamplifier System
F. Arams and M. Wang (AIL)
Proc. IEEE, Vol. 53, p. 329, March 1965 (3.39μ HeNe laser
preamp, 32 db net improvement over PbS detector)

11.24 Multimode Effects in the Gain of Raman Amplifiers and Oscil-
lators II. Amplifiers
P. Lallemand and N. Bloembergen (Harvard U.)
Appl. Phys. Letters, Vol. 6, pp. 212–213, May 15, 1965

11.25 Gain Narrowing in a Laser Amplifier
D. F. Hotz (Hughes)
Appl. Optics, Vol. 4, pp. 527–530, May 1965 (3.39μ He-Ne
oscillator and Zeeman tunable amplifier)

11.26 Low-Level Single-Resonator Laser Amplifiers
A. D. Jacobson and T. R. O'Meara (Hughes)
Proc. IEEE, Vol. 53, p. 529, May 1965 (reflection amplifier,
3-dB bandwidth, gain sensitivity)

11.27 A Limit Upon Laser Amplifier Pump Level
A. C. Scott (U. Wisconsin)
Proc. IEEE, Vol. 53, p. 537, May 1965 (suggests instability
at high pumping levels)

11.28 Correction: A Limit Upon Laser Amplifier Pump Level
A. C. Scott (U. of Wisconsin)
Proc. IEEE, Vol. 53, p. 1806, November 1965 (Proc. IEEE, Vol.
53, p. 537, May 1965)

11.29 Amplification by Reflection from an Active Interferometer
Vern N. Smiley, David K. Forbes, and Adolph L. Lewis
(U.S. Navy Electronics Lab.)
Appl. Phys. Letters, Vol. 7, pp. 1–2, July 1, 1965 (2.03-μ He-Xe
laser, gain up to 500 just below oscillation threshold)

11.30 Quantum Statistical Dynamics of Laser Amplifiers
A. E. Glassgold (New York U.) and D. Holliday (RAND)
Physics Letters, Vol. 17, pp. 249–250, July 15, 1965

11.31 Theory of Optical Maser Amplifiers
F. T. Arecchi and R. Bonifacio (U. of Milan, Italy)
IEEE J. Quantum Electronics, Vol. QE-1, pp. 169–178, July 1965
(electromagnetic wave and medium interaction)

11.32 Erratum: Theory of Optical Maser Amplifiers
F. T. Arecchi and R. Bonifacio (Istituto di Fisica dell'Univer-
sità Milan, Italy)
IEEE J. Quantum Electronics, Vol. QE-2, p. 105, May 1966
(IEEE J. Quantum Electronics, Vol. QE-1, pp. 169–178, July
1965)

11.33 Generalized Solutions for Optical Maser Amplifiers
N. Kumagai and H. Yamamoto (Osaka U., Japan)
IEEE Trans. on Microwave Theory and Techniques,
Vol. MTT-13, pp. 445–451, July 1965 (Laplace transforms, transient terms, nonlinear above threshold)

11.34 Amplification of Light by Four-Level Quantum Systems
Yu. A. Anan'ev, A. A. Mak, and B. M. Sedov (Vavilov State Institute of Optics, U.S.S.R.)
Soviet Physics JETP, Vol. 21, pp. 4–7, July 1965 (experiments with $CaF_2:Sm^{2+}$ oscillator and amplifier, varying inversion)

11.35 Saturation Effects in High-Gain Lasers
W. W. Rigrod (BTL)
J. Appl. Phys., Vol. 36, pp. 2487–2490, August 1965 (high loss in cavity)

11.36 Quantum Statistical Dynamics of Laser Amplifiers
A. E. Glassgold (New York U.) and Dennis Holliday (RAND)
Phys. Rev., Vol. 139, pp. A1717–A1734, September 13, 1965 (Heisenberg picture, density matrix, damping)

11.37 Gain in a Diffusely Pumped Raman Amplifier
D. P. Bortfeld and W. R. Sooy (Hughes)
Appl. Phys. Letters, Vol. 7, pp. 283–285, November 15, 1965 (about 8 percent gain)

11.38 Study of Helium-Neon Laser Amplification at 3.39 μ
G. K. Moeller and T. K. McCubbin, Jr. (Pennsylvania State U.)
Appl. Optics, Vol. 4, pp. 1412–1415, November 1965 (power gain, input signal power, discharge current, mixture ratio, total pressure)

11.39 Gain and Bandwidth Narrowing in a Regenerative He–Xe Laser Amplifier
Vern N. Smiley, Adolph L. Lewis, and David K. Forbes (U.S. Navy Electronics Lab., San Diego, Calif.)
J. Opt. Soc. Am., Vol. 55, pp. 1552–1553, November 1965 (2.03 μ wavelength, 27 dB gain, magnetostrictively tuned amplifier)

11.40 Saturation Operation and Gain Coefficient of a Neodymium-Glass Amplifier
C. G. Young and J. W. Kantorski (American Optical)
Appl. Optics, Vol. 4, pp. 1675–1677, December 1965 (106 dB small-signal round-trip gain, spectral narrowing, amplified spontaneous emission, saturation onset at 13 J/cm^2, 7.5 percent gain cm^{-1}/Jcm^{-3}, 140 J output)

11.41 Noise Properties of Pulsed Ruby Laser Amplifiers
I. J. D'Haenens and C. R. Giuliano (Hughes)
IEEE J. Quantum Electronics, Vol. QE-1, pp. 393–397, December 1965 (spontaneous fluorescence has broader beam and wider bandwidth than laser beam)

11.42 Experimental and Theoretical Ruby Laser Amplifier Dynamics
Petras V. Avizonis and Ronald L. Grotbeck (Air Force Weapons Laboratory)
J. Appl. Phys., Vol. 37, pp. 687–693, February 1966

11.43 Regenerative Ruby Laser Amplifiers
H. Jacobs, J. Castro, F. A. Brand, C. LoCascio, S. Weitz, and G. Novick (U.S. Army Electronics Command)
J. Opt. Soc. Am., Vol. 56, pp. 149–156, February 1966

11.44 Laser with Nonresonant Feedback
R. V. Ambartsumyan, N. G. Basov, P. G. Kryukov, and V. S. Letokhov (Lebedev Physics Institute, U.S.S.R.)
JETP Letters, Vol. 3, pp. 167–169, March 15, 1966 (diffuse and specular mirrors in laser cavity, two ruby amplifiers)

11.45 Etude Optique du Faisceau Émis par un Laser de Grande Intensité
J. de Metz, A. Terneaud, et P. Veyrie (AEC, France)

Appl. Optics, Vol. 5, pp. 819–822, May 1966 (5% Nd glass, Q-sw oscillator, 4 amplifiers, 30 J in 25 ns)

11.46 Scanning Active Interferometer Employing Linear Drive Excitation and Reflectance Monitor
Vern N. Smiley, Adolph L. Lewis, and David K. Forbes (U.S. Navy Electronics Lab.)
Appl. Optics, Vol. 5, pp. 827–829, May 1966 (regenerative gas laser amplifiers)

11.47 Backward Wave Optical Amplification by an Asymmetric Active Interference Filter
Vern N. Smiley (U.S. Navy Electronics Lab.)
Appl. Optics, Vol. 5, pp. 977–980, June 1966 (gain, bandwidth)

11.48 Nonlinear Negative Absorption of Resonance Light in Ruby and Neodymium Glass
V. I. Borodulin, N. A. Ermakova, L. A. Rivlin, V. V. Tsvetkov, and V. S. Shil'Dyaev
Soviet Phys. JETP, Vol. 22, pp. 1174–1176, June 1966 (pulsed amplifier experiments)

11.49 Propagation of a Light Pulse in a Nonlinearly Amplifying and Absorbing Medium
R. V. Ambartsumyan, N. G. Basov, V. S. Zuev, P. G. Kryukov, and V. S. Letokhov (Lebedev Phys. Inst., Acad. of Sci., USSR)
JETP Letters, Vol. 4, pp. 12–14, July 1, 1966 (Q-sw ruby laser oscillator, 3 ruby amplifiers, eliminated transverse mode structure, amplified pulse shortened to 2 ns, 3 GW/cm^2 power density)

11.50 Laser Amplifier Noise at 3.5 Microns in Helium-Xenon
J. W. Klüver (BTL)
J. Appl. Phys., Vol. 37, pp. 2987–2999, July 1966 (unsaturated gain of 17 dB, saturated gain of 1 dB, cone angle of noise, good noise properties of amplifier)

11.51 Nonlinear Amplification of Light Pulses
N. G. Basov, R. V. Ambartsumyan, V. S. Zuev, P. G. Kryukov, and V. S. Letokhov (Lebedev Phys. Inst., Acad. of Sci., USSR)
Soviet Phys. JETP, Vol. 23, pp. 16–22, July 1966 (propagation rate, pulse shortening)

11.52 Effect of Spectral Hole-Burning and Cross Relaxation on the Gain Saturation of Laser Amplifiers
Amado Y. Cabezas and Richard P. Treat (Hughes Aircraft)
J. Appl. Phys., Vol. 37, pp. 3556–3563, August 1966 (Nd laser experiments)

11.53 Amplification of Spontaneous Radiation in an Inversely Populated Medium
V. N. Morozov
Optics and Spectroscopy, Vol. 21, pp. 133–134, August 1966 (highest gain from segmented laser crystal)

11.54 Theoretical and Experimental Investigation of Regenerative Laser Amplifiers and Their Applications
H. Boersch and G. Herziger (Tech. Univ. Berlin, Germany)
IEEE J. Quantum Electronics, Vol. QE-2, pp. 549–552, September 1966 (oscillator and resonator in series, measure electron densities of 10^{13} to 10^{15} cm^{-3})

11.55 Competition between Two Types of Oscillation in a Traveling Wave Laser (TWL)
S. G. Zeiger and E. E. Fradkin
Optics and Spectroscopy, Vol. 21, pp. 217–219, September 1966

11.56 Change of Light-Pulse Shape by Nonlinear Amplification
N. G. Basov and V. S. Letokhov (Lebedev Phys. Inst., Acad. of Sci., USSR)
Soviet Phys. Doklady, Vol. 11, pp. 222–224, September 1966

11.57 Optical Avalanche Laser
C. G. Young, J. W. Kantorski, and E. O. Dixon (American Optical)
J. Appl. Phys., Vol. 37, pp. 4319–4324, November 1966 (Nd-doped glass, five 20-cm and one 100-cm long rods, amplified spontaneous emission, 185-dB gain, spike free, 70-ns 1 GW output pulse, 1-mrad full-angle beam, unfocused air breakdown)

12. LASER RESONATORS AND MODE CONSIDERATIONS

12.1 Splitting of Fabry–Perot Rings by Microwave Modulation of Light
I. P. Kaminow (BTL)
Appl. Phys. Letters, Vol. 2, pp. 41–42, January 15, 1963

12.2 Isolation of Axi-symmetrical Optical-Resonator Modes
W. W. Rigrod (BTL)
Appl. Phys. Letters, Vol. 2, pp. 51–53, February 1, 1963 (Note: Erratum on p. 121)

12.3 Generation of Laser Axial Mode Difference Frequencies in a Nonlinear Dielectric
Kenneth E. Niebuhr (IBM)
Appl. Phys. Letters, Vol. 2, pp. 136–137, April 1, 1963

12.4 Resonances of the Fabry–Perot Laser
S. R. Barone (TRG)
J. Appl. Phys., Vol. 34, pp. 831–843, Part I, April 1963

12.5 Diffraction Studies with Plane-Parallel Maser Interferometer
W. W. Rigrod and A. J. Rustako, Jr. (BTL)
J. Appl. Phys., Vol. 34, pp. 967–968, Part I, April 1963

12.6 X-Band Microwave Phototube for Demodulation of Laser Beams
M. D. Petroff, H. A. Spetzler, and E. K. Bjørnerud (Calif. Inst. of Tech.)
Proc. IEEE, Vol. 51, pp. 614–615, April 1963

12.7 Light Scattering from Dielectric Film Laser Mirrors
Walter A. Specht, Jr. (Calif. Inst. of Tech.)
Proc. IEEE, Vol. 51, pp. 615–616, April 1963

12.8 Observation of Beats between Transverse Modes in Ruby Lasers
C. M. Stickley (AFCRL)
Proc. IEEE, Vol. 51, pp. 848–849, May 1963

12.9 Optical Lattice Filters from the Wave Field of Laser Radiation
R. Gerharz
Proc. IEEE, Vol. 51, pp. 862–863, May 1963

12.10 Regular Spiking and Single-Mode Operation of Ruby Laser
C. L. Tang, H. Statz, and G. DeMars (Raytheon)
Appl. Phys. Letters, Vol. 2, pp. 222–224, June 1, 1963

12.11 Multireflector Optical Resonators
Peter O. Clark (Calif. Inst. of Tech.)
Proc. IEEE, Vol. 51, pp. 949–950, June 1963

12.12 Off-Axial Mode Rejection of Ruby Laser Using Ball Mirror
Akira Okaya (Westinghouse)
Proc. IEEE, Vol. 51, pp. 1033–1034, July 1963

12.13 Tilted-Plate Interferometry with Large Plate Separations
H. W. Moos, G. F. Imbusch, L. F. Mollenauer, and A. L. Schawlow (Stanford U.)
Appl. Optics, Vol. 2, pp. 817–822, August 1963

12.14 Note on the paper, "Tilted-Plate Interferometry with Large Plate Separations"
H. W. Moos, G. F. Imbusch, L. F. Mollenauer, and A. L. Schawlow (Stanford U.)
Appl Optics, Vol. 2, pp. 1330–1331, December 1963

12.15 Spherical-Mirror Oscillating Interferometer
Donald R. Herriott (BTL)
Appl. Optics, Vol. 2, pp. 865–866, August 1963

12.16 Mode Selection in an Aperture-Limited Concentric Maser Interferometer
Tingye Li (BTL)
Bell Sys. Tech. J., Vol. 42, pp. 2609–2620, November 1963

12.17 Selection of Axial Modes in Optical Masers
H. Manger and H. Rothe (Technische Hochschule Karlsruhe, Germany)
Physics Letters, Vol. 7, pp. 330–331, December 15, 1963

12.18 An Experimental Verification of Transverse Modes in Concave Optical Resonators
J. T. Verdeyen, J. B. Gerardo, and E. P. Bialecke (U. of Illinois)
Proc. IEEE, Vol. 51, pp. 1775–1776, December 1963

12.19 Zeeman Effect and Nonlinear Interactions Between Oscillating Modes in Masers
H. Statz and C. L. Tang (Raytheon)
Proc. Third Int. Conf. Quantum Electronics, Columbia U. Press, N. Y., Vol. 1, pp. 469–480, 1964

12.20 Modes in Confocal Geometries
G. D. Boyd (BTL)
Proc. Third Int. Conf. Quantum Electronics, Columbia U. Press, N. Y., Vol. 2, pp. 1173–1186, 1964

12.21 Design of Resonators
J. M. Burch (Natl. Physical Lab., Teddington, England)
Proc. Third Int. Conf. Quantum Electronics, Columbia U. Press, N. Y., Vol. 2, pp. 1187–1202, 1964

12.22 New Modes of Optical Oscillation in Closed Resonators
R. J. Collins (Inst. for Defense Analyses) and J. A. Giordmaine (BTL)
Proc. Third Int. Conf. Quantum Electronics, Columbia U. Press, N. Y., Vol. 2, pp. 1239–1246, 1964

12.23 Etude des Modes d'Oscillations des Lasers à Miroirs Termineaux Sphériques
M. Pauthier (Lab. Central des Telecommunications, France)
Proc. Third Int. Conf. Quantum Electronics, Columbia U. Press, N. Y., Vol. 2, pp. 1253–1262, 1964

12.24 Modes in a Maser Interferometer with Curved Mirrors
A. G. Fox and T. Li (BTL)
Proc. Third Int. Conf. Quantum Electronics, Columbia U. Press, N. Y., Vol. 2, pp. 1263–1270, 1964

12.25 Isolation of Axi-Symmetrical Optical-Resonator Modes
W. W. Rigrod (BTL)
Proc. Third Int. Conf. Quantum Electronics, Columbia U. Press, N. Y., Vol. 2, pp. 1285–1290, 1964

12.26 Interferometer Laser Mode Selector
S. A. Collins and G. R. White (Sperry Gyroscope)
Proc. Third Int. Conf. Quantum Electronics, Columbia U. Press, N. Y., Vol. 2, pp. 1291–1300, 1964

12.27 Mode Spectrum in the He-Ne Maser
D. Rosenberger (Siemens and Halske, Germany)
Proc. Third Int. Conf. Quantum Electronics, Columbia U. Press, N. Y., Vol. 2, pp. 1301–1304, 1964

12.28   Characteristics of Ruby Laser Modes in a Nominally Plane Parallel Resonator
        V. Evtuhov and J. K. Neeland (Hughes)
        Proc. Third Int. Conf. Quantum Electronics, Columbia U. Press, N. Y., Vol. 2, pp. 1405–1414, 1964

12.29   A Diffraction Limited Oscillator
        J. G. Skinner and J. E. Geusic (BTL)
        Proc. Third Int. Conf. Quantum Electronics, Columbia U. Press, N. Y., Vol. 2, pp. 1437–1444, 1964

12.30   Transverse Mode Patterns in Neodymium Glass and Ruby
        E. S. Dayhoff (U. S. Naval Ordnance Lab., White Oak, Silver Spring, Md.)
        Proc. Third Int. Conf. Quantum Electronics, Columbia U. Press, N. Y., Vol. 2, pp. 1445–1451, 1964

12.31   An Internally Reflecting Optical Resonator with Confocal Properties
        D. F. Holshouser (U. of Illinois)
        Proc. Third Int. Conf. Quantum Electronics, Columbia U. Press, N. Y., Vol. 2, pp. 1453–1458, 1964

12.32   The Effects of Temperature on Ruby Optical Maser Mode Sequences
        M. C. Adamson, T. P. Hughes, and K. M. Young (A.E.I. Res. Lab., England)
        Proc. Third Int. Conf. Quantum Electronics, Columbia U. Press, N. Y., Vol. 2, pp. 1459–1467, 1964

12.33   Etudes des Variations Temporelles de la Composition Spectrale et de la Répartition Angulaire du Rayonnement Emis par les Lasers
        M. P. Vanukov, V. I. Issayenko, and V. V. Lubimov (Academy of Sciences, U.S.S.R.)
        Proc. Third Int. Conf. Quantum Electronics, Columbia U. Press, N. Y., Vol. 2, pp. 1477–1482, 1964

12.34   Matching of Optical Modes
        H. Kogelnik (BTL)
        Bell Sys. Tech. J., Vol. 43, pp. 334–337, Part I, January 1964

12.35   Beat Frequencies between Modes of a Concave-Mirror Optical Resonator
        John P. Goldsborough (IBM)
        Appl. Optics, Vol. 3, pp. 267–275, February 1964

12.36   Ruby Laser with Mode-Selective Etalon Reflector
        Dieter Roess (Siemen & Halske, Germany)
        Proc. IEEE, Vol. 52, pp. 196–197, February 1964

12.37   Reflecting Prisms for Dispersive Optical Maser Cavities
        A. D. White (BTL)
        Appl. Optics, Vol. 3, pp. 431–432, March 1964

12.38   The Diffraction Loss Curve for Nonconfocal Spherical Mirrors
        Takeo Omori (N. T. & T., Japan)
        IEEE Trans. on Microwave Theory and Techniques, Vol. MTT-12, p. 258, March 1964

12.39   Calculated Divergence of Laser Beam from Generalized Spherical Mirror Cavities
        Norio Karube (Matsushita, Japan)
        Proc. IEEE, Vol. 52, pp. 327–328, March 1964

12.40   Off-Axis Paths in Spherical Mirror Interferometers
        D. Herriott, H. Kogelnik, and R. Kompfner (BTL)
        Appl. Optics, Vol. 3, pp. 523–526, April 1964

12.41   Resonator Influence on Temporal Character of Laser Output
        M. Katzman and J. W. Strozyk (U.S. Army)
        Proc. IEEE, Vol. 52, pp. 433–434, April 1964

12.42   Capillary Quartz Cell for Liquid Laser Research
        Erhard J. Schimitschek and Edward R. Schumacher (U.S. Naval Electr. Lab.)
        Rev. Sci. Instr., Vol. 35, p. 521, April 1964 (mirrors on piston)

12.43   Precise Method for Measuring the Absolute Phase Change on Reflection
        Jean M. Bennett (Michelson Lab.)
        J. Opt. Soc. Am., Vol. 54, pp. 612–624, May 1964

12.44   Competition between Internal and External Modes of a Ruby Laser
        A. Korpel and J. Free (Zenith)
        Proc. IEEE, Vol. 52, pp. 619–620, May 1964 (0.6° between internal and external cavity axes, spikes from two cavities generally coincident)

12.45   Analysis of Spherical Sector Resonators for the Production of a Focused Laser Beam
        C. Y. She and H. Heffner (Stanford U.)
        Appl. Optics, Vol. 3, pp. 703–708, June 1964 (both Fabry-Perot reflectors have common curvature center)

12.46   Frustrated Total Internal Reflection and Application of its Principle to Laser Cavity Design
        I. N. Court and F. K. von Willisen (G. E.)
        Appl. Optics, Vol. 3, pp. 719–726, June 1964 (variable output coupling structure)

12.47   He-Ne Laser with Perpendicular End Windows
        K. D. Mielenz, K. F. Nefflen, and K. E. Gillilland (Natl. Bur. Standards)
        Appl. Optics, Vol. 3, pp. 785–786, June 1964 (confocal Fabry-Perot, near-field mode patterns, 1.15 microns)

12.48   Beat Frequency Between Two Traveling Waves in a Fabry-Perot Square Cavity
        P. K. Cheo and C. V. Heer (Ohio State U.)
        Appl. Optics, Vol. 3, pp. 788–789, June 1964 (He-Ne ring laser, 3.39 μ, sharp edge near beam produces nonreciprocal frequency bias)

12.49   The Laser Interferometer: Application to Plasma Diagnostics
        J. B. Gerardo and J. T. Verdeyen (U. of Illinois)
        Proc. IEEE, Vol. 52, pp. 690–697, June 1964 (He-Ne gas laser as both source and detector, helium plasma afterglow study)

12.50   Triangular Injection Lasers
        J. C. Marinace, A. E. Michel, and M. I. Nathan (IBM)
        Proc. IEEE, Vol. 52, pp. 722–723, June 1964 (equilateral triangle)

12.51   Messung der Güte von Laser-Resonatoren
        H. Boersch, G. Herziger, and H. Lindner (Tech. U., Berlin)
        Physics Letters, Vol. 11, pp. 38–39, July 1, 1964

12.52   Locking of He-Ne Laser Modes Induced by Synchronous Intracavity Modulation
        L. E. Hargrove, R. L. Fork, and M. A. Pollack (BTL)
        Appl. Phys. Letters, Vol. 5, pp. 4–5, July 1, 1964 (56 Mc/s synchronous modulation)

12.53   Resonant Modes of Optic Interferometer Cavities. I. Plane-Parallel End Reflectors
        L. Bergstein and H. Schachter (P.I.B.)
        J. Opt. Soc. Am., Vol. 54, pp. 887–903, July 1964

12.54   Comments on "Resonant Modes of Optic Interferometric Cavities"
        S. P. Morgan and T. Li (BTL)
        J. Opt. Soc. Am., Vol. 55, pp. 1128–1132, September 1965 (L. Bergstein and H. Schachter, Jr., J. Opt. Soc. Am., Vol. 54, p. 887, 1964)

12.55   Modes in Coupled Optical Resonators with Active Media
        J. R. Fontana (U. of Minnesota)
        IEEE Transactions on Microwave Theory and Techniques, Vol. MTT-12, pp. 400–405, July 1964 (multiresonator systems, analytical)

12.56   The Confocal Resonator System with a Large Fresnel Number (V-Type Eigenmodes)
        H. Lotsch (Stanford U.)
        IEEE Transactions on Microwave Theory and Techniques, Vol. MTT-12, pp. 482–483, July 1964

12.57 A Technique for Measuring Small Optical Loss using an Oscillating Spherical Mirror Interferometer
A. J. Rack and M. R. Biazzo (BTL)
Bell Sys. Tech. J., Vol. 43, pp. 1563–1579, July 1964, Pt. 2 (accurate high reflectance or small transmission measurements)

12.58 A Circle Diagram for Optical Resonators
J. P. Gordon (BTL)
Bell Sys. Tech. J., Vol. 43, pp. 1826–1827, July 1964, Pt. 2 (spherical mirror resonators)

12.59 A Note on the Resonant Modes and Spatial Coherency of a Fabry-Perot Maser Interferometer
G. J. Fan (IBM)
IBM J. Res. and Dev., Vol. 8, pp. 335–337, July 1964

12.60 Note Concerning the Analogy Between the Fabry-Perot Interferometer and the Iris-Type Beam Waveguide
H. K. V. Lotsch (Stanford U.)
Physics Letters, Vol. 11, pp. 221–222, August 1, 1964

12.61 A Study of Transverse Modes of Ruby Lasers using Beat-Frequency Detection and Fast Photography
C. M. Stickley (AFCRL)
Appl. Optics, Vol. 3, pp. 967–979, August 1964 (low beat frequencies, different resonators)

12.62 Improved Laser Angular Brightness Through Diffraction Coupling
J. T. LaTourette, S. F. Jacobs, and P. Rabinowitz (TRG)
Appl. Optics, Vol. 3, pp. 981–982, August 1964 (single-mode output)

12.63 A Rigid Low-Loss Polarizing Device for Giant Pulse Lasers using Brewster Angle Polarization
G. Weiser (Institut für Plasmaphysik, Germany)
Proc. IEEE, Vol. 52, p. 966, August 1964 (polarizer set at Brewster's angle)

12.64 On the Physical Formulation of the Fabry-Perot Interferometer Used as a Laser Resonator
H. K. V. Lotsch (Stanford U.)
Physics Letters, Vol. 12, pp. 99–101, September 15, 1964

12.65 Choice of Mirror Curvatures for Gas Laser Cavities
D. C. Sinclair (U. of Rochester)
Appl. Optics, Vol. 3, pp. 1067–1071, September 1964 (near-hemispherical cavity easiest to align, relative output power dependent on mode volume)

12.66 Continuous Operation of a Ruby Laser During Pumping Pulse
R. V. Pole and H. Wieder (IBM)
Appl. Optics, Vol. 3, pp. 1086–1087, September 1964 (conjugate-concentric resonator, helical lamp)

12.67 Single and Double Axial Mode Operation of a Nd-Optical Maser
H. Manger and H. Rothe (Tech. Hoch., Karlsruhe, Germany)
Physics Letters, Vol. 12, pp. 182–183, October 1, 1964 ($CaWO_4$: Nd, 5.9 and 11.95-mm thick etalons)

12.68 A Simple Way of Demonstrating the Phase Reversals in the TEM_{10}, TEM_{20}, TEM_{30} Modes of a Gas Laser Source
M. V. R. K. Murty (U. of Rochester)
Appl. Optics, Vol. 3, pp. 1192–1193, October 1964

12.69 Fabry-Perot Resonances at Small Fresnel Numbers
S. R. Barone and M. C. Newstein (TRG)
Appl. Optics, Vol. 3, p. 1194, October 1964

12.70 Eigenfrequencies and Quality Factors of Multi-Mirror Etalons
G. Bouwhuis (Philips)
Philips Res. Repts., Vol. 19, pp. 422–428, October 1964 (up to four-mirror etalons analyzed)

12.71 Eigenstates of Polarization in Lasers
H. de Lang (Philips)
Philips Res. Repts., Vol. 19, pp. 429–440, October 1964 (multiple-pass interferometer, multiple Eigenfrequencies)

12.72 Analysis of Optical Resonators Involving Focusing Elements
S. A. Collins, Jr. (Sperry Gyroscope)
Appl. Optics, Vol. 3, pp. 1263–1275, November 1964 (universal chart of phase front curvature and spot size)

12.73 Modes in a Triangular Ring Optical Resonator
S. A. Collins, Jr., and D. T. M. Davis, Jr. (Sperry Gyroscope)
Appl. Optics, Vol. 3, pp. 1314–1315, November 1964

12.74 Dual Forms of the Gaussian Beam Chart
T. Li (BTL)
Appl. Optics, Vol. 3, pp. 1315–1316, November 1964 (discussion of Collin's chart and its dual)

12.75 Diffraction Loss and Beam Size in Lasers with Spherical Mirrors
M. J. Taylor, G. R. Hanes, and K. M. Baird (Natl. Res. Council, Canada)
J. Opt. Soc. Am., Vol. 54, pp. 1310–1314, November 1964 (TEM_{00} mode losses determined experimentally)

12.76 Nonresonant Cavity Sources of Well-Defined Frequency
M. S. Cook (Laser, Inc.)
Proc. IEEE, Vol. 52, pp. 1351–1353, November 1964

12.77 Selection of Discrete Modes in Toroidal Lasers
D. Roess and G. Gehrer (Siemens and Halske, Germany)
Proc. IEEE, Vol. 52, pp. 1359–1360, November 1964 (ruby toroid, ruby cube, near field patterns)

12.78 Planar Dielectric-Waveguide Modes
A. H. Luther (U. S. Navy Ordnance Lab.)
Proc. IEEE, Vol. 52, p. 1386, November 1964 (applicable to p-n junctions)

12.79 Equivalence Relations Among Spherical Mirror Optical Resonators
J. P. Gordon and H. Kogelink (BTL)
Bell Sys. Tech. J., Vol. 43, pp. 2873–2886, November 1964

12.80 Modes of Optical Maser Cavities with Roof-Top and Corner Cube Reflectors
D. L. Bobroff (Raytheon)
Appl. Optics, Vol. 3, pp. 1485–1487, December 1964 (both traveling and standing wave modes, rotated reflectors)

12.81 Mode-Selecting Prism Reflectors for Optical Masers
J. A. Giordmaine and W. Kaiser (BTL)
J. Appl. Phys., Vol. 35, pp. 3446–3451, December 1964 (high reflectivity over 1 minute angular range)

12.82 Geometry of the Radiation Field for a Laser Interferometer
M. J. Offerhaus (Philips, The Netherlands)
Philips Res. Repts., Vol. 19, pp. 520–523, December 1964 (like Collin's charts in Applied Optics)

12.83 Mode Suppression on Lasers by Metal Wires
A. Okaya (Electro-Optical Systems)
Proc. IEEE, Vol. 52, p. 1741, December 1964 (gas laser, wire grid)

12.84 Coupling and Conversion Coefficients for Optical Modes
H. Kogelnik (BTL)
Proc. Symp. on Quasi-Optics, Polytech. Inst. of Bklyn., New York, MRI Symp. Series, Vol. 14, pp. 333–349, 1964

12.85 On the Schmidt Expansion for Optical Resonator Modes
W. Streifer and H. Gamo (U. of Rochester)
Proc. Symp. on Quasi-Optics, Polytech. Inst. of Bklyn., New York, MRI Symp. Series, Vol. 14, pp. 351–365, 1964

12.86 Hyperspheroidal Functions—Optical Resonators with Circular Mirrors
J. C. Heurtley (U. of Rochester)
Proc. Symp. on Quasi-Optics, Polytech. Inst. of Bklyn., New York, MRI Symp. Series, Vol. 14, pp. 367–377, 1964

12.87 Beam Tracing and Applications
G. A. Deschamps and P. E. Mast (U. of Illinois)
Proc. Symp. on Quasi-Optics, Polytech. Inst. of Bklyn., New York, MRI Symp. Series, Vol. 14, pp. 379–395, 1964 (resonators and iterative structures)

12.88 Round Table Discussion on Resonators
W. K. Kahn, Moderator, (PIB), L. Bergstein (PIB), H. Gamo

(U. of Rochester), G. J. E. Goubau (U. S. Army Electr. Labs.),
J. T. LaTourrette (TRG) and G. Toraldo Di Francia (U. of Florence, Italy)
Proc. Symp. on Quasi-Optics, Polytech. Inst. of Bklyn., New York, MRI Symp. Series, Vol. 14, pp. 397–414, 1964

12.89 Laser Mode Locking by an External Doppler Cell
L. C. Foster, M. D. Ewy, and C. B. Crumly (Zenith)
Appl. Phys. Letters, Vol. 6, pp. 6–8, January 1, 1965 (He–Ne, 57 Mc/s sonic quartz cell)

12.90 Measurements of Cavity Loss in a Pulsed Ruby Laser
D. Chen (Honeywell)
Nature, Vol. 205, pp. 271–272; January 16, 1965 (lower threshold yields narrower R_1 line width)

12.91 Microwave Models to Study the Performance of Laser Resonators
P. F. Checcacci, A. M. Scheggi, and G. Toraldo Di Francia (Centro Microonde del C.N.R., Florence, Italy)
Z. angewandte Mathematik und Physik, Vol. 16, No. 1, pp. 170–172; January 25, 1965

12.92 A "Twisted-Mode" Technique for Obtaining Axially Uniform Energy Density in a Laser Cavity
V. Evtuhov (Hughes) and A. E. Siegman (Stanford U.)
Appl. Optics, Vol. 4, pp. 142–143, January 1965 (birefringent plates adjacent to cavity mirrors)

12.93 Active Image Formation in Lasers
W. A. Hardy (IBM)
IBM J. Res. & Dev., Vol. 9, pp. 31–46, January 1965 (modes controlled by imaging masks, can be projected)

12.94 Self-Consistent Field Analysis of Multireflector Optical Resonators
P. O. Clark (Calif. Inst. of Tech.)
J. Appl. Phys., Vol. 36, pp. 66–72, January 1965

12.95 A Self-Consistent Field Analysis of Spherical-Mirror Fabry-Perot Resonators
P. O. Clark (Hughes)
Proc. IEEE, Vol. 53, pp. 36–41, January 1965

12.96 Broadband Dielectric Mirrors for Multiple Wavelength Laser Operation in the Visible
D. L. Perry (BTL)
Proc. IEEE, Vol. 53, pp. 76–77, January 1965 (over 30 layers, 99.5% reflectance, 0.43 to 0.74 μ)

12.97 The Observation of Mode Impurity in Gas Lasers Apparently Resonating in the TEM-00 Mode
H. M. Heinemann and H. W. Redlien, Jr. (Wheeler)
Proc. IEEE, Vol. 53, pp. 77–78, January 1965

12.98 Multimode Resonators with a Small Fresnel Number (Lowest-Order Eigenmodes)
Helmut K. V. Lotsch (Stanford U.)
Z. Naturforschung, Vol. 20a, No. 1, pp. 38–48, January 1965

12.99 Equivalence of Different Integral Equations for the Fabry-Perot Interferometer used as a Laser Resonator
A. G. Fox and T. Li (BTL)
Physics Letters, Vol. 14, pp. 187–188, February 1, 1965

12.100 Eigenmodes of a Symmetric Cylindrical Confocal Laser Resonator and Their Perturbation by Output-Coupling Apertures
D. E. McCumber (BTL)
Bell Sys. Tech. J., Vol. 44, pp. 333–363, February 1965 (diffraction loss, field distribution)

12.101 GaAs Injection Laser with Novel Mode Control and Switching Properties
M. I. Nathan, J. C. Marinace, R. F. Rutz, A. E. Michel, and G. J. Lasher (IBM)
J. Appl. Phys., Vol. 36, pp. 473–480, February 1965 (control by current distribution, dual units, bistable)

12.102 Internal Reflection Barriers as Reflectors in a Modified Fabry-Perot Interferometer

H. A. Daw and J. R. Izatt (New Mexico State U.)
J. Opt. Soc. Am., Vol. 55, pp. 201–202, February 1965

12.103 A Circular Ring Laser
I. Itzkan (Sperry)
Proc. IEEE, Vol. 53, p. 164, February 1965 (toroid, 45°-inclined optical coupler, conjecture)

12.104 Regular Emission from a Many-Element Laser During the Pumping Pulse
R. Pratesi and G. Toraldo di Francia (U. di Firenze, Italy)
Proc. IEEE, Vol. 53, pp. 196–197, February 1965 (ten-element ruby laser, regular spiking)

12.105 The Axicon-Scanned Fabry-Perot Spectrometer
J. Katzenstein (Culham Lab., England)
Appl. Optics, Vol. 4, pp. 263–266, March 1965 (uses prism of revolution)

12.106 A Spatial Spectral Scanning Technique for the Fabry-Perot Spectrometer
G. G. Shepherd, C. W. Lake, J. R. Müller, and L. L. Cogger (U. of Saskatchewan, Canada)
Appl. Optics, Vol. 4, pp. 267–272, March 1965 (angle-sensitive spectrometer, multi-channel output)

12.107 Imaging of Optical Modes—Resonators with Internal Lenses
H. Kogelnik (BTL)
Bell Sys. Tech. J., Vol. 44, pp. 455–494, March 1965 (Gaussian beams of light)

12.108 Off-Axial Modes in Cylindrical Glass Laser Rods
R. Yokota and H. Imagawa (Toshiba, Japan)
Japanese J. Appl. Phys., Vol. 4, pp. 231–232, March 1965 (6% Nd-doped glass, far field patterns)

12.109 Conjugate-Concentric Laser Resonator
R. V. Pole (IBM)
J. Opt. Soc. Am., Vol. 55, pp. 254–260, March 1965 (reflective and refractive surfaces)

12.110 Making of Spacers for Fabry-Perot Etalons
F. M. Phelps, III (U. of Michigan)
J. Opt. Soc. Am., Vol. 55, pp. 293–295, March 1965 (inexpensive, Invar spacer, fabrication and testing techniques)

12.111 Unstable Optical Resonators for Laser Applications
A. E. Siegman (Stanford U.)
Proc. IEEE, Vol. 53, pp. 277–287, March 1965 (Cassegrainian type, large diffraction losses, useful for high-gain lasers, ruby experiments)

12.112 A Proposed Novel Method for Obtaining a Nonspiking Pulsed Laser
R. H. Pantell and H. E. Puthoff (Stanford U.)
Proc. IEEE, Vol. 53, p. 295, March 1965 (segmented rod, unequal pumping)

12.113 Single Mode Differential Efficiency for Circular and Rectangular Laser Diodes
A. C. Scott (U. of Wisconsin)
Proc. IEEE, Vol. 53, pp. 315–316, March 1965 (omni-directional circular, bi-directional rectangular)

12.114 Locking of He-Ne Laser Modes by Intracavity Acoustic Modulation in Coupled Interferometers
M. DiDomenico, Jr., and V. Czarniewski (BTL)
Appl. Phys. Letters, Vol. 6, pp. 150–152, April 15, 1965 (triple mirror Fabry-Perot)

12.115 Study of the Output Spectra of Ruby Lasers
V. Evtuhov and J. K. Neeland (Hughes)
IEEE J. Quantum Electronics, Vol. QE-1, pp. 7–12, April 1965 (frequency separation between transverse modes, mode hopping to lower frequencies, mode selection)

12.116 Characteristics of Mode-Coupled Lasers
M. H. Crowell (BTL)
IEEE J. Quantum Electronics, Vol. QE-1, pp. 12–20, April 1965 (modes in a gas laser, time-varying loss, operates like a pulse regenerative oscillator, He–Ne and argon ion lasers)

12.117 Modes in Spherical-Mirror Resonators
W. A. Specht, Jr. (Calif. Inst. of Tech.)
J. Appl. Phys., Vol. 36, pp. 1306–1313, April 1965

12.118 Theory of Deformed Fabry-Perot Resonator
H. Ogura, Y. Yoshida and J. Ikenoue (Kyoto U., Japan)
J. Phys. Soc. Japan, Vol. 20, pp. 598–609, April 1965 (tilted mirror, curved mirror)

12.119 A Scalar Resonator Theory for Optical Frequencies
Helmut K. V. Lotsch (Northrop)
Optica Acta, Vol. 12, pp. 113–136, April 1965

12.120 The Fabry-Perot Interferometer as an Antenna Problem
H. K. V. Lotsch (Northrop)
Proc. IEEE, Vol. 53, pp. 398–399, April 1965

12.121 Mode Selection and Mode Volume Enhancement in a Gas Laser with Internal Lens
T. Li and P. W. Smith (BTL)
Proc. IEEE, Vol. 53, pp. 399–400, April 1965 ("cat's eye" configuration, HeNe laser)

12.122 Comments on the Equivalence of Different Integral Equations Formulated to Describe the Fabry-Perot Interferometer
H. K. V. Lotsch (Northrop)
Physics Letters, Vol. 16, pp. 45–46, May 1, 1965

12.123 Spectral Characteristics of a Gas Laser with Traveling Wave
S. N. Bagaev, V. S. Kuznetsov, Yu. V. Troitskii, and B. I. Troshin (Inst. of Semiconductor Physics, Siberian Div., Academy of Sciences, U.S.S.R.)
JETP Letters, Vol. 1, pp. 114–116, May 15, 1965 (0.63 μ wavelength, triangular resonator)

12.124 Design Considerations for Mode Selective Fabry-Perot Laser Resonator
N. Kumagai, M. Matsuhara, and H. Mori (Osaka U., Japan)
IEEE J. Quantum Electronics, Vol. QE-1, pp. 85–94, May 1965 (2-, 3-, and 4-reflector resonators)

12.125 Longitudinal Mode Control in Giant Pulse Lasers
F. J. McClung and D. Weiner (Hughes)
IEEE J. Quantum Electronics, Vol. QE-1, pp. 94–99, May 1965 (three control techniques, measurement of mode structure, ruby laser, single mode possible)

12.126 A Traveling-Wave Type Resonator Theory for Optical Frequencies
Helmut K. V. Lotsch (Stanford U.)
Physica, Vol. 31, pp. 629–644, May 1965

12.127 On the Field Distributions in a Fabry-Perot Interferometer
H. K. V. Lotsch (Northrop)
Proc. IEEE, Vol. 53, pp. 489–490, May 1965

12.128 Axial Modes of a Ruby Laser with External Reflectors
S. Singh, R. G. Smith, and M. Di Domenico, Jr. (BTL)
Proc. IEEE, Vol. 53, pp. 507–508, May 1965 (minimum fundamental beat with rod at cavity center)

12.129 Geometric Optical Derivation of Formula for the Variation of the Spot Size in a Spherical Mirror Resonator
W. K. Kahn (P.I.B.)
Appl. Optics, Vol. 4, pp. 758–759, June 1965

12.130 The Optical Ring Resonator
W. W. Rigrod (BTL)
Bell Sys. Tech. J., Vol. 44, pp. 907–916, May-June 1965 (stability parameter, spot size, wavefront curvature, up to four mirrors)

12.131 Diffraction Loss and Selection of Modes in Maser Resonators with Circular Mirrors
T. Li (BTL)
Bell Sys. Tech. J., Vol. 44, pp. 917–932, May-June 1965 (good mode-selective property with confocal configuration)

12.132 Resonant Waves in a Fabry-Perot Interferometer
Helmut K. V. Lotsch (Northrop)
Japanese J. Appl. Phys., Vol. 4, pp. 435–444, June 1965 (plane-parallel strip mirrors considered, aperture-limited operation)

12.133 A Ruby Laser with External Mirrors of Large Spacing
Tadao Shimizu (Inst. of Physical and Chemical Research, Tokyo, Japan) and Fujio Shimizu, Minato Kawaguti, and Koichi Shimoda (U. of Tokyo, Japan)
Japanese J. App. Phys., Vol. 4, pp. 445–451, June 1965 (up to 24.5 meters, low and high Q modes)

12.134 Problem of Mode Deformation in Optical Masers
H. Statz and C. L. Tang (Raytheon)
J. Appl. Phys., Vol. 36, pp. 1816–1819, June 1965 (nonuniform gain distribution within cavity)

12.135 The Confocal Resonator System with a Large Fresnel Number (V-Type Eigenmodes)
Helmut K. V. Lotsch (Stanford U.)
Z. angewandte Physik, Vol. 19, No. 3, pp. 162–168, June 1965

12.136 Laser with Radiation Diagram of Diffraction Width
A. L. Mikaelyan, A. V. Korovitsyn, and L. V. Naumova
JETP Letters, Vol. 2, pp. 22–24, July 1, 1965 (lowest loss near true confocal conditions)

12.137 The Natural Selection of Modes in a Passive Q-Switched Laser
W. R. Sooy (Hughes)
Appl. Phys. Letters, Vol. 7, pp. 36–37, July 15, 1965 (slow buildup enhances mode selection)

12.138 Single-Mode Operation of a Q-Switched Ruby Laser
M. Hercher (U. of Rochester)
Appl. Phys. Letters, Vol. 7, pp. 39–41, July 15, 1965 (saturable filter and resonant reflector, less than 0.001 Å width, over 5 MW peak power)

12.139 Mode Degeneracy-Dips on Output of Gas Laser
Masanobu Yamanaka, Mamoru Nakasuji, Yoshihiro Ohtsuka, and Hiroshi Yoshinaga (Osaka U., Japan)
Japanese J. Appl. Phys., Vol. 4, pp. 548–549, July 1965 (He-Ne laser, transverse mode interactions)

12.140 Optical Resonator Modes—Rectangular Reflectors of Spherical Curvature
William Streifer (U. of Rochester)
J. Opt. Soc. Am., Vol. 55, pp. 868–877, July 1965 (integral equation, numerical results)

12.141 Reduction of Infrared Maser Losses by Cavity Tubing
J. A. Little and C. V. Heer (Ohio State U.)
Rev. Sci. Instr., Vol. 36, p. 1061, July 1965 (92 $\times$ 92-cm-square cavity, 3.39 μ He-Ne, internal reflection from tubes)

12.142 Excitation of Auxiliary Off-Axis Laser Modes
M. P. Vanyukov, V. I. Isaenko, L. A. Luizova, and O. A. Shorokhov (Vavilov State Inst. of Optics, U.S.S.R.)
Soviet Physics JETP, Vol. 21, pp. 1–3, July 1965 (rod axis inclined to Fabry-Perot cavity axis, far-field pattern)

12.143 Excitation of Modes and Oscillation Kinetics in a Ruby Laser with a Concentric Resonator
V. V. Korobkin, A. M. Leontovich, and M. N. Smirnova (Lebedev Physics Inst., Academy of Sciences, U.S.S.R.)
Soviet Physics JETP, Vol. 21, pp. 53–58, July 1965 (regular and irregular spiking, emission spectra, explanations)

12.144 Changes in the Resonator of a Ruby Laser when Heated by Pumping
A. P. Veduta, A. M. Leontovich, and V. N. Smorchkov (Lebedev Physics Inst., Academy of Sciences, U.S.S.R.)
Soviet Physics JETP, Vol. 21, pp. 59–63, July 1965 (0.63-μ

laser interferometer probe, 4.8°C temp rise, nonuniform across rod diameter, distortion during 3.5 ms pump pulse)·

12.145 Laser Pulse-Shaping and Mode-Locking with Acoustic Waves
A. J. DeMaria and D. A. Stetser (United Aircraft)
Appl. Phys. Letters, Vol. 7, pp. 71–73, August 1, 1965 (27 Mc/s acoustic waves in quartz within argon ion laser cavity)

12.146 Mode Competition and Frequency Splitting in Magnetic-Field-Tuned Optical Masers
R. L. Fork (BTL) and M. Sargent, III (Yale U.)
Phys. Rev., Vol. 139, pp. A617–A618, August 2, 1965 (similar to hole burning)

12.147 Mode-Locking Effects in an Internally Modulated Ruby Laser
Thomas Deutsch (Raytheon)
Appl. Phys. Letters, Vol. 7, pp. 80–82, August 15, 1965 (KDP, 50 and 148 Mc/s, varied cavity length)

12.148 On the Integral Equations for Fabry-Perot Interferometer Modes
W. Streifer (U. of Rochester)
Physics Letters, Vol. 18, pp. 118–119, August 15, 1965 (Lotsch's formulation proved equivalent to that of Fox and Li)

12.149 Mode Competition and Collision Effects in Gaseous Optical Masers
R. L. Fork and M. A. Pollack (BTL)
Phys. Rev., Vol. 139, pp. A1408–A1414, August 30, 1965 (computer results, He–Ne laser experiments)

12.150 Low-Loss Multilayer Dielectric Mirrors
D. L. Perry (BTL)
Appl. Optics, Vol. 4, pp. 987–991, August 1965 (preparation, 99.8 percent from 25 layers, broadband)

12.151 Observation of the Transverse Modes of a Laser with a Scanning Interferometer
P. W. Smith (BTL)
Appl. Optics, Vol. 4, pp. 1038–1039, August 1965

12.152 Oscillating Modes in Ruby Lasers with Nonuniform Pumping Energy Distribution
Tingye Li and J. G. Skinner (BTL)
J. Appl. Phys., Vol. 36, pp. 2595–2596, August 1965

12.153 Mode Locking Behavior of Gas Lasers in Long Cavities
R. E. McClure (Sperry Gyroscope)
Appl. Phys. Letters, Vol. 7, pp. 148–150, September 15, 1965 (up to 5-m-long cavities, He–Ne laser)

12.154 On the Filaments in a Typical Laser with Plane Parallel Mirrors
Helmut K. V. Lotsch (Northrop)
Z. angewandte Mathematik und Physik, Vol. 16, pp. 704–706, September 25, 1965 (laser material imperfections)

12.155 The Fabry-Perot Interferometer with a Large Fresnel Number
Helmut K. V. Lotsch (Stanford U.)
Annalen der Physik, Series 7, Vol. 16, pp. 7–16, September 1965

12.156 Comparison of Focused and Parallel-Plane Fabry-Perot Resonators at Small Fresnel Numbers
K. E. Lonngren and J. B. Beyer (U. of Wisconsin)
Appl. Optics, Vol. 4, pp. 1206–1207, September 1965 (loss per iteration)

12.157 On Mode Losses in Confocal Resonator and Transmission Systems
J. C. Heurtley and W. Streifer (U. of Rochester)
IEEE Trans. on Microwave Theory and Techniques, Vol. MTT-13, pp. 711–712, September 1965

12.158 Open Resonators with Mirrors Having Variable Reflection Coefficients
N. G. Vakhimov
Radio Eng. Electronic Physics, Vol. 9, pp. 1439–1446, September 1965 (parallel plane and concave mirrors)

12.159 Resonators with Mirrors Having a Variable Reflection Coefficient
S. N. Vlasov
Radio Eng. Electronic Physics, Vol. 9, pp. 1473–1476, September 1965

12.160 Flat-Roof Resonators
G. Toraldo di Francia (U. of Florence, Italy)
Appl. Optics, Vol. 4, pp. 1267–1270, October 1965 (inside angle greater than 90°)

12.161 Bistable Traveling-Wave Oscillations of Ion Ring Laser
W. W. Rigrod and T. J. Bridges (BTL)
IEEE J. Quantum Electronics, Vol. QE-1, pp. 298–303, October 1965 (4879.9 Å from argon, spontaneous and controlled one-way oscillation)

12.162 Properties of an Anisotropic Fabry-Perot Resonator
Walter M. Doyle and Matthew B. White (Philco)
J. Opt. Soc. Am., Vol. 55, pp. 1221–1225, October 1965 (birefringent plate, optical rotator, gas laser experiment, frequency splitting)

12.163 Resonant Modes of Optic Cavities of Small Fresnel Numbers
Leonard Bergstein and Harry Schachter (Polytechnic Inst. of Brooklyn)
J. Opt. Soc. Am., Vol. 55, pp. 1226–1233, October 1965 (power loss, intensity distribution over mirror)

12.164 Mode Competition and Self-Locking Effects in a Q-Switched Ruby Laser
Hans W. Mocker (Honeywell) and R. J. Collins (U. of Minnesota)
Appl. Phys. Letters, Vol. 7, pp. 270–273, November 15, 1965 (passive Q switch, varied cavity, few axial modes)

12.165 Transmission Characteristics of Fabry-Perot Interferometers and a Related Electrooptic Modulator
V. N. Del Piano, Jr., and A. F. Quesada (Baird-Atomic)
Appl. Optics, Vol. 4, pp. 1386–1390, November 1965 (effects of maladjustments and surface imperfections)

12.166 The Optical Whispering Mode of Polished Cylinders and its Implications in Laser Technology
Franklin G. Reick (ITT)
Appl. Optics, Vol. 4, pp. 1395–1399, November 1965 (clockwise and counterclockwise trapped light in cylinder)

12.167 Stabilized, Single-Frequency Output from a Long Laser Cavity
P. W. Smith (BTL)
IEEE J. Quantum Electronics, Vol. QE-1, pp. 343–348, November 1965 (external and internal mode selectors, 1.5-m-long laser cavity, 15 mW, 0.63 μ from He–Ne laser)

12.168 On the Regular and Irregular Spiking Behaviour of Solid-State Lasers
Helmut K. V. Lotsch (Stanford U.)
International J. of Electronics, Vol. 19, pp. 453–467, November 1965 (parabolic cylinder functions in Fabry-Perot cavity, filaments, ruby laser experiments, up to 20 meter long cavity, aperture limited)

12.169 Optical Resonator Modes—Circular Reflectors of Spherical Curvature
John C. Heurtley and William Streifer (U. of Rochester)
J. Opt. Soc. Am., Vol. 55, pp. 1472–1479, November 1965 (phase and intensity of modes, large Fresnel number configurations)

12.170 A Generalized Diffraction Formula
H. K. V. Lotsch (Northrop)
Optik, Vol. 23, No. 2, pp. 189–196, November 1965 (illuminated aperture, discussion of open-walled resonator)

12.171 A Multiple Internal-Reflection Folded-Path Optical Maser Geometry
A. J. DeMaria (United Aircraft Corp. Research Labs.)
Proc. IEEE, Vol. 53, pp. 1757–1758, November 1965 (rectangular configuration, ruby immersed in liquid nitrogen)

12.172 Microwave Models of Optical Resonators
P. F. Checcacci and A. M. Scheggi (Centro Microonde, Consiglio Nazionale delle Ricerche, Florence, Italy)
Appl. Optics, Vol. 4, pp. 1529–1532, December 1965 (open X-band models, flatness within λ/100 of plane mirrors)

12.173 On "Microwave Models of Optical Resonators"
Helmut K. V. Lotsch (Northrop)
Appl. Optics, Vol. 5, pp. 673–674, April 1966

12.174 On "Microwave Models of Optical Resonators"
A. G. Fox and Tingye Li (BTL)
Appl. Optics, Vol. 5, p. 1341, August 1966

12.175 Phase Locking of Modes in Lasers
H. Statz and C. L. Tang (Raytheon)
J. Appl. Phys., Vol. 36, pp. 3923–3927, December 1965 (important with multimode lasers, some data using ruby, analyses)

12.176 The Field Distributions in a Fabry-Perot Interferometer with a Very Small Fresnel Number
H. K. V. Lotsch (Stanford U.)
Physica, Vol. 31, pp. 1796–1808, December 1965

12.177 Reflection Gratings as Elements in Far Infrared Masers
Eric Brannen (U. of Western Ontario, Canada)
Proc. IEEE, Vol. 53, pp. 2134–2135, December 1965 (resonator uses first-order diffracted ray, zero-order reflection is coupled out)

12.178 Mode Locking of a Nd^{3+}-Doped Glass Laser
A. J. DeMaria, C. M. Ferrar, and G. E. Danielson, Jr. (United Aircraft Res. Labs.)
Appl. Phys. Letters, Vol. 8, pp. 22–24, January 1, 1966 (pulsed, 12.2 cm long rod, 47 Mc/s acoustic modulator within laser cavity, 0.5 ns pulses at 94 Mc/s rate)

12.179 Interaction of Traveling Waves in a Ring Laser
E. M. Belenov, E. P. Markin, V. N. Morozov, and A. N. Oraevskii (Lebedev Physics Institute, Academy of Sciences, U.S.S.R.)
JETP Letters, Vol. 3, pp. 32–35, January 1, 1966 (3.39-μ He–Ne laser, varied output mirror reflectivity)

12.180 Spectrum of a Fabry-Perot Resonant Cavity Containing an Active Medium
B. F. Hochheimer (Johns Hopkins U.)
Appl. Optics, Vol. 5, pp. 113–120, January 1966 (below laser threshold, He–Ne discharge cell in Michelson interferometer, line shape)

12.181 A Method for Evaluating the Cavity Loss and an Optimum Reflectivity of the Output Mirror in a Ruby Laser with an External Mirror
Yoshihiro Ohtsuka (Osaka U., Japan)
Japanese J. Appl. Phys., Vol 5, pp. 74–78, January 1966 (mirror tilt, reflectivity, insert glass plate)

12.182 Angular Spectra of Optic Cavities
Leonard Bergstein and Emanuel Marom (Polytechnic Inst. of Brooklyn)
J. Opt. Soc. Am., Vol. 56, pp. 16–32, January 1966 (plane-parallel cavities, infinite strip and circular reflectors, relative power losses)

12.183 Comments on "Angular Spectra of Optic Cavities"
S. P. Morgan and T. Li (BTL)

J. Opt. Soc. Am., Vol. 56, pp. 1138–1139, August 1966 (J. Opt. Soc. Am., Vol. 56, p. 16, 1966, and reply by L. Bergstein and E. Marom)

12.184 A Laser End Reflector with Spectral Tuning Capability
Robert M. Zoot (Hughes)
Appl. Optics, Vol. 5, pp. 349–350, February 1966 (Pellin-Broca type prism with reflective interface)

12.185 Laser Cavities with Increased Axial Mode Separation
A. D. White
Bell Sys. Tech. J., Vol. 45, pp. 339–343, February 1966 (split beam for bicavity arrangement)

12.186 Experimental Verification of Fox and Li Patterns in Open Resonators
P. F. Checcacci, A. M. Scheggi, and G. Toraldo Di Francia (Centro Microonde, Italy)
Electronics Letters, Vol. 2, pp. 63–64, February 1966 (3 cm wavelength experiments, Fresnel numbers up to 6.25, good check on Fox and Li)

12.187 On the Diffraction Loss of Optical Resonator
Ryuichi Hioki and Takeomi Suzuki (U. of Tokyo, Japan)
Japanese J. Appl. Phys., Vol. 5, pp. 160–166, February 1966 (aperture effects, mirror spacing, confocal losses)

12.188 Mode Selection Properties of Segmented-Rod Giant Pulse Lasers
Milton Birnbaum and Tom L. Stocker (Aerospace)
J. Appl. Phys., Vol. 37, pp. 531–534, February 1966

12.189 Influence of Mode Number and Mode Degeneracy on the Output of a Ruby Laser
Harold Wieder (IBM)
J. Appl. Phys., Vol. 37, pp. 615–621, February 1966

12.190 Formulation of the Integral Equation for the Resonant Modes of Optic Cavities
Emanuel Marom (Polytechnic Institute of Brooklyn)
J. Appl. Phys., Vol. 37, pp. 1942–1943, March 15, 1966

12.191 Unstable Optical Resonators
Walter K. Kahn (Polytechnic Inst. of Brooklyn)
Appl. Optics, Vol. 5, pp. 407–413, March 1966 (ray analysis)

12.192 Experimental Confirmation of Standing Waves in Laser Resonators
A. M. Ledger (Canadian Westinghouse, Canada)
Appl. Optics, Vol. 5, pp. 476–477, March 1966 (evaporated pattern on thin metallic film at small angle within laser cavity)

12.193 Slight Deformation of Confocal Fabry-Perot Resonator
Hisanao Ogura, Yasuo Yoshida, Yoji Furuhama and Jun-ichi Ikenoue (Kyoto U., Japan)
Japanese J. Appl. Phys., Vol. 5, pp. 225–233, March 1966 (tilt)

12.194 Alignment Characteristics of a Roof Prism Optical Maser
Norio Karube, Yasuhide Sakai and Eiso Yamaka (Matsushita Res. Inst. Tokyo, Japan)
Japanese J. Appl. Phys., Vol. 5, pp. 257–258, March 1966

12.195 Technique for Aligning Laser Mirrors Using Gas Laser
P. N. Everett (Mitre Corp.)
Rev. Sci. Instr., Vol. 37, p. 375, March 1966

12.196 Nonlinear Mode Interaction in Lasers
N. G. Basov, V. N. Morozov and A. N. Oraevskii (Lebedev Physics Institute, U.S.S.R.)
Soviet Phys. JETP, Vol. 22, pp. 622–628, March 1966

12.197 Operating Features of the Ring Laser
I. L. Bershtein and Yu. I. Zaĭtsev (Radiophysics Institute, Gorkiĭ State U.)
Soviet Phys. JETP, Vol. 22, pp. 663–667, March 1966 (3-mirrors, 0.63 μ)

12.198 Investigation of Spatial Coherence and Phase Change in the Modes of a He–Ne Laser
L. Csillag, M. Jánossy and K. Kántor (Central Res. Inst. of Physics, Hungary)
Physics Letters, Vol. 20, pp. 636–637, April 1, 1966 (180° change across TEM$_{10}$ mode)

12.199 Some Effects of Nonuniform Pumping on the Mode Structure of Solid State Lasers
Peter J. Warter, Jr., and Ramon U. Martinelli (Princeton U.)
J. Appl. Phys., Vol. 37, pp. 2103–2111, April 1966

12.200 Autoresonant Feedback in Lasers
V. S. Letokhov (Lebedev Physics Inst., U.S.S.R.)
JETP Letters, Vol. 3, pp. 269–271, May 15, 1966 (reflection from three-dimensional phase lattice)

12.201 Modes of a Tilted-Mirror Optical Resonator for the Far Infrared
W. H. Wells (Jet Propulsion Lab.)
IEEE J. Quantum Electronics, Vol. QE-2, pp. 94–102, May 1966 (output coupled by spillover)

12.202 Interferometry and Laser Control with Solid Fabry-Perot Etalons
Don G. Peterson (Lockheed) and Amnon Yariv (Calif. Inst. of Tech.)
Appl. Optics, Vol. 5, pp. 985–991, June 1966 (quartz etalon, thermal effects, scattering)

12.203 Laser Mode Control by Internal Modulation Using the Transverse Electrooptic Effect in Quartz
G. A. Massey (Sylvania)
Appl. Optics, Vol. 5, pp. 999–1001, June 1966 (0.63-μ He–Ne laser, negligible reduction in output power)

12.204 Rays and Ray Envelopes within Stable Optical Resonators Containing Focusing Media
Noritaka Kurauchi (Sumitomo Electric Industries, Japan) and Walter K. Kahn (Polytechnic Inst. of Brooklyn)
Appl. Optics, Vol. 5, pp. 1023–1029, June 1966

12.205 On the Model of the "Equivalent" Confocal Resonator System
Helmut K. V. Lotsch (Northrop)
Optica Acta, Vol. 13, pp. 229–239, July 1966

12.206 Stable, Easily Fabricated, Tunable, Passive Optical Cavity
A. D. White (BTL)
Rev. Sci. Instr., Vol. 37, pp. 968–969, July 1966 (piezoelectric cylinder for tuning, thermally insulated cavity)

12.207 Method for Decoupling Laser Mirror Transducers from Mechanical Resonances of Laser Cavity
A. D. White (BTL)
Rev. Sci. Instr., Vol. 37, pp. 976–977, July 1966 (symmetrical transducer)

12.208 Interaction Between Axial Modes of a Zeeman Laser
J. Kannelaud and W. Culshaw (Lockheed)
Appl. Phys. Letters, Vol. 9, pp. 120–123, August 1, 1966 (2.65-μ transition of Xe)

12.209 Direct Spectroscopic Detection of Ruby Laser Giant Pulse Off-Axial Mode Structure
Daniel J. Bradley, Malcolm S. Engwell, and A. W. McCullough (U. of London, England), and George Magyar and Martin C. Richardson (UKAEA, England)
Appl. Phys. Letters, Vol. 9, pp. 150–152, August 15, 1966

12.210 Single Transverse and Longitudinal Mode Q-Switched Ruby Laser
V. Daneu, C. A. Sacchi, and O. Svelto (Milano Politecnico, Italy)
IEEE J. Quantum Electronics, Vol. QE-2, pp. 290–293, August 1966 (spherical mirrors, vanadium phthalocyanine saturable absorber)

12.211 Analysis of a Room-Temperature CW Ruby Laser of 10-mm Resonator Length: The Ruby Laser as a Thermal Lens
Dieter Roess (Siemens and Halske, Germany)
J. Appl. Phys., Vol. 37, pp. 3587–3594, August 1966 (axial and transverse modes)

12.212 Minimizing the Divergence of Laser Beams
V. V. Lyubimov
Optics and Spectroscopy, Vol. 21, pp. 129–130, August 1966 (considers resonator distortions)

12.213 The Output Power of a Gas Laser Using Nearly Confocal Resonators
N. I. Kaliteevskii, M. M. Popov, Yu. A. Rymarchuk, T. B. Tolchinskaya, and M. P. Chaika
Optics and Spectroscopy, Vol. 21, pp. 152–154, August 1966 (varied resonator length)

12.214 Discrimination of Axial Oscillation Modes in a Laser with External Mirrors
V. I. Malyshev and A. S. Markin (Lebedev Phys. Inst., Acad. of Sci., USSR)
Soviet Phys. JETP, Vol. 23, pp. 225–227, August 1966 (Nd-glass laser, rod position relative to mirrors, bleachable liquid, four discrete beat frequencies)

12.215 New Property of the Stability Diagram for Curved-Mirror Resonators
S. A. Harrison and W. K. Kahn (Poly. Inst. of Brooklyn)
Electronics Letters, Vol. 2, pp. 326–327, September 1966 (insertion of inhomogeneous focusing medium)

12.216 Effect of Axial Modes on Doppler Experiments with Gas Lasers
P. T. Bolwijn, Th. H. Peek, and C. Th. J. Alkemade (Utrecht State U., Netherlands)
Physics Letters, Vol. 23, pp. 88–90, October 3, 1966 (moving mirror, power output modulation, He–Ne laser)

12.217 Laser Beams and Resonators
H. Kogelnik and T. Li (BTL)
Appl. Optics, Vol. 5, pp. 1550–1567, October 1966 (review, stability criteria, modes, losses)

12.218 Modes, Phase Shifts, and Losses of Flat-Roof Open Resonators
P. F. Checcacci, Anna Consortini, and Annamaria Scheggi (Centro Microonde, Italy)
Appl. Optics, Vol. 5, pp. 1567–1572, October 1966 (Fox and Li iterative method, microwave models)

12.219 Fabry-Perot Resonators in Uniaxially Anisotropic Media
L. Bergstein (Poly. Inst. of Brooklyn) and T. H. Zachos (BTL)
IEEE J. Quantum Electronics, Vol. QE-2, pp. 677–690, October 1966 (TE and TM modes)

12.220 Optical Resonators
G. Toraldo di Francia (U. of Firenze, Italy)
Optica Acta, Vol. 13, pp. 323–342, October 1966 (several resonators, review)

12.221 Laser Beams and Resonators
H. Kogelnik and T. Li (BTL)
Proc. IEEE, Vol. 54, pp. 1312–1329, October 1966 (review, stability criteria, modes, losses)

12.222 Modes, Phase Shifts, and Losses of Flat-Roof Open Resonators
P. F. Checcacci, Anna Consortini, and Annamaria Scheggi (Centro Microonde, Italy)
Proc. IEEE, Vol. 54, pp. 1329–1334, October 1966 (Fox and Li iterative method, microwave models)

13. GIANT PULSE TECHNIQUES (Q-SWITCHING)

A. Miller (RCA)
J. Appl. Phys., Vol. 36, pp. 334–335; January 1965

13.24 Use of Dielectric Etalon as a Reflector for Q-Switched Laser Operation
H. Pawel, J. R. Sanford, J. H. Wenzel, and G. J. Wolga (G. E.)
Proc. IEEE, Vol. 52, pp. 1048–1049, September 1964 (26 per cent reflectance from sapphire etalon at 0.69 micron)

13.25 An Improved Method of Mechanical Q-Switching Using Total Internal Reflection
R. Daly and S. D. Sims (TRG)
Appl. Optics, Vol. 3, pp. 1063–1066, September 1964 (12 reflection Lummer-Gehrcke plate provides faster switching)

13.26 Generation of Giant Pulses from a Neodymium Laser by a Reversibly Bleachable Absorber
B. H. Soffer and R. H. Hoskins (Korad)
Nature, Vol. 204, p. 276, October 17, 1964

13.27 Giant Pulse Lasers and Stimulated Raman Scattering
R. W. Hellwarth, F. J. McClung, W. G. Wagner, and D. Weiner (Hughes)
Z. angewandte Mathematik und Physik, Vol. 16, No. 1, pp. 27–32, January 25, 1965

13.28 Study of Giant Laser Pulse Formation, Utilizing Ultrasonic Q Switching
V. A. Suprynowicz (U. of Connecticut)
Bull. Am. Phys. Soc., Series II, Vol. 10, No. 1, p. 97–HG6, January 1965

13.29 Giant Pulses from a Laser: Optimum Conditions
M. Menat (Israel AEC, Israel)
J. Appl. Phys., Vol. 36, pp. 73–76, January 1965

13.30 Erratum: Giant Pulses from a Laser: Optimum Conditions
M. Menat (Israel Atomic Energy Commission, Israel)
J. Appl. Phys., Vol. 37, p. 936, February 1966 (J. Appl. Phys., Vol. 36, p. 73, 1965)

13.31 Comparison of Passive Q-Switch Components and Observation of Scattering Effects
J. I. Masters and E. M. E. Murray (Tech. Op. Res.)
Proc. IEEE, Vol. 53, p. 76, January 1965 (Beckman & Whitley thin film gives highest peak power)

13.32 Saturable Optical Absorption in Phthalocyanine Dyes
J. A. Armstrong (IBM)
J. Appl. Phys., Vol. 36, pp. 471–473, February 1965 (with ruby laser)

13.33 Production of Giant Laser Emission Pulses in Neodymium Activated Glass with the Aid of Translucent Solutions
O. L. Lebedev, V. N. Gavrilov, Yu. M. Gryaznov, and A. A. Chastov
JETP Letters, Vol. 1, pp. 47–48, April 15, 1965 (polymethine dye in quinoline)

13.34 A Repetitively Q-Switched, Continuously Pumped YAG:Nd Laser
J. E. Geusic, M. L. Hensel, and R. G. Smith (BTL)
Appl. Phys. Letters, Vol. 6, pp. 175–177, May 1, 1965 (1000 watts pump, 50 to 600 rps rotating mirror)

13.35 Q-Modulation of a Neodymium-Glass Laser with the Aid of a Passive Shutter
V. I. Malyshev, A. S. Markin, and V. S. Petrov (Lebedev Physics Inst., Academy of Sciences, U.S.S.R.)
JETP Letters, Vol. 1, pp. 99–101, May 1, 1965 (polymethine dye, 3000 J threshold, transmission loss through breakdown region)

13.36 Theory of Laser Giant Pulsing by a Saturable Absorber
A. Szabo (Nat. Res. Council, Canada) and R. A. Stein (U. Alberta, Canada)
J. Appl. Phys., Vol. 36, pp. 1562–1566, May 1965 (optimum output mirror reflectivity for maximum power)

13.37 Optical Amplification of the Apparent Rate of Rotation of a Reflector in Q-switching a Laser Resonator
J. W. Gates and R. G. N. Hall (Natl. Physical Lab., England)
Nature, Vol. 206, p. 1141, June 12, 1965 (afocal system to increase beam diameter, 3000 rpm rotating prism, about 1.2 joules output from 10-cm long ruby)

13.38 Study on Saturation Process by Anomalous Dispersion of Ruby Laser
Pil Hyon Kim and Susumu Namba (Inst. of Physical and Chemical Research, Tokyo, Japan)
Japanese J. Appl. Phys., Vol. 4, pp. 469–470, June 1965 (kryptocyanine bleachable dye, laser wavelength depends on concentration and intensity)

13.39 The Natural Selection of Modes in a Passive Q-Switched Laser
W. R. Sooy (Hughes)
Appl. Phys. Letters, Vol. 7, pp. 36–37, July 15, 1965 (slow buildup enhances mode selection)

13.40 Single-Mode Operation of a Q-Switched Ruby Laser
M. Hercher (U. of Rochester)
Appl. Phys. Letters, Vol. 7, pp. 39–41, July 15, 1965 (saturable filter and resonant reflector, less than 0.001 Å width, over 5-MW peak power)

13.41 Characteristics of a Q-Switched Ruby Laser
T. V. Gvaladze, I. K. Krasyuk, P. P. Pashinin, A. V. Prokhindeev, and A. M. Prokhorov (Lebedev Physics Inst., Academy of Sciences, U.S.S.R.)
Soviet Physics JETP, Vol. 21, pp. 72–74, July 1965 (rotating TIR prism, single mode with 3-mm-thick mirror)

13.42 Output Power and Energy in a Q-Switched Ruby Laser with a Saturable Absorber
G. Potenza and A. Sona (Laboratori C.I.S.E., Milan, Italy)
Nuovo Cimento, Vol. 38, pp. 1438–1440, August 1, 1965 (multiple spikes, higher peak power and lower energy output with increasing concentration of phthalocyanine in nitrobenzene, absorber loss of 10^{-4})

13.43 The Theory of Q-switching Applied to Slow Switching and Pulse Shaping for Solid State Lasers
J. E. Midwinter (Royal Radar Estab., England)
British J. Appl. Phys., Vol. 16, pp. 1125–1133, August 1965

13.44 The Change in the Emission Characteristics of a Ruby Laser Caused by Phthalocyanine Solutions in the Laser Cavity
V. N. Gavrilov, Yu. M. Gryaznov, O. L. Lebedev, and A. A. Chastov
Soviet Physics JETP, Vol. 21, pp. 510–511, August 1965 (several phthalocyanines in pyridine)

13.45 Narrow Spectral Emission from a Passively Q-Spoiled Neodymium-Glass Laser
B. B. McFarland, R. H. Hoskins, and B. H. Soffer (Korad)
Nature, Vol. 207, pp. 1180–1181, September 11, 1965 (bleachable polymethine dye, < 0.02 Å)

13.46 A Kerr Cell with Roof Prism
Milton Laikin (Electro-Optical Systems)
Appl. Optics, Vol. 4, pp. 1177–1178, September 1965 (prism apex parallel or perpendicular to Kerr cell plates, phase shift in prism)

13.47 Laser Q-Switch Using a Roof Prism End Reflector and Electro-Optical Retarder
W. W. Buchman (Korad)
IEEE J. Quantum Electronics, Vol. QE-1, pp. 280–281, September 1965 (phase retardation plate for polarized laser and arbitrary prism roof angle)

13.48 Emission of Monopulses of Coherent Light by a Two-Component Medium with Negative Absorption
V. I. Borodulin, N. A. Ermakova, L. A. Rivlin, and V. S. Shil'dyaev
Soviet Physics JETP, Vol. 21, pp. 563–566, September 1965 (ruby, KS-19 bleachable glass plate, air breakdown)

13.49 Giant Pulses from Neodymium Doped Calcium Tungstate by Gain-Switching
R. A. Clay and D. Findlay (Royal Radar Estab., England)
Physics Letters, Vol. 19, pp. 212–213, October 15, 1965 (rotating prism between two laser rods, 2.5 MW peak power)

13.50 More Power to the Laser—with Q Switching
Joseph I. Masters (Technical Operations)
Electronics, Vol. 38, No. 21, pp. 91–95, October 18, 1965 (general, active, and passive switching)

13.51 Laser Oscillations and Self Q-Switching in Triply Activated Glass
H. W. Gandy, R. J. Ginther, and J. F. Weller (U.S. Naval Research Lab.)
Appl. Phys. Letters, Vol. 7, pp. 233–236, November 1, 1965 (UO_2^{2+}, Nd^{3+} and Yb^{3+}; filtered flash lamp)

13.52 Quasicontinuous Ruby Giant Pulse Laser using a Saturable Absorber as a Q Switch
Dieter Roess and Günter Zeidler (Siemens & Halske A. G., Munich, Germany)
Appl. Phys. Letters, Vol. 8, pp. 10–12, January 1, 1966 (ellipsoidal pump cavity, room temp., very low saturation level in methylene blue absorber)

13.53 Lifetimes of Saturable Absorbers
P. W. A. Bowe, W. E. K. Gibbs, and J. Tregellas-Williams (Defence Standards Labs., Maribyrnong, Victoria, Australia)
Nature, Vol. 209, pp. 65–66, January 1, 1966 (aluminum phthalocyanine, Q-sw ruby laser, about 5 ns)

13.54 Synchronization of Several Q-Switched Optical Masers
Nguyen van Tran and D. Kehl (Compagnie Générale d'Electricité, France)
Appl. Optics, Vol. 5, pp. 168–169, January 1966 (one common rotating prism)

13.55 Bleaching Waves in Two-Level Systems
V. M. Ovchinnikov and V. E. Khartsiev (A. F. Ioffe Physicotechnical Institute, U.S.S.R.)
Soviet Physics JETP, Vol. 22, pp. 221–222, January 1966 (predict 42 ns bleaching time in cryptocyanine by ruby)

13.56 Saturation Effects in the Absorption of the Laser Light by Organic Dyes
V. Degiorgio and G. Potenza (Laboratori CISE, Italy)
Nuovo Cimento, Vol. 41 B, pp. 254–257, February 11, 1966 (about 10^8 W/cm², Q-sw ruby, decay times)

13.57 Giant Laser Pulse Formation using Ultrasonic Q-Spoiling
V. A. Suprynowicz (U. of Connecticut)
J. Appl. Phys., Vol. 37, pp. 778–784, February 1966

13.58 The Effect of Absorber Relaxation on Passive Q-Switch Laser Performance
R. McLeary and P. W. Bowe (Defence Standards Laboratories, Australia)
Appl. Phys. Letters, Vol. 8, pp. 116–117, March 1, 1966

13.59 Frequency Locking and Dye Spectral Hole Burning in Q-Spoiled Lasers
B. H. Soffer and B. B. McFarland (Korad)
Appl. Phys. Letters, Vol. 8, pp. 166–169, April 1, 1966 (two laser beams intersecting in common cell containing dye)

13.60 Self Mode-Locking of Lasers with Saturable Absorbers
A. J. DeMaria, D. A. Stetser, and H. Heynau (United Aircraft)
Appl. Phys. Letters, Vol. 8, pp. 174–176, April 1, 1966 (recirculating loop, Nd glass amplifier, 1 ns pulses at 100 Mc/s rate)

13.61 Generation of Ultrashort Optical Pulses by Mode Locking the YAlG: Nd Laser
M. DiDomenico, Jr., J. E. Geusic, H. M. Marcos, and R. G. Smith (B.T.L.)
Appl. Phys. Letters, Vol. 8, pp. 180–183, April 1, 1966 (acoustic modulator in laser cavity, 8×10^{-11} s pulse length)

13.62 Dynamics of the Field and Generation Frequency in a Giant Pulse of a Laser with Passive Shutter
V. V. Korobkin, A. M. Leontovich, M. N. Popova, and M. Ya. Shchelev (Lebedev Physics Institute, USSR)
JETP Letters, Vol. 3, pp. 194–196, April 1, 1966 (cryptocyanine, near field, streak pictures of slit emission on ruby rod end, far zone, emission wavelengths)

13.63 Nonlinear Polarizers as λ-Independent, Variable Q-Switches for Lasers
Gy. Farkas and I. Kertesz (Central Res. Inst. of Physics, Hungary)
Physics Letters, Vol. 20, pp. 634–635, April 1, 1966 (polarized ruby laser, HN polarizer sheet, comparison with 3 bleachable glass filters)

13.64 Giant Pulse Shortening by Resonator Transients
Dieter Roess (Siemens & Halske A. G., Germany)
J. Appl. Phys., Vol. 37, pp. 2004–2006, April 1966 (Q-sw ruby laser pumps Nd^{3+} and Dy^{3+} lasers, add saturated amplifier, calculations, 10^{-11} s risetime)

13.65 Self Q-Switched Nd^{3+} Glass Laser
W. Shiner, E. Snitzer, and R. Woodcock (American Optical)
Physics Letters, Vol. 21, pp. 412–413, June 1, 1966 (unfiltered pump light, color centers formed, regular spikes at about 100 kc/s)

13.66 A Faraday-Switched Ruby Laser
U. Ascoli-Bartoli, G. Benedetti-Michelangeli, L. Lovisetto (Laboratorio Gas Ionizzati, EURATOM-C.N.E.N., Italy
Appl. Phys. Letters, Vol. 8, pp. 332–333, June 15, 1966 (10-cm long Schott SFS 09 glass)

13.67 Optical Properties and Applications of Photochromic Glass
G. K. Megla (Corning Glass Works)
Appl. Optics, Vol. 5, pp. 945–960, June 1966 (usable as Q switch)

13.68 Faraday Rotators for High Power Laser Cavities
Nicholas George (California Inst. of Tech.), and R. W. Waniek (Advanced Kinetics)
Appl. Optics, Vol. 5, pp. 1183–1185, July 1966 (quartz, up to 400 kOe magnetic field, cascaded with cryptocyanine cell, ruby laser)

13.69 Giant Pulse Laser Operation with Semiconductor Mirrors
M. Birnbaum and T. L. Stocker (Aerospace)
IEEE J. Quantum Electronics, Vol. QE-2, pp. 184–185, July 1966 (broad range of wavelengths, 1 to 7 MW/cm², 25 to 50 ns pulses)

13.70 Giant-Pulse Laser Activity in Neodymium-Doped Silicate Glass: The Energy Conversion Process
J. H. Wenzel (GE)
J. Appl. Phys., Vol. 37, pp. 3100–3110, July 1966 (standing waves in resonator, radial variation in gain)

13.71 Laser Cavity Dumping Using Time Variable Reflection
W. R. Hook, R. H. Dishington, and R. P. Hilberg (TRW/Systems)

Appl. Phys. Letters, Vol. 9, pp. 125–127, August 1, 1966 (Nd lasers, time variable reflector in a laser resonator, 4 ns laser output pulse)

13.72 Giant Optical Pulse Shortening Through Pulse-Transmission Mode Operation of a Ruby Laser
J. Ernest, M. Michon, and J. DeBrie (CGE, France)
Physics Letters, Vol. 22, pp. 147–149, August 1, 1966 (Kerr cell within laser cavity, 3.5 ns output pulse)

13.73 Combination Laser Q-Switch Using a Spinning Mirror and Saturable Dye
D. Hull (Sandia)
Appl. Optics, Vol. 5 pp. 1342–1343, August 1966 (1.5 cm diam and 15.2 cm long ruby, cryptocyanine, 10 ns pulse, 10^9 W)

13.74 Narrow Spectrum Giant Pulse Laser
L. M. Frantz and R. S. Witte (TRW/Systems)
IEEE J. Quantum Electronics, Vol. QE-2, pp. 333–334, August 1966 (inject radiation from low power non Q-switched laser)

13.75 Optical Properties of Cryptocyanine
Thomas F. Deutsch and Marvin J. Weber (Raytheon)
J. Appl. Phys., Vol. 37, p. 3629, August 1966 (wideband transmission, concentration, fluorescence, and excitation spectra)

13.76 Etude de l'Emission d'un Laser au Néodyme, Déclenché par Effet Pockels
P. Wurtz
Philips Research Repts., Vol. 21, pp. 213–245, August 1966 (fluorescence intensity, KDP Q-sw cell)

13.77 Giant Pulse Laser Action at 77°K
R. Williamson and D. Walsh (Oxford U., England)
Proc. IEEE, Vol. 54, pp. 1122–1123, August 1966 (ruby, chloroaluminum phthalocyanine, 30 ns pulse half width, relatively low peak power)

13.78 Investigation of Spectral Bleaching in Passive Q-Switch Dyes
C. R. Giuliano and L. D. Hess (Hughes Research)
Appl. Phys. Letters, Vol. 9, pp. 196–198, September 1, 1966 (vary spectral width of Q-sw ruby laser)

13.79 The Effect of Absorber Concentration on a Pulsed Laser System
C. Y. She and Ang-Tiek Tan (U. of Minnesota)
Appl. Phys. Letters., Vol. 9, pp. 198–200, September 1, 1966 (more dye, higher threshold, more delay for first pulse)

13.80 Giant Superluminescence Pulses
V. S. Zuev, V. S. Letokhov, and Yu. V. Senatskii (Lebedev Phys. Inst., Acad. of Sci., USSR)
JETP Letters, Vol. 4, pp. 125–127, September 1, 1966 (Nd glass rods 90 cm long, 40-dB single pass gain, Kerr cell shutter, 500 MW/cm² output, 9 to 12 ns pulse)

13.81 The Operation of a Neodymium Glass Laser Using a Saturable Liquid Q-Switch
C. H. Skeen (TRW/Systems) and C. M. York (U. of California)
Appl. Optics, Vol. 5, pp. 1463–1464, September 1966 (Kodak 9740 bleachable liquid, 6 ns output pulses)

13.82 Short-Pulse Q-Switched Laser with Variable Pulse Length
R. V. Ambartsumyan, N. G. Basov, V. S. Zuev, P. G. Kryukov, and V. S. Letokhov (Lebedev Phys. Inst., Acad. of Sci., USSR)
IEEE J. Quantum Electronics, Vol. QE-2, pp. 436–441, September 1966 (Q-sw laser, Kerr cell second shutter, nonlinear laser amplifiers)

13.83 The Influence of Nd^{3+} Ion Properties in a Glass Matrix on the Dynamics of a Q-Spoiled Laser
Maurice Michon (CGE France)
IEEE J. Quantum Electronics, Vol. QE-2, pp. 612–616, September 1966 (20 Å wide spectral output, laser terminal level lifetime of 25 to 90 ns)

13.84 Experimental Studies of Saturable Optical Absorption
F. Gires (CSF/CEPCA, France)
IEEE J. Quantum Electronics, Vol. QE-2, pp. 624–626, September 1966 (several dyes and glasses, 0.69 μ experiments, residual absorption, lifetimes)

13.85 Saturable Absorption of Color Centers in Nd^{3+} and Nd^{3+}-Yb^{3+} Laser Glass
E. Snitzer and R. Woodcock (American Optical)
IEEE J. Quantum Electronics, Vol. QE-2, pp. 627–632, September 1966 (self Q-switching, limit-cycle output, progressive solarization)

13.86 Fast Liquid Shutter for Q-Modulation of a Neodymium Glass Laser
M. P. Vanyukov, O. D. Dmitrievskii, V. I. Isaenko, and V. A. Serebryakov
Soviet Phys. Doklady, Vol. 11, pp. 233–234, September 1966 (cyanic dye 3, 3-diethyl-9, 11, 15, 17-dineopentylene-thiapentacarbocyaniniodide, 20 to 30 ns 1.5 J output, narrow spectrum, concentration, output mirror reflectivity)

13.87 PTM Single-Pulse Selection from a Mode-Locked Nd^{3+}-Glass Laser Using a Bleachable Dye
A. W. Penney, Jr., and H. A. Heynau (United Aircraft)
Appl. Phys. Letters, Vol. 9, pp. 257–258, October 1, 1966 (pulse-transmission mode, 25 MW peak power, 0.2 J output, 0.8 ns duration)

13.88 Internal Q Switching of Ho^{3+}-Stimulated Emission in Iron-Containing Glasses
H. W. Gandy, R. J. Ginther, and J. F. Weller (NRL)
Appl. Phys. Letters, Vol. 9, pp. 277–279, October 15, 1966 (2.08 μ output, 77°K)

13.89 Magnetic Q Spoiling of Cooled Ruby
R. C. Eckardt, J. N. Bradford, and J. W. Tucker (NRL)
Appl. Phys. Letters, Vol. 9, pp. 285–287, October 15, 1966 (77°K, 30 000 gauss in 2.5 ms)

13.90 Observation of Saturable Filter Action in a UO_2^{++} Doped Glass
L. A. Cross (U. of Michigan) and L. G. Cross (Lear Siegler)
Proc. IEEE, Vol. 54, pp. 1460–1461, October 1966 (Corning #3-79 filter glass, optically pumped, Q-sw ruby, $1.5 \times 10^9 \lambda$ W 20 ns pulse output)

13.91 Generation Dynamics of a Giant Coherent Light Pulse
V. S. Letokhov and A. F. Suchkov (Lebedev Phys. Inst., Acad. of Sci., USSR)
Soviet Phys. JETP, Vol. 23, pp. 763–767, October 1966 (fine structure of giant pulse)

13.92 Characteristics of a $\lambda/2$ Kerr Cell Ruby Oscillator for Use as an Optical Radar (Lidar)
Hans W. Mocker (Honeywell)
Appl. Optics, Vol. 5, pp. 1829–1831, November 1966 (2–3 ns buildup time, 1 μs recovery time, up to 3×10^7 W peak power with 15 ns pulse width)

13.93 Saturation of Absorption and Fluorescence in Solutions of Phthalocyanines
Walter F. Kosonocky and Sol E. Harrison (RCA)
J. Appl. Phys., Vol. 37, pp. 4789–4797, December 1966

13.94 Effects of Saturable Absorber Lifetime on the Performance of Giant-Pulse Lasers
L. E. Erickson and A. Szabo (Nat'l Research Council, Canada)
J. Appl. Phys., Vol. 37, pp. 4953–4961, December 1966

13.95 A TVR Laser Oscillator Using a Single Pockels Cell for Q-Switching and Synchronized Cavity Dumping
W. R. Hook, R. P. Hilberg, and R. H. Dishington (TRW/

Systems)
Proc. IEEE, Vol. 54, pp. 1954–1955, December 1966 (time variable reflection technique within laser cavity, predict extraction of 80 percent of stored energy)

13.96 On the Theory of Single-Pulse Laser Operation
Yu. A. Anan'ev, I. F. Balashov, and A. A. Mak
Soviet Phys. Doklady, Vol. 11, pp. 124–126, December 1966 (Q-switching)

14. PUMPING, LIGHT SOURCES, SENSITIZATION, AND POPULATION INVERSION

14.1 Radiant Energies and Irradiances of Capacitor Discharge Lamps
N. A. Kuebler and L. S. Nelson (BTL)
J. Opt. Soc. Am., Vol. 51, pp. 1411–1416, December 1961

14.2 Light Source System for Ruby Laser
Gunji Shinoda, Tatsuro Suzuki, and Masataka Umeno (Osaka U., Japan)
Japanese J. Appl. Phys., Vol. 1, pp. 364–365, December 1962

14.3 Strong Blue Emission in ZnS:Tm
S. Ibuki (U. of Dayton) and D. W. Langer (Wright-Patterson AFB)
Appl. Phys. Letters, Vol. 2, pp. 95–97, March 1, 1963

14.4 High-Index-of-Refraction Spherical Sheath Composite-Rod Optical Masers
O. Svelto and M. Di Domenico, Jr. (Stanford U.)
Appl. Optics, Vol. 2, pp. 431–439, April 1963

14.5 Spectral Characteristics of Tubular Pulsed Discharge Lamps
I. M. Gurevich and F. A. Charnaya
Optics and Spectroscopy, Vol. 14, pp. 297–300, April 1963

14.6 High Intensity Cesium Lamp for Optical Pumping
F. A. Franz (U. of Illinois)
Rev. Sci. Instr., Vol. 34, pp. 589–590, May 1963

14.7 Enhanced Ultraviolet Output from Double-Pulsed Flash Lamps
J. L. Emmett and A. L. Schawlow (Stanford U.)
Appl. Phys. Letters, Vol. 2, pp. 204–206, June 1, 1963

14.8 DC Electroluminescence in Thin Films of ZnS
P. Goldberg and J. W. Nickerson (G. T. & E. Labs.)
J. Appl. Phys., Vol. 34, pp. 1601–1608, June 1963

14.9 Energy Density Distribution in a Polished Cylinder of Laser Material
W. R. Sooy and M. L. Stitch (Hughes)
J. Appl. Phys., Vol. 34, pp. 1719–1723, June 1963

14.10 Design and Operation of Xenon Flashtubes
Arden Buck (U. of Colorado), Roger Erickson (Minneapolis–Honeywell), and Frank Barnes (U. of Colorado)
J. Appl. Phys., Vol. 34, pp. 2115–2116, July 1963

14.11 Optimization of the Parameters of Multi-elliptical Laser Head Configurations
J. A. Ackerman (Aircraft Armaments)
Proc. IEEE, Vol. 51, pp. 1032–1033, July 1963

14.12 Limiting Parameters and Generalized Characteristics of Xenon Lamps
I. S. Marshak (Electric Lamps Works, Moscow, U.S.S.R.)
Appl. Optics, Vol. 2, pp. 827–831, August 1963

14.13 A New Condenser for a Sun-Powered Continuous Laser
P. H. Keck, J. J. Redmann, C. E. White, and R. E. DeKinder, Jr. (Texas Instruments)
Appl. Optics, Vol. 2, pp. 827–831, August 1963

14.14 Variation of Arc Resistance and Arc Power with Current in Pulsed Xenon Optical Pump Lamps
H. G. Heard (Radiation at Stanford)
Proc. IEEE, Vol. 51, pp. 1234–1235, September 1963

14.15 Plasma Pinch Excitation of a Ruby Laser
Richard A. Brandewie, Joe S. Hitt, and J. M. Feldman (Carnegie Inst. of Tech.)
J. Appl. Phys., Vol. 34, pp. 3415–3416, November 1963

14.16 Effect of Pre-excitation on Flash Lamp Optical Quality
H. G. Heard (Energy Systems)
Appl. Optics, Vol. 2, p. 1329, December 1963

14.17 Nondestructive Laser Pumping by High Explosives
John K. Crosby and R. C. Honey (Stanford Res. Inst.)
Appl. Optics, Vol. 2, pp. 1339–1340, December 1963

14.18 A Theory of Pumping by Incoherent Waves
H. Dormont (Labs. d'Electronique et de Physique Appliquée, Paris)
Proc. Third Int. Conf. Quantum Electronics, Columbia U. Press, N. Y., Vol. 1, pp. 139–149, 1964

14.19 Recent Progress in Optical Pumping
J. Brossel (Lab. de Physique, Ecole Normale Supérieure, Paris)
Proc. Third Int. Conf. Quantum Electronics, Columbia U. Press, N. Y., Vol. 1, pp. 201–212, 1964

14.20 Pompage par Résonance Magnétique dans les Solides et les Liquides
J. Uebersfeld (CNRS, France)
Proc. Third Int. Conf. Quantum Electronics, Columbia U. Press, N. Y., Vol. 1, pp. 213–230, 1964

14.21 Selective Excitation by Photodissociation of Molecules
G. Gould (TRG)
Proc. Third Int. Conf. Quantum Electronics, Columbia U. Press, N. Y., Vol. 1, pp. 459–467, 1964

14.22 On Fast Electron Pumping of Ruby
R. P. de Figueiredo (Purdue U.)
Proc. Third Int. Conf. Quantum Electronics, Columbia U. Press, N. Y., Vol. 2, pp. 1353–1371, 1964

14.23 An Investigation of Confined Arc Discharges in Xenon
A. Buck, F. Barnes (U. of Colorado), and R. Erickson (Minneapolis–Honeywell)
Proc. Third Int. Conf. Quantum Electronics, Columbia U. Press, N. Y., Vol. 2, pp. 1379–1396, 1964

14.24 Comparison of Excitation Geometries for Ruby Lasers
R. S. Congleton, W. R. Sooy, D. R. Dewhirst, and L. D. Riley

(Hughes)
Proc. Third Int. Conf. Quantum Electronics, Columbia U. Press, N. Y., Vol. 2, pp. 1415–1425, 1964

14.25 Persistent Enhanced UV Radiation from Double-Pulsed Flash Lamps
H. W. Gandy, A. C. Kolb, W. H. Lupton, and J. F. Weller (U.S. Naval Res. Lab.)
Appl. Phys. Letters, Vol. 4, pp. 11–13, January 1, 1964

14.26 Irradiance Linearity Corrections for Multiplier Phototubes
D. J. Baker and C. L. Wyatt (Utah State U.)
Appl. Optics, Vol. 3, pp. 89–91, January 1964

14.27 Argon Tube Pumped Laser
Leslie T. Long and Robert L. Conger (U.S. Naval Ordnance Lab.)
Appl. Optics, Vol. 3, p. 156, January 1964

14.28 Shock Waves in Xenon Flashtubes and Tube Deterioration
Kenneth R. Lang and Frank S. Barnes (U. of Colorado)
J. Appl. Phys., Vol. 35, pp. 107–110, January 1964

14.29 A High Repetition Rate Laser System
W. T. Haswell, III, J. S. Hitt, and J. M. Feldman (Carnegie Inst. of Tech.)
Proc. IEEE, Vol. 52, p. 93, January 1964

14.30 Injection Luminescent Pumping of $CaF_2:U^{3+}$ with GaAs Diode Lasers
R. J. Keyes and T. M. Quist (MIT)
Appl. Phys. Letters, Vol. 4, pp. 50–52, February 1, 1964

14.31 Exfocal Pumping of Optical Masers in Elliptical Mirrors
Dieter Roess (Siemens & Halske, Germany)
Appl. Optics, Vol. 3, pp. 259–265, February 1964

14.32 Data Processing System for the Automatic Transformation of Observed Plasma Intensities into Their Radial Distribution
D. R. Paquette and W. L. Wiese (Nat. Bur. Stds.)
Appl. Optics, Vol. 3, pp. 291–295, February 1964

14.33 Optical Absorption and Diffuse Reflectance of Powders
Peter D. Johnson (G.E.)
J. Appl. Phys., Vol. 35, pp. 334–336, February 1964

14.34 Thermal Radiation Characteristics of Xenon Flashtubes
Kenneth A. Lincoln (U.S. Naval Radiological Defense Lab.)
Appl. Optics, Vol. 3, pp. 405–412, March 1964

14.35 Fluorescent Filters to Increase the Efficiency of Optical Pumping: Potential Uses in Flash Photolysis and Laser Work
B. Holmstrom and B. Stening (U. of Uppsala, Sweden)
Photochemistry and Photobiology, Vol. 3, pp. 55–59, March 1964 (fluorescein in alkaline aqueous solution, 3000j into 2 flash lamps, enhanced spectral radiation)

14.36 Cathodoluminescence for $CaWO_4$: Nd Laser Pumping
J. W. Ogland, C. W. Baugh, and W. E. Horn (Westinghouse)
Appl. Phys. Letters, Vol. 4, pp. 133–134, April 1, 1964 (laser action reported, 25 Kv, 1A onto phosphor)

14.37 Some Characteristic of Metal Iodide–Mercury Arcs
T. H. Rautenberg, Jr., and P. D. Johnson (G.E.)
Appl. Optics, Vol. 3, pp. 487–492, April 1964

14.38 Mechanism of Afterglow in Neon Flashtubes
P. D. Johnson, T. H. Rautenberg, Jr., and B. Harris (G.E.)
J. Appl. Phys., Vol. 35, pp. 1128–1130, April 1964

14.39 Characteristics of Mercury Vapor–Metallic Iodide Arc Lamps
G. H. Reiling (G.E.)
J. Opt. Soc. Am., Vol. 54, pp. 532–540, April 1964

14.40 Spectral Characteristics for a Neodymium Laser Liquid Filter System
K. F. Tittel (G.E.)
Rev. Sci. Instr., Vol. 35, pp. 522–524, April 1964

14.41 Superconducting Pulse Power Supply
P. R. Wiederhold (Ion Physics Corp.)
IEEE Trans. on Microwave Theory and Techniques, Vol. MTT-12, p. 386, May 1964 (1200-J experiment, superconducting-to-normal transition)

14.42 Parallel Theta-Pinch Pumping of a Laser Oscillator-Amplifier
J. S. Hitt and J. M. Feldman (Carnegie Inst. of Tech.)
Proc. IEEE, Vol. 52, pp. 616–617, May 1964 (20 torr argon, 800 j, 1 Mcps ring, 0.1 μs rise time of Nd laser)

14.43 Radiationless Resonance Energy Transfer from UO_2^{2+} to Nd^{3+} in Coactivated Barium Crown Glass
H. W. Gandy, R. J. Ginther, and J. F. Weller (U. S. NRL)
Appl. Phys. Letters, Vol. 4, pp. 188–190, June 1, 1964 (green-UV absorption in UO_2^{++} transfers energy to Nd)

14.44 Continuous Sun-Pumped Room Temperature Glass Laser Operation
G. R. Simpson (American Optical)
Appl. Optics, Vol. 3, pp. 783–784, June 1964 (neodymium, parabolic mirror and two aplanatic refractors, 15 per cent over threshold)

14.45 Injection-Luminescence Pumping of a $CaF_2:Dy^{2+}$ Laser
S. A. Ochs and J. I. Pankove (RCA)
Proc. IEEE, Vol. 52, pp. 713–714, June 1964 (He temp $GaAs_x$ P_{1-x} diodes for max. 0.2-second pulse, 7200 Å pump, 2.36-μ output, cylindrical cluster of diodes)

14.46 Elliptical Head for Liquid Laser Research
E. J. Schimitschek and A. L. Lewis (USN Electronics Lab.)
Rev. Sci. Instr., Vol. 35, pp. 911–912, July 1964 (hinged halves)

14.47 Distribution of Absorbed Power in a Side-Pumped Ruby Rod
C. H. Cooke, J. McKenna, and J. G. Skinner (BTL)
Appl. Optics, Vol. 3, pp. 957–961, August 1964 (polychromatic pump, three-dimensional model, numerical analyses)

14.48 Pumping Energy Distribution in Ruby Rods
J. G. Skinner (BTL)
Appl. Optics, Vol. 3, pp. 963–965, August 1964 (fluorescence experiments on side-pumped thin ruby disk)

14.49 Cast Plastic Laser Pump Cavities
H. A. Daw and J. R. Izatt (New Mexico State U.)
Appl. Optics, Vol. 3, pp. 984–985, August 1964 (10.2-cm diameter spherical cavity)

14.50 High Voltage Flash Photolysis Lamp
L. Lindqvist (U. S. Army Natick Labs.)
Rev. Sci. Instr., Vol. 35, pp. 993–995, August 1964 (3 μs at 1450 J, 40 kV, four 20-cm linear lamps)

14.51 Ultraviolet Lamp for the Generation of Intense, Constant-Shape Pulses in the Subnanosecond Region
J. T. D'Alessio, P. K. Ludwig, and M. Burton (U. of Notre Dame)
Rev. Sci. Instr., Vol. 35, pp. 1015–1017, August 1964 (modified mercury contact relay, 0.5-ns uV pulse)

14.52 Energy Transfer in Silicate Glass Coactivated with Cerium and Neodymium
H. W. Gandy, R. J. Ginther, and J. F. Weller (NRL)
Physics Letters, Vol. 11, pp. 213–214, August 1, 1964 (from Ce to Nd)

14.53 Direct Measurement of Xenon Flashtube Opacity
J. L. Emmett, A. L. Schawlow (Stanford U.), and E. H. Weinberg (ONR)
J. Appl. Phys., Vol. 35, pp. 2601–2604, September 1964 (2500 to 10,000 Å data, up to 5000 A/cm²)

14.54   Light Scattering in a Solution of Europium Benzoylacetonate
During Optical Pumping
E. P. Riedel (Westinghouse)
Appl. Phys. Letters, Vol. 5, pp. 162–165, October 15, 1964 (opti-
cal homogeneity strongly temperature dependent)

14.55   A New Optical Laser Pump
G. J. Fan, C. B. Smoyer, and J. Nuñez (IBM)
Appl. Optics, Vol. 3, pp. 1277–1279, November 1964 (linear
lamp, two reflecting axicons and cylinder, high efficiency, 25 J
for Nd-glass laser)

14.56   Xenon Flash Lamp for Laser Pumping in Liquid Nitrogen
J. W. Tucker and J. N. Bradford (U. S. NRL)
Rev. Sci. Instr., Vol. 35, pp. 1615–1616, November 1964 (mixture
of xenon and argon, spectral output like pure xenon)

14.57   Large Energy Transfer from Uranyl to Europium Ions in Glass
L. G. DeShazer and A. Y. Cabezas (Hughes)
Proc. IEEE, Vol. 52, p. 1355, November 1964 (uranyl fluoresces
at 5250 Å)

14.58   Correction to "Large Energy Transfer from Uranyl to Europium
Ions in Glass"
L. G. DeShazer and A. Y. Cabezas (Hughes)
Proc. IEEE, Vol. 53, p. 305, March 1965 (Proc. IEEE, Vol. 52,
p. 1355; November 1964)

14.59   Dynamics of Energy Transfer from 3d to 4f Electrons in $LaAlO_3$:
$Cr^{3+}Nd^{3+}$
Z. J. Kiss (RCA)
Phys. Rev. Letters, Vol. 13, pp. 654–656, November 30, 1964 (red
fluorescence of Cr pumps Nd, disproves an earlier paper)

14.60   Energy Transfer in Silicate Glass Coactivated with Cerium and
Ytterbium
H. W. Gandy, R. J. Ginther, and J. Weller (U. S. Naval Res.
Lab.)
Appl. Phys. Letters, Vol. 5, pp. 220–222, December 1, 1964 (from
Ce to Yb, fluorescence at 1.015 μ)

14.61   Nonradiative Energy Exchange from Gd^{3+} to Tb^{3+} in Borate Glass
A. D. Pearson and G. E. Peterson (BTL)
Appl. Phys. Letters, Vol. 5, pp. 222–223, December 1, 1964

14.62   Pump Energy Absorption in a Ruby Rod
G. Lampis, C. A. Sacchi, and O. Svelto (Inst. Fis. Politec., Italy)
Appl. Optics, Vol. 3, pp. 1467–1470, December 1964 (helical
lamp, immersed ruby rod, distribution deduced from fluorescence)

14.63   Model for the Angular Distribution of Light Rays from a Linear
Gas Discharge Source
J. Wenzel (G. E.)
Appl. Optics, Vol. 3, pp. 1503–1504, December 1964 (slit on lamp
wall, gas radiation density uniform in volume, almost fills tube
bore)

14.64   Some Characteristics of Pulsed Na- and Tl-Iodide Monatomic
Gas Arcs
P. D. Johnson and T. H. Rautenberg, Jr. (G. E.)
J. Opt. Soc. Am., Vol. 54, pp. 1425–1428, December 1964 (spec-
tral distribution, intensity and excitation temperature)

14.65   Stable High-Pressure Xenon Arc Lamp
Gerald S. Wasserman (MIT)
J. Opt. Soc. Am., Vol. 54, pp. 1492–1493, December 1964 (150
watt, 1.4 mm long arc, much greater stabilities)

14.66   Performance of the Plasma Theta-Pinch for Laser Pumping
S. Aisenberg, D. V. Missio, and P. A. Silberg (Raytheon)
J. Appl. Phys., Vol. 35, pp. 3625–3626, December 1964 (noble
gases, 3.2-J threshold in $Nd:CaWO_4$)

14.67   Microwave Electron Cyclotron Resonance Pumping of a Gas
Laser
S. A. Ahmed and R. Kocher (RCA)
Proc. IEEE, Vol. 52, pp. 1737–1738, December 1964 (He-Ne,
2.45-Gc magnetron)

14.68   Evidence of Stimulated Emission in Ruby-Laser-Pumped GaAs
J. J. Schlickman, M. E. Fitzgerald, and R. H. Kingston (MIT)
Proc. IEEE, Vol. 52, pp. 1739–1740, December 1964 (threshold
ruby power of $10^4W/cm^2$)

14.69   Charge Transfer as a Possible Laser Pumping Mechanism
J. William McGowan and R. F. Stebbings (General Dynamics)
Appl. Optics, Suppl. 2: Chemical Lasers, pp. 68–72, 1965 (be-
tween positive ions and neutral particles)

14.70   Nonequilibrium Chemical Excitation and Chemical Pumping of
Lasers
K. E. Shuler and T. Carrington (NBS) and J. C. Light (U. of
Chicago)
Appl. Optics, Suppl. 2: Chemical Lasers, pp. 81–104, 1965
(process appears feasible)

14.71   Inverted Population Distributions Produced by Chemical Reac-
tions
H. P. Broida (U. of California)
Appl. Optics, Suppl. 2: Chemical Lasers, pp. 105–108, 1965
(some low-pressure atomic gas inversions reported)

14.72   Vibrational-Rotational Population Inversion
J. C. Polanyi (U. of Toronto, Canada)
Appl. Optics, Suppl. 2: Chemical Lasers, pp. 109–127, 1965

14.73   Measurement of Energy Transfer in Molecular Collisions
A. B. Callear (U. of Cambridge, England)
Appl. Optics, Suppl. 2: Chemical Lasers, pp. 145–170, 1965
(transfer of vibrational, electronic and translational energies)

14.74   Laser Possibilities of Chemically Excited Molecules Formed with
Atomic Species
T. T. Kikuchi (General Motors) and H. P. Broida (U. of
California)
Appl. Optics, Suppl. 2: Chemical Lasers, pp. 171–178, 1965
(CN* may lase at infrared and red wavelengths)

14.75   Flame Laser: Model and Some Preliminary Experimental Results
R. Bleekrode and W. C. Nieuwpoort (Philips, The Netherlands)
Appl. Optics, Suppl. 2: Chemical Lasers, pp. 179–180, 1965 (low
pressure oxyacetylene flame is a possibility)

14.76   Population Inversions Produced by Chemical Depletion of
Ground States
M. L. Seman (Xerox)
Appl. Optics, Suppl. 2: Chemical Lasers, pp. 181–183, 1965

14.77   Explosion Flame Emission
J. A. Howe (BTL)
Appl. Optics, Suppl. 2: Chemical Lasers, pp. 184–186, 1965
(negative results of selective emission from various gaseous
explosions)

14.78   Exploratory Research on Population Inversions in Gaseous Explosions
I. Wieder, R. R. Neiman, and A. P. Rodgers (Interphase)
Appl. Optics, Suppl. 2: Chemical Lasers, pp. 187–192, 1965
(low pressure oxyacetylene)

14.79   Atomic and Molecular Fluorescence Excited by Photo-dissociation
R. N. Zare (Joint Institute for Laboratory Astrophysics) and
D. R. Herschbach (Harvard U.)
Appl. Optics, Suppl. 2: Chemical Lasers, pp. 193–200, 1965
(many references, di-, tri-, and polyatomic molecules)

14.80   Photodissociation of Thallium Bromide and Cesium Bromide
W. T. Walter and S. M. Jarrett (TRG)
Appl. Optics, Suppl. 2: Chemical Lasers, pp. 201–204, 1965
(possible laser outputs at 5350 Å and 2.9μ in Tl and Cs, respectively)

14.81   Investigations on Plasma-Pinch Pump-Lamps
V. Met (Watkins-Johnson)
Z. angewandte Mathematik und Physik, Vol. 16, No. 1, pp.
63–64, January 25, 1965 (argon and helium, 5 μsec discharge)

14.82   On the Efficiency of Single and Multiple Elliptical Laser Cavities
C. Bowness (Raytheon)

Appl. Optics, Vol. 4, pp. 103–107, January 1965 (highest efficiency from single cavity, higher outputs from multiple cavities)

14.83 The Effect of Optical Pump Pulse Shape on Ruby Inversion
P. N. Mace and G. McCall (Los Alamos)
Proc. IEEE, Vol. 53, p. 74, January 1965 (little effect calculated)

14.84 Quasi-Continuous Operation of a CaWO₄:Nd³⁺ Maser Using Long Duration Pumping Pulses
H. Manger (Technische Hochschule Karlsruhe, Germany)
Proc. IEEE, Vol. 53, pp. 83–84, January 1965 (100 m sec possible)

14.85 Cathodoluminescence of Thin Films Containing Rare-Earth Oxides
W. W. Hansen and R. E. Myers (SRI)
Appl. Phys. Letters, Vol. 6, pp. 58–59, February 1, 1965 (rise and decay times, down to 10^{-7} secs)

14.86 Energy Transfer from Copper and Silver to Rare Earths in II-VI Compounds
J. D. Kingsley, J. S. Prener, and M. Aven (G.E.)
Phys. Rev. Letters, Vol. 14, pp. 136–138, February 1, 1965 (many lanthanides, powder samples)

14.87 An Injection Laser Pump for Nd⁺³ Doped Hosts
R. H. Harada and C. K. Suzuki (Autonetics)
Appl. Optics, Vol. 4, pp. 225–227, February 1965 (possible pumping at 0.87μ)

14.88 Microwave Electron-Cyclotron Resonance Pumping of a Gas Laser
S. A. Ahmed (RCA)
Bull. Am. Phys. Soc., Series II, Vol. 10, No. 2, p. 227–Q12, February 1965 (12 eV electron temperature, 5-fold increase in 3.39μ He–Ne output)

14.89 Constricted Discharges in the Rare Gases. I. Spectroscopic and Electrical Measurements
J. T. Massey and S. M. Cannon (Johns Hopkins U.)
J. Appl. Phys., Vol. 36, pp. 361–372, February 1965

14.90 Constricted Discharges in the Rare Gases. II. Analysis of the Macroscopic Properties of the Discharge
J. T. Massey (Johns Hopkins U.)
J. Appl. Phys., Vol. 36, pp. 373–380, February 1965

14.91 Shock-Induced Luminescence
N. L. Coleburn, M. Solow, and R. C. Wiley (NOL)
J. Appl. Phys., Vol. 36, pp. 507–510, February 1965 (blue luminescence from aluminum oxide surface)

14.92 Sensitization of Nd³⁺ Luminescence by Mn²⁺ and Ce³⁺ in Glasses
S. Shionoya and E. Nakazawa (U. of Tokyo, Japan)
Appl. Phys. Letters, Vol. 6, pp. 117–118, March 15, 1965 (calcium phosphate glass, 1.06μ luminescence)

14.93 Sensitization of Tb³⁺ Luminescence by Ce³⁺ and Cu⁺ in Glasses
S. Shionoya and E. Nakazawa (U. of Tokyo, Japan)
Appl. Phys. Letters, Vol. 6, pp. 118–120, March 15, 1965 (calcium phosphate glass, 0.54μ luminescence)

14.94 The Spectrum of the Argon Bomb
R. L. Conger, L. T. Long, J. A. Parks, and J. H. Johnson (NOL)
Appl. Optics, Vol. 4, pp. 273–276, March 1965 (explosive charges of argon, visible continuum, 20,000°K blackbody)

14.95 Resistivity of Xenon Plasma
J. H. Goncz (Edgerton, Germeshausen & Grier)
J. Appl. Phys., Vol. 36, pp. 742–743, Part 1, March 1965 (resistivity proportional to inverse square root of current density, 10^{-2} ohm-cm at 10^4 amp/cm²)

14.96 Series Triggering of Xenon Flash Lamps
J. P. Chernoch and K. F. Tittel (G.E.)
Rev. Sci. Instr., Vol. 36, p. 392, March 1965

14.97 Wall Deterioration in Flash Lamps
J. H. Rosolowski and R. J. Charles (G.E.)

J. Appl. Phys., Vol. 36, pp. 1792–1793, May 1965 (silica smoke deposits, circumferential hairline cracks)

14.98 Energy Exchange from Nd³⁺ to Yb³⁺ in Calibo Glass
G. E. Peterson, A. D. Pearson and P. M. Bridenbaugh (BTL)
J. Appl. Phys., Vol. 36, pp. 1962–1966, June 1965 (nonradiative, 0.6 quantum efficiency, time resolved spectroscopic technique)

14.99 Electron Beam Pumped Lasers of PbS, PbSe, and PbTe
C. E. Hurwitz, A. R. Calawa, and R. H. Rediker (Lincoln Lab., M.I.T.)
IEEE J. Quantum Electronics, Vol. QE-1, pp. 102–103, May 1965 (liquid He temp)

14.100 Energy Transfer and Ho³⁺ Laser Action in Silicate Glass Coactivated with Yb³⁺ and Ho³⁺
H. W. Gandy, R. J. Ginther, and J. F. Weller (U.S. Naval Research Lab.)
Appl. Phys. Letters, Vol. 6, pp. 237–239, June 15, 1965 (laser output wavelength $> 1.9\ \mu$, 150 J threshold)

14.101 Use of the Pinch Effect for Optical Laser Pumping
M. R. Bedilov, V. M. Likhachev, G. V. Mikhailov, and M. S. Rabinovich (Lebedev Physics Inst., Academy of Sciences, U.S.S.R.)
JETP Letters, Vol. 2, pp. 59–60, July 15, 1965 (Kr discharge, 35 000°K black-body temp, Nd-doped glass and ruby lasers)

14.102 Design of Efficient Explosively Driven Electromechanical Energy Converters
Shimshon Frankenthal, Oscar P. Manley, and Yvain M. Treve (American Science and Engineering, Inc.)
J. Appl. Phys., Vol. 36, pp. 2137–2139, July 1965 (possibly for laser flash tube)

14.103 Energy Distribution in a Glass:Nd³⁺ Laser Rod
N. F. Borrelli and M. L. Charter (Corning Glass Works)
J. Appl. Phys., Vol. 36, pp. 2172–2174, July 1965 (gain and heating are nonuniform across diameter)

14.104 Energy Transfer from UO₂²⁺ to Er³⁺ in Glass
Akio Kitamura (Matsushita Communication Ind. Co., Japan)
J. Phys. Soc. Japan, Vol. 20, pp. 1283–1284, July 1965 (1.6 μ fluorescence emission)

14.105 Laser Action in Neodymium-Doped Glass Produced Through Energy Transfer
N. T. Melamed, C. Hirayama, and E. K. Davis (Westinghouse)
Appl. Phys. Letters, Vol. 7, pp. 170–172, September 15, 1965 (0.2 ms transfer time, pulsed operation, Mn²⁺ sensitized phosphate glass)

14.106 Stabilized Operation of Xenon Flashlamps Using Ultraviolet Illumination
R. J. Pressley and H. Y. S. Tang (RCA)
Rev. Sci. Instr., Vol. 36, p. 1381, September 1965 (2537 Å irradiation, lower trigger voltage)

14.107 Electrical and Spectrographic Study of a Double-Pulsed Flashtube as Applied to Laser Pumping
Petras V. Avizonis and Tony Legato (Air Force Weapons Lab.)
J. Appl. Phys., Vol. 36, pp. 3302–3307, October 1965 (helical lamp explodes at 1.22×10^4 J/cm³, max 5600 Å emission at 25 000 A/cm², highest temperature of 12 000°K, ruby experiments)

14.108 Reflectance of Nonperfect Surfaces in the Integrating Sphere
Bjarne J. Hisdal (Norwegian Council for Scientific and Industrial Research, Norway)
J. Opt. Soc. Am., Vol. 55, pp. 1255–1260, October 1965 (also nonperfect samples in sphere)

14.109 Population Inversion Obtained by Thermal Dissociation of Molecules in a Shock Wave
A. N. Oraevskiĭ (Lebedev Physics Inst., Academy of Sciences, U.S.S.R.)
Soviet Physics JETP, Vol. 21, pp. 768–770, October 1965 (example of N_2O molecule)

14.110 Laser Oscillations and Self Q-Switching in Triply Activated Glass
H. W. Gandy, R. J. Ginther, and J. F. Weller (U.S. Naval Research Lab.)
Appl. Phys. Letters, Vol. 7, pp. 233–236, November 1, 1965 (UO_2^{2+}, Nd^{3+} and Yb^{3+}; filtered flash lamp)

14.111 Performance Study of a Vortex-Stabilized Arc Radiation Source
J. E. Anderson, R. C. Eschenbach, and H. H. Troue (Union Carbide)
Appl. Optics, Vol. 4, pp. 1435–1441, November 1965 (highest total radiance of 5250 W cm^{-2} sr^{-1} from argon, high spectral radiance between 0.4 and 1.0 μ, other gases)

14.112 Optimum Designs of Elliptical Cavities Compared with Cylindrical Ones
K. Kamiryo, T. Kano, H. Matsuzawa, and M. Yoshida (Tohoku U., Sendai, Japan)
Proc. IEEE, Vol. 53, pp. 1750–1751, November 1965 (elliptical cavities much better than cylindrical)

14.113 Method for Determining the Peak Irradiance of an Optical Pump and Total Pump Energy That is Incident upon a Laser Crystal
Thomas J. Negrelli (Systems Research Labs.)
Rev. Sci. Instr., Vol. 36, pp. 1755–1759, December 1965 (Chromel-Alumel thermocouple)

14.114 Energy Transfer from Cr^{3+} to Nd^{3+} in Solids
R. C. Ohlmann and R. Mazelsky (Westinghouse)
Proc. Physics of Quantum Electronics Conf., McGraw-Hill Book Co., New York, pp. 322–331, 1966

14.115 Electron Beam Excitation in Laser Crystals
W. W. Anderson (Stanford U.)
Appl. Optics, Vol. 5, pp. 167–168, January 1966 (depth of active region, method proposed)

14.116 Efficient High Energy Laser Radiation Utilizing a Coaxial Optical Pump
J. P. Lesnick and C. H. Church (Westinghouse)
IEEE J. Quantum Electronics, Vol. QE-2, pp. 16–17, January 1966 (94-cm long Nd-doped glass, 5.1% slope, 800 J output, 150 kJ max input to lamp)

14.117 New Developments in Electronic Flashtubes
John H. Goncz (Edgerton, Germeshausen, & Grier)
ISA Trans., Vol. 5, pp. 28–36, January 1966 (ultimate limits, flashtube life, spectral emission)

14.118 Spectra of Pulsed and Continuous Xenon Discharges
John H. Goncz and P. Bruce Newell (Edgerton, Germeshausen & Grier)
J. Opt. Soc. Am., Vol. 56, pp. 87–92, January 1966 (spectral range of 0.35 to 1.1 μ, several intense IR lines with dc, almost blackbody for pulsed)

14.119 Effect of Ultraviolet Pumping on Ruby Laser Output
R. L. Greene, J. L. Emmett, and A. L. Schawlow (Stanford U.)
Appl. Optics, Vol. 5, pp. 350–351, February 1966 (detrimental, highest output with Pyrex filter)

14.120 Quantitative Measurements of the Optical Emission from Dense Theta-Pinch Discharges
J. M. Feldman (Carnegie Institute of Technology)
J. Appl. Phys., Vol. 37, pp. 674–681, February 1966

14.121 Tb^{3+} Fluorescence and Nonradiative Energy Transfer from Gd^{3+} to Tb^{3+} in Borate Glass
A. David Pearson, G. E. Peterson, and W. R. Northover (B.T.L.)
J. Appl. Phys., Vol. 37, pp. 729–734, February 1966

14.122 Characteristics and Operation of Xenon Filled Linear Flashlamps
David E. Perlman (Eastman Kodak)
Rev. Sci. Instr., Vol. 37, pp. 340–343, March 1966 (experimental parametric investigation)

14.123 Laser Pumping by Intense Discharges in z-Pinch Geometry
Rudolf G. Buser and Dietolf Ramm (U. S. Army Electronics Command)
Appl. Optics, Vol. 5, pp. 627–631, April 1966 (Nd laser)

14.124 Effect of Multiple Reflections on the Design of an Elliptic Cavity for Solid State Lasers
D. Fekete (Grumman)
Appl. Optics, Vol. 5, pp. 643–646, April 1966

14.125 Anomalous Surface Heating Rates
R. E. Harrington (Carbon Products Div., Union Carbide)
J. Appl. Phys., Vol. 37, pp. 2028–2034, April 1966 (apparently lower thermal conductivity, metals, semiconductors and insulators, MgO reflectivity saturates with xenon flashtube, hysteresis effect)

14.126 Some Efficiency Measurements of the Theta-Pinch
P. A. Silberg (Raytheon)
J. Appl. Phys., Vol. 37, pp. 2155–2161, April 1966 (59% transfer efficiency, argon at 2 and 5 torr)

14.127 Duration and Intensity of the Radiation Pulse from a Flash Tube
E. R. Wooding (U. of Sheffield, England)
Appl. Optics, Vol. 5, pp. 777–785, May 1966 (linear lamps)

14.128 Lithium Spectral Lamp for Optical Pumping
Paolo Minguzzi (Scuola Normale Superiore, Pisa, Italy) and Franco Strumia and Paolo Violino (Istituto di Fisica dell'Università, Pisa, Italy)
J. Opt. Soc. Am., Vol. 56, pp. 707–708, May 1966 (6707 Å emission)

14.129 Population Inversion in Adiabatic Expansion of a Gas Mixture
V. K. Konyukhov and A. M. Prokhorov (Lebedev Phys. Inst., Acad. of Sci., USSR)
JETP Letters, Vol. 3, pp. 286–288, June 1, 1966 (molecular gases, vibrational levels, different relaxation times, proposed)

14.130 An Equation for the Local Thermal Emissivity at the Vertex of a Diffuse Conical or V-Groove Cavity
Francis J. Kelly (U. S. Naval Ordnance Lab.)
Appl. Optics, Vol. 5, pp. 925–927, June 1966

14.131 Effective Emissivity of a Spherical Cavity
P. Campanaro and T. Ricolfi (Istituto Termometrico Italiano, Italy)
Appl. Optics, Vol. 5, pp. 929–932, June 1966

14.132 Transfer Efficiency Formula for Diffusely Reflecting Laser Pumping Cavities
J. Whittle and D. R. Skinner (Defence Stds. Labs., Australia)
Appl. Optics, Vol. 5, pp. 1179–1182, July 1966

14.133 Argon Bomb Pumping of Ruby Laser
R. L. Conger, J. H. Johnson, L. T. Long, and J. A. Parks (USN Ordnance Lab.)
Appl. Optics, Vol. 5, pp. 1240–1241, July 1966 (chemically powered laser pump)

14.134 Radiant Emission Characteristics of a Nonisothermal Spherical Cavity
P. Campanaro and T. Ricolfi (Istituto Termometrico Italiano, Italy)
Appl. Optics, Vol. 5, pp. 1271–1273, August 1966

15. MODULATION

15.15 Measurement of the Direct Electro-Optic Effect in Quartz at UHF
D. D. Eden and G. H. Thiess (Texas Instruments)
Appl. Optics, Vol. 2, pp. 868–869, August 1963

15.16 Pulse Modulation of Gallium Arsenide Injection Luminescent Diode Laser
J. P. Quine, K. Tomiyasu, and C. Younger (G.E.)
Proc. IEEE, Vol. 51, pp. 1141–1142, August 1963

15.17 Optical Modulation by Light Bunching
Carl F. Buhrer (G. T. & E. Labs.)
Proc. IEEE, Vol. 51, p. 1151, August 1963

15.18 Low-Power Microwave Modulation of a 0.63 μ He–Ne Laser
G. Grau and D. Rosenberger (Siemens Res. Lab., Germany)
Physics Letters, Vol. 6, pp. 129–131, September 1, 1963 (2 Gc, 10-mW microwave power, 10 volts on KDP in Fabry-Perot; also 1-kc sq wave mod)

15.19 Single Sideband Modulation of Coherent Light by Bragg Reflection from Acoustical Waves
H. Z. Cummins and N. Knable (Columbia U.)
Proc. IEEE, Vol. 51, p. 1246, September 1963

15.20 Electro-Optic Interference Filter Light Modulator
X. De Angelis and W. Niblack (Sylvania)
Proc. IEEE, Vol. 51, p. 1258, September 1963

15.21 Ultrasonic-Diffraction Shutters for Optical Maser Oscillators
A. J. DeMaria (United Aircraft)
J. Appl. Phys., Vol. 34, pp. 2984–2988, October 1963 (Note: Erratum in Vol. 35, p. 465, February 1964)

15.22 Solid-State Modulation and Direct Demodulation of Gas Laser Light at a Microwave Frequency
Kenneth M. Johnson (Texas Instruments)
Proc. IEEE, Vol. 51, pp. 1368–1369, October 1963

15.23 Zeeman Tuning and Internal Modulation of the $CaF_2:Dy^{2+}$ Laser
Zoltan J. Kiss (RCA)
Appl. Phys. Letters, Vol. 3, pp. 145–148, November 1, 1963

15.24 The Stressed-Plate Shutter, a New Moderate-Speed Electro-Optical Light Modulator
S. M. Hauser, L. S. Smith, D. G. Marlowe, and P. R. Yoder, Jr.
Appl. Optics, Vol. 2, pp. 1175–1179, November 1963

15.25 "Push-Pull" Optical Modulators and Demodulators
Fred Sterzer (RCA)
Appl. Optics, Vol. 2, pp. 1197–1198, November 1963

15.26 Broadband Electro-Optic Traveling-Wave Light Modulators
M. DiDomenico, Jr., and L. K. Anderson
Bell Sys. Tech. J., Vol. 42, pp. 2621–2678, November 1963

15.27 Voice Modulation of an Electroacoustically Deflected Light Beam
A. Reich and S. S. Verner (Lockheed)
Proc. IEEE, Vol. 51, pp. 1661–1662, November 1963

15.28 Index of Refraction of KDP
Jane H. Dennis and R. H. Kingston (MIT Lincoln Lab.)
Appl. Optics, Vol. 2, pp. 1334–1335, December 1963

15.29 Harmonic Structure of Modulated Light Beams
James E. Hopson (Arthur D. Little)
IEEE Trans. on Communications Systems, Vol. CS-11, pp. 464–469, December 1963 (theoretical comparison of modulators)

15.30 Proposal for Microwave Modulation of Light Employing the Shift of Optical Absorption Edge with Applied Electric Field
R. C. Eden and P. D. Coleman (Stanford U.)
Proc. IEEE, Vol. 51, pp. 1776–1777, December 1963

15.31 Zeeman Effect and Nonlinear Interactions Between Oscillating Modes in Masers
H. Statz and C. L. Tang (Raytheon)
Proc. Third Int. Conf. Quantum Electronics, Columbia U. Press, N. Y., Vol. 1, pp. 469–480, 1964

15.32 Zeeman Effects in Helium-Neon Lasers
W. Culshaw and J. Kannelaud (G. T. & E. Labs.)
Proc. Third Int. Conf. Quantum Electronics, Columbia U. Press, N. Y., Vol. 1, pp. 523–535, 1964

15.33 The Operation of Optical Masers in Uniform Magnetic Fields
W. A. Runciman (MIT)
Proc. Third Int. Conf. Quantum Electronics, Columbia U. Press, N. Y., Vol. 1, pp. 673–676, 1964

15.34 Beats and Modulation in Optical Ruby-Masers
K. Gürs (Siemens and Halske, Germany)
Proc. Third Int. Conf. Quantum Electronics, Columbia U. Press, N. Y., Vol. 2, pp. 1113–1119, 1964

15.35 Interferometric Processing of a Phase Modulated Optical Carrier
W. F. Davison (Texas Instruments)
Proc. Third Int. Conf. Quantum Electronics, Columbia U. Press, N. Y., Vol. 2, pp. 1277–1283, 1964

15.36 Frequency Shifts and Modulation of Light Beams
C. F. Buhrer, L. R. Bloom, V. J. Fowler, D. H. Baird, and E. M. Conwell (GT&E Labs.)
Proc. Third Int. Conf. Quantum Electronics, Columbia U. Press, N. Y., Vol. 2, pp. 1609–1617, 1964

15.37 Temperature Dependence of the Complex Dielectric Constant in KH_2PO_4-Type Crystals and the Design of Microwave Light Modulators
I. P. Kaminow (BTL)
Proc. Third Int. Conf. Quantum Electronics, Columbia U. Press, N. Y., Vol. 2, pp. 1659–1665, 1964

15.38 Modulation of an Optical Source by Microwave Using an Optical Resonator
C. C. Eaglesfield and M. M. Ramsay (Standard Telecommunication Labs., Harlow, Essex, England)
Proc. Third Int. Conf. Quantum Electronics, Columbia U. Press, N. Y., Vol. 2, pp. 1667–1670, 1964

15.39 Microwave and Electro-Optical Properties of Carbon Disulfide
O. L. Gaddy, D. F. Holshouser, and R. E. Stanfield (U. of Illinois)
Proc. Third Int. Conf. Quantum Electronics, Columbia U. Press, N. Y., Vol. 2, pp. 1679–1686, 1964

15.40 Anomalous Dispersion Optical Modulator and Demodulator
P. Parzen (RCA)
Proc. Third Int. Conf. Quantum Electronics, Columbia U. Press, N. Y., Vol. 2, pp. 1711–1715, 1964

15.41 Modulation Transfer Function Associated with Image Transmission through Turbulent Media
R. E. Hufnagel and N. R. Stanley (Perkin-Elmer)
J. Opt. Soc. Am., Vol. 54, pp. 52–61, January 1964

15.42 Cuprous Chloride Light Modulators
F. Sterzer, D. Blattner, and S. Miniter (RCA)
J. Opt. Soc. Am., Vol. 54, pp. 62–68, January 1964

15.43 Amplitude Modulation of Light by Reverse Biased p–n Junctions
C. A. Renton (U. of Pennsylvania)
Proc. IEEE, Vol. 52, pp. 93–94, January 1964

15.44 Polarization Modulation and Demodulation of Light
W. Niblack and E. Wolf (Sylvania)
Appl. Optics, Vol. 3, pp. 277–279, February 1964

15.45 Electro-Optic Effect in $(NH_4)_2Cd_2(SO_4)_3$ and $(NH_4)_2Mn_2(SO_4)_3$
Carl F. Buhrer and Lily Ho (G. T. & E. Labs.)
Appl. Optics, Vol. 3, p. 314, February 1964

15.46 Linear Electro-Optic Effect in CdS
Denis J. A. Gainon (Clevite)
J. Opt. Soc. Am., Vol. 54, pp. 270–271, February 1964

15.47 Transmitter for Coherent Light Communication System
M. Ito (Nippon Electric)
IEEE Int. Conv. Record, Vol. 12, Pt. 2, pp. 59–66, 1964 (discussion of components, cooled KDP suggested)

15.48 Electro-Optic Properties of Some ABO_3 Perovskites in the Paraelectric Phase
J. E. Geusic, S. K. Kurtz, L. G. VanUitert, and S. H. Wemple (BTL)
Appl. Phys. Letters, Vol. 4, pp. 141–143, April 15, 1964 (highest electro-optical effect in $KTa_xNb_{1-x}O_3$ (called KTN), low modulation voltages, Gc possibilities good, dielectric constant $\doteq$ 20,000)

15.49 Strain Effects in Electro Optic Light Modulators
I. P. Kaminow (BTL)
Appl. Optics, Vol. 3, pp. 511–515, April 1964

15.50 Electro Optic Effect in Optically Active Crystals
C. F. Buhrer, L. Ho, and J. Zucker (G. T. & E. Labs.)
Appl. Optics, Vol. 3, pp. 517–521, April 1964

15.51 Modulation of Infrared by Free Carrier Absorption
R. B. McQuistan and J. W. Schultz (Honeywell)
J. Appl. Phys., Vol. 35, pp. 1243–1248, April 1964 (germanium, 10^5 to 10^6 cps bandwidth)

15.52 A Microwave Electro-Optic Modulator Which Overcomes Transit Time Limitation
Samuel M. Stone (G. T. & E. Labs.)
Proc. IEEE, Vol. 52, pp. 409–410, April 1964

15.53 Verdet Constant of the "Active Medium" in a Laser Cavity
I. Tobias (Rutgers U.) and R. A. Wallace (G. T. & E.)
Phys. Rev., Vol. 134, pp. A549–A552, May 4, 1964 (axial magnetic field, 0.633-μ He-Ne laser, 2.8 kc/s·Oe)

15.54 Cross Modulation of Optical, RF, and Microwave Signals with Optical Pumping
B. W. Harned, L. B. Leder, and M. E. Lasser (Philco) and D. L. Carter (U. of Pennsylvania)
Proc. IEEE, Vol. 52, p. 632, May 1964 (cesium vapor)

15.55 2000-A Pulse Generator
N. A. Sullivan (Lincoln Lab., M.I.T.)
Rev. Sci. Instr., Vol. 35, pp. 639–640, May 1964 (for low impedance GaAs diode, KN-6 cold cathode trigger tube, 50-ns pulses)

15.56 An Interferometric Optical Modulator
D. L. Fried, W. S. Read, and D. B. Pollock (North American Aviation)
Appl. Optics, Vol. 3, pp. 697–701, June 1964 (Twyman-Green type, optical path length modulated up to 2 Mc/s, about one milliwatt dissipated power)

15.57 Microwave Modulation of a GaAs Injection Laser
B. S. Goldstein and J. D. Welch (Lincoln Lab., M.I.T.)
Proc. IEEE, Vol. 52, p. 715, June 1964 (0.1-μs 2-Gc pulse, 77°K)

15.58 Absorption Edge Modulator Utilizing a p-n Junction
G. Racette (Philco)
Proc. IEEE, Vol. 52, p. 716, June 1964 (light through thin 0.030 inches dia. GaAs junction layer, few per cent mod up to 100 kc)

15.59 Novel Laser Q-Switching Mechanism
J. L. Wentz (Westinghouse)

Proc. IEEE, Vol. 52, pp. 716–717, June 1964 (pair of KDP crystals, transverse 400 volts, over 400-kc mod rate)

15.60 Electro-Optic Frequency Modulation in Optical Resonators
A. Yariv (Lockheed)
Proc. IEEE, Vol. 52, pp. 719–720, June 1964 (electro-optical modulator in cavity, frequency stabilization)

15.61 Frequency Translation of an He-Ne Laser's Output Frequency by Acoustic Output Coupling Inside the Resonant Cavity
A. E. Siegman, C. F. Quate, J. Bjorkholm, and G. Francois (Stanford U.)
Appl. Phys. Letters, Vol. 5, pp. 1–2, July 1, 1964 (800-Mc/s phonons in quartz)

15.62 Locking of He-Ne Laser Modes Induced by Synchronous Intracavity Modulation
L. E. Hargrove, R. L. Fork, and M. A. Pollack (BTL)
Appl. Phys. Letters, Vol. 5, pp. 4–5, July 1, 1964 (56-Mc/s synchronous modulation)

15.63 Shift of Optical Absorption Edge by an Electric Field: Modulation of Light in the Space-Charge Region of a Ge p-n Junction
A. Frova and P. Handler (U. of Illinois)
Appl. Phys. Letters, Vol. 5, pp. 11–13, July 1, 1964 (90 percent modulation at 1.56 μ, no degradation up to 10 Mc/s)

15.64 Electro-Optic Amplitude Modulation of Laser-Generated Second Harmonics in KH_2PO_4 (KDP)
J. P. van der Ziel (Harvard U.)
Appl. Phys. Letters, Vol. 5, pp. 27–29, July 15, 1964 (bias voltage changes harmonic amplitude)

15.65 Frequency Modulation of an He-Ne Laser Beam via Ultrasonic Waves in Quartz
W. J. Thaler (Georgetown U.)
Appl. Phys. Letters, Vol. 5, pp. 29–31, July 15, 1964 (10 Mc with 50 mW)

15.66 An Electro-Optical Light Intensity Modulator
G. H. Ujhelyi and S. T. Ribeiro (U. of Illinois)
Proc. IEEE, Vol 52, p. 845, July 1964 (vary reflection between glass and nitrobenzene, 9.2-kV, 1-μs pulses)

15.67 Laser Experiments Involving In-Cavity Modulation with Electro-Optic Crystals
A.D. Rugari and P. E. Nordborg (RADC)
Proc. IEEE, Vol. 52, p. 852, July 1964 (He-Ne laser medium, KDP, 50-kc modulation)

15.68 Electro-Optic Properties of $LiNbO_3$
G. E. Peterson, A. A. Ballman, P. V. Lenzo, and P. M. Bridenbaugh (BTL)
Appl. Phys. Letters, Vol. 5, pp. 62–64, August 1, 1964 (temperatures up to 100°C)

15.69 Control of Laser Radiation with Birefringent Crystals: the Microwave Circuit Viewpoint
M. R. Wohlers and K. G. Leib (Grumman)
J. Appl. Phys., Vol. 35, pp. 2311-2312, August 1964

15.70 Single Sideband Microwave Light Modulation
Carl F. Buhrer (GT&E)
Proc. IEEE, Vol. 52, pp. 969–970, August 1964 (3-Gc TM_{010} cavity)

15.71 Wideband Optical Modulator
C. E. Enderby (G.E.)
Proc. IEEE, Vol. 52, pp. 981-982, August 1964 (ring-plane slow wave circuit, 0.23 W/Mc bandwidth)

15.72 Internal Gating of Optically Pumped, High-Gain, Solid-State Lasers
G. E. Danielson, Jr. and A. J. DeMaria (United Aircraft)
Appl. Phys. Letters, Vol. 5, pp. 123-125, September 15 1964 (split cylindrical transducer produces ultrasonic modulation in 53 cm long Nd-doped glass laser)

15.73 Polarization as Modulation Control
K. G. Leib (Grumman)

Appl. Optics, Vol. 3, pp. 1088-1089; September 1964 (combined modulation proposed)

15.74 A Technique for Optical Frequency Translation Utilizing the Quadratic Electro-Optic Effect in Cubic Crystals
S. E. Harris and A. E. Siegman (Stanford U.)
Appl. Optics, Vol. 3, pp. 1089-1090, September 1964

15.75 Traveling-Wave Phase Modulation
W. W. Buchman and M. Laikin (Korad)
Rebuttal by C. J. Peters (Sylvania)
Proc. IEEE, Vol. 52, pp. 1054-1056, September 1964 (comments and rebuttal)

15.76 Conversion of Frequency-Modulated Light to Space-Modulated Light
S. E. Harris (Stanford U.)
J. Opt. Soc. Am., Vol. 54, pp. 1147-1151, September 1964

15.77 Light Modulation by the Electro-Optic Effect in Reverse-Biased GaP p-n Junctions
D. F. Nelson and F. K. Reinhart (BTL)
Appl. Phys. Letters, Vol. 5, pp. 148-150, October 1, 1964 (index change about 10^{-3} near 0.6μ)

15.78 Small-Signal Analysis of Internal (Coupling-Type) Modulation of Lasers
M. DiDomenico, Jr. (BTL)
J. Appl. Phys., Vol. 35, pp. 2870-2876, October 1964 (analytical, vary cavity losses, bandwidth limited by axial mode separation)

15.79 Laser Frequency Translation by Means of Electro-Optic Coupling Control
R. Targ, G. A. Massey (Sylvania), and S. E. Harris (Stanford U.)
Proc. IEEE, Vol. 52, pp. 1247-1248, October 1964 (KDP inside cavity, one-watt S-band modulation power, SSB output)

15.80 The Use of Perovskite Paraelectrics in Beam Deflectors and Light Modulators
F. S. Chen, J. E. Geusic, S. K. Kurtz, J. G. Skinner, and S. H. Wemple (BTL)
Proc. IEEE, Vol. 52, pp. 1258-1259, October 1964 (KTN prism beam deflector, 17 volts for π shift in modulator)

15.81 The Dielectric and Electro-Optical Properties of a Molecular Crystal-Hexamine
G. H. Heilmeier (RCA)
Appl. Optics, Vol. 3, pp. 1281-1287, November 1964 (compared with other materials, cubic, 3.2 dielectric constant, 10^{-3} loss tangent)

15.82 **Linear Electro-Optic Effects in Class 32, 6, 3m, and $\overline{4}3m$ Crystals**
T. R. Sliker (Clevite)
J. Opt. Soc. Am., Vol. 54, pp. 1348–1351, November 1964 (KD_2PO_4 has highest electro-optic coefficient, $LiKSO_4$ looks promising)

15.83 Internal Frequency Modulation of GaAs Junction Lasers by Changing the Index of Refraction through Electron Injection
G. E. Fenner (G.E.)
Appl. Phys. Letters, Vol. 5, pp. 198-199, November 15, 1964 (pulsed dual diode, different currents, continuous mode shift up to 1 Å or 45 Gc)

15.84 The Pockels Effect of Hexamethylenetetramine
C. F. Buhrer and L. Ho (GT&E)
Appl. Optics, Vol. 3, pp. 1500-1501, December 1964 (90-kV half-wave retardation voltage at 5893Å)

15.85 **Linear Electro-Optic Effects in KH_2PO_4 and Its Isomorphs**
J. H. Ott and T. R. Sliker (Clevite)
J. Opt. Soc. Am., Vol. 54, pp. 1442–1444, December 1964 (r_{63} and r_{41} coefficients given, 45 volts rms for proposed transverse modulator based on 45° Y-cut crystal)

15.86 Cross Modulation of Gaseous Lasers by High-Intensity Light
B. Pariser and T. C. Marshall (Columbia U.)
Proc. IEEE, Vol. 52, pp. 1740-1741, December 1964 (flash lamp with filters increases or decreases output)

15.87 **A Proposal for a Magnetically-Controlled Light-Beam Modulator and Shutter**
R. M. Zilberstein (Sharon, Mass.)
Proc. IEEE, Vol. 52, pp. 1751-1752, December 1964

15.88 Generation of Single-Frequency Light Using the FM Laser
G. A. Massey, M. K. Oshman and R. Targ (Sylvania)
Appl. Phys. Letters, Vol. 6, pp. 10–11, January 1, 1965 (He–Ne, KDP phase modulator at 104 Mc/s)

15.89 Laser Applications to Communication
D. Sette (U. of Rome, Rome, Italy)
Z. angewandte Mathematik und Physik, Vol. 16, No. 1, pp. 156–169, January 25, 1965 (general discussion, comparison of modulation and detection methods)

15.90 Internal Modulation of Optical Masers (Bandwidth Limitations)
I. P. Kaminow (BTL)
Appl. Optics, Vol. 4, pp. 123–127, January 1965

15.91 Light Modulation Experiments at 16 Gc/sec
R. A. Myers and P. S. Pershan (Harvard U.)
J. Appl. Phys., Vol. 36, pp. 22–28, January 1965 (cavity and traveling-wave types, Pockels effect)

15.92 Amplitude Modulation of Infrared and Sub-mm Wave Radiation by Reverse-Biased Junction Diodes
S. Deb and P. K. Chaudhuri (Calcutta, India)
Proc. IEEE, Vol. 53, pp. 81–82, January 1965

15.93 Frequency Modulation of a Gas Laser by Pressure Waves
R. L. Carter and T. K. Lewin (U. of Missouri)
Bull. Am. Phys. Soc., Series II, Vol. 10, No. 2, p. 164–G1, February 1965 (5-cm air column inside Fabry-Perot)

15.94 Internal Modulation in Multimode Laser Oscillators
A. Yariv (Calif. Inst. of Tech.)
J. Appl. Phys., Vol. 36, pp. 388–391, February 1965 (reactive and loss modulation)

15.95 GaAs Injection Laser with Novel Mode Control and Switching Properties
M. I. Nathan, J. C. Marinace, R. F. Rutz, A. E. Michel, and G. J. Lasher (IBM)
J. Appl. Phys., Vol. 36, pp. 473–480, February 1965 (control by current distribution, dual units, bistable)

15.96 Modulation of the Reflectivity of Semiconductors
M. Birnbaum (Aerospace)
J. Appl. Phys., Vol. 36, pp. 657–658, February 1965 (reflectivity increases with ruby laser irradiation)

15.97 Piezo-Optic Resonances in Crystals of the Dihydrogen Phosphate Type
J. F. Stephany (General Dynamics)
J. Opt. Soc. Am., Vol. 55, pp. 136–142, February 1965 (very low voltages required at resonance, typically at 63 kc/s)

15.98 Transverse Traveling-Wave Light Modulator
R. A. Myers (IBM)
Proc. IEEE, Vol. 53, p. 159, February 1965 (electro-optic prisms, microwave example)

15.99 Electro-Optic TV Communication System
W. J. Hannan, J. Bordogna, and T. E. Penn (RCA)
Proc. IEEE, Vol. 53, pp. 171–172, February 1965 (GaAs electro-optic birefringent modulator, 400 volts for 13% modulation, Nd:YAG laser oscillator)

15.100 Internal Modulation of a He–Ne Laser with a Television Signal
G. Schiffner and O. Hintringer (Technische Hochschule Wien, Austria)
Proc. IEEE, Vol. 53, pp. 172–173, February 1965 (KDP inside HeNe laser cavity, 120 volts for modulator)

15.101 X-Band Modulation of GaAs Lasers
B. S. Goldstein and R. M. Weigand (MIT)
Proc. IEEE, Vol. 53, p. 195, February 1965 (re-entrant microwave cavity, laser on post, 4.2°K)

15.102 Regular Emission from a Many-Element Laser During the Pumping Pulse
R. Pratesi and G. Toraldo di Francia (U. di Firenze, Italy)
Proc. IEEE, Vol. 53, pp. 196–197, February 1965 (ten-element ruby laser, regular spiking)

15.103 Ruby Laser Oscillations Modulated by Ultrasonic Vibration
Y. Sakai (Matsushita, Japan)
Proc. IEEE, Vol. 53, pp. 204–205, February 1965 (nickel transducer, 24 kc, superimposed modulation)

15.104 FM Laser Oscillation—Theory and Experiment
S. E. Harris and O. P. McDuff (Stanford U.) and R. Targ (Sylvania)
IEEE Int. Conv. Record, Vol. 13, Part 5, pp. 21–26, March 1965 (may convert FM light to single frequency light)

15.105 Microwave Modulation of Light in Ferromagnetic Resonance
J. T. Hanlon and J. F. Dillon, Jr. (BTL)
J. Appl. Phys., Vol. 36, pp. 1269–1270, Part 2, March 1965 (chromium tribromide crystal, 4810 Å beam, 23.5 Gc/s)

15.106 Amplitude Modulation of Infrared Light by GaAs p-n Junctions
M. Migitaka (Hitachi, Japan)
Proc. IEEE, Vol. 53, pp. 326–327, March 1965 (GaAs emitter and modulator, 10^3 c/s)

15.107 Modification of Devices Normally Operating between Input and Output Polarizers to Allow their Use with Arbitrarily Polarized Light
E. O. Ammann (Sylvania)
J. Opt. Soc. Am., Vol. 55, pp. 412–417, April 1965 (electrooptic shutter, modulator, KDP in 2.08 Gc/sec resonant cavity)

15.108 Microwave Dielectric Properties of $NH_4H_2PO_4$, KH_2AsO_4, and Partially Deuterated KH_2PO_4
I. P. Kaminow (BTL)
Phys. Rev., Vol. 138, pp. A1539–A1543, May 31, 1965 (temperature effects, 9.2 Gc/s data)

15.109 Wide-Band Electrooptic Light Modulation Utilizing an Asynchronous Traveling-Wave Interaction
Carl F. Buhrer (G.T. & E.)
Appl. Optics, Vol. 4, pp. 545–550, May 1965 (bandwidth from 7.5 to 12 Gc/s)

15.110 The Electrooptic Effect in Calcium Pyroniobate
C. H. Holmes (U. of Nottingham, England), E. G. Spencer, A. A. Ballman and P. V. Lenzo (BTL)
Appl. Optics, Vol. 4, pp. 551–553, May 1965 (about 5 kv for $\lambda/2$ phase retardation, experiments at 3 Gc/s)

15.111 Gigacycle-Bandwidth Coherent-Light Traveling-Wave Amplitude Modulator
Charles J. Peters (Sylvania)
Proc. IEEE, Vol. 53, pp. 455–460, May 1965 (16 inch length of KDP, 50 volts for 100% modulation)

15.112 Wide-Band Optical Frequency Translation
J. R. Kerr (h nu)
Proc. IEEE, Vol. 53, pp. 496–497, May 1965 (proposed, gas laser, several Gc/s, few watts drive)

15.113 Microwave Modulation of a Ruby Laser Output by Absorption
Di Chen (Honeywell)
IEEE J. Quantum Electronics, Vol. QE-1, pp. 125–131, June 1965 (ruby absorber in 3.2 Gc/s resonant cavity, Zeeman splitting, dc bias of 2.5 kOe)

15.114 Comments on "Bandwidth Limitations of In-Cavity Laser Modulation"

D. W. Jackson and W..H. Huntley, Jr. (Stanford Electronics Lab.)
Proc. IEEE, Vol. 53, pp. 616–617, June 1965 (also rebuttals)

15.115 A Coherent Light Modulator
J. T. Ruscio (BTL)
IEEE J. Quantum Electronics, Vol. QE-1, pp. 182–183, July 1965 (KDP between mirrors, 3 W modulation power at 70 Mc/s)

15.116 Optical Network Synthesis Using Birefringent Crystals. II. Synthesis of Networks Containing One Crystal, Optical Compensator, and Polarizer per Stage
E. O. Ammann (Sylvania) and I. C. Chang (Stanford U.)
J. Opt. Soc. Am., Vol. 55, pp. 835–841, July 1965

15.117 Laser Pulse-Shaping and Mode-Locking with Acoustic Waves
A. J. DeMaria and D. A. Stetser (United Aircraft)
Appl. Phys. Letters, Vol. 7, pp. 71–73, August 1, 1965 (27 Mc/s acoustic waves in quartz within argon ion laser cavity)

15.118 Mode-Locking Effects in an Internally Modulated Ruby Laser
Thomas Deutsch (Raytheon)
Appl. Phys. Letters, Vol. 7, pp. 80–82, August 15, 1965 (KDP, 50 and 148 Mc/s, varied cavity length)

15.119 Electro-Optic Diffraction Grating for Light Beam Modulation
E. I. Gordon and M. G. Cohen (BTL)
IEEE J. Quantum Electronics, Vol. QE-1, pp. 191–198, August 1965 (5 W of microwave power, one cm interaction length, also beam deflector)

15.120 Crystal Growth and Electro-Optic Effect of Bismuth Germanate, $Bi_4(GeO_4)_3$
Rudolf Nitsche (RCA, Switzerland)
J. Appl. Phys., Vol. 36, pp. 2358–2360, August 1965

15.121 Dielectric Constant and Magneto-Optical Kerr Rotation of Ferromagnetic Chromium Tribromide above the Absorption Band Edge
Wun Jung (BTL)
J. Appl. Phys., Vol. 36, pp. 2422–2426, August 1965

15.122 Infrared Modulator Using Multiple Internal Reflections and Induced Conductivity
D. W. Peters (SRI)
Proc. IEEE, Vol. 53, pp. 1148–1149, August 1965 (TlBr-TlI semiconductor crystal, modulated 8 to 9 μ wavelength region)

15.123 Electro-Optic Effect of Ferroelectric Microcrystals in a Glass Matrix
N. F. Borelli, Andrew Herczog, and R. D. Maurer (Corning Glass Works)
Appl. Phys. Letters, Vol. 7, pp. 117–118, September 1, 1965 (sodium niobate crystals in high silica glass)

15.124 Barium Titanate Light Phase Modulator
I. P. Kaminow (BTL)
Appl. Phys. Letters, Vol. 7, pp. 123–125, September 1, 1965 (beam incident on edge of 0.004-in. thick platelet, 70 Mc/s, 2 dB optical loss)

15.125 Electron-Beam Pumping of Noble-Gas Ion Lasers
J. M. Hammer and C. P. Wen (RCA)
Appl. Phys. Letters, Vol. 7, pp. 159–161, September 15, 1965 (electron tube transverse to laser beam, both inside Fabry-Perot cavity, modulation)

15.126 Traveling-Wave Light Modulator
W. A. Scanga (Aircraft Armaments)
Appl. Optics, Vol. 4, pp. 1103–1106, September 1965 (62 crystals, zig-zag path, 0.66 watts, 3.2 dB optical loss, broad bandwidth)

15.127 A Kerr Cell with Roof Prism
Milton Laikin (Electro-Optical Systems)

Appl. Optics, Vol. 4, pp. 1177–1178, September 1965 (prism apex parallel or perpendicular to Kerr cell plates, phase shift in prism)

15.128 Some Dielectric and Electro-Optic Properties of BaTiO₃ Single Crystals
C. J. Johnson (M.I.T.)
Appl. Phys. Letters, Vol. 7, pp. 221–223, October 15, 1965

15.129 Magnetostrictive Light Modulator
C. C. Aleksoff (U. of Michigan) and N. J. Harrick (Philips Labs., N.Y.)
Proc. IEEE, Vol. 53, pp. 1636–1637, October 1965 (movable ferrite absorber near TIR surface of Ge element, 12–15 kc/s modulation, 2–7 μ wavelength)

15.130 Transmission Characteristics of Fabry-Perot Interferometers and a Related Electrooptic Modulator
V. N. Del Piano, Jr., and A. F. Quesada (Baird-Atomic)
Appl. Optics, Vol. 4, pp. 1386–1390, November 1965 (effects of maladjustments and surface imperfections)

15.131 Design of a Wideband Electro-optic Light-Modulating System for CW Operations
G. W. Hong (Lawrence Radiation Lab., U. of California)
Appl. Optics, Vol. 4, pp. 1391–1394, November 1965 (network techniques for broadbanding)

15.132 Direct Modulation of Gas Lasers
Teiji Uchida (Nippon Electric Co., Japan)
IEEE J. Quantum Electronics, Vol. QE-1, pp. 336–343, November 1965 (KDP inside laser cavity, up to 10 Mc/s modulation)

15.133 Optical Transmission in Multidomained KH₂PO₄: Polarization Scattering
R. M. Hill, G. F. Herrmann, and S. K. Ichiki (Lockheed)
J. Appl. Phys., Vol. 36, pp. 3672–3677, November 1965

15.134 The Electro-Optic Effect of Sodium Uranyl Acetate
J. Warner, D. S. Robertson, and H. T. Parfit (Royal Radar Estab., England)
Physics Letters, Vol. 19, pp. 479–480, December 1, 1965 (cubic crystal, very small effect)

15.135 Doppler Shift of Ruby Laser Light by Means of a Kerr Cell Traveling-Wave Line
F. P. Küpper and E. Fünfer (Institut für Plasmaphysik, Germany)
Physics Letters, Vol. 19, pp. 486–487, December 1, 1965 (60-cm-long cell, theoretical shift of 2.3×10^{10} c/s)

15.136 Electro-Optic Effect in NH₄Cl
M. Vassell and E. M. Conwell (G.T. and E. Labs.)
Phys. Rev., Vol. 140, pp. A2110–A2116, December 13, 1965 (Pockels effect, 45 kV for half-wave shift at 0.589 μ, 77°K)

15.137 Time Resolution of Acoustic Mode Patterns in KDP Crystals
G. E. Peterson and P. M. Bridenbaugh (BTL)
Appl. Optics, Vol. 4, pp. 1655–1659, December 1965 (strobe flash lamp, photographic technique, 530 kc/s patterns)

15.138 An Improved Mode of Kerr Cell Operation
H. A. Heynau (United Aircraft Research Labs.)
Proc. IEEE, Vol. 53, pp. 2145–2156, December 1965 (add $\lambda/4$ quartz plate within laser cavity)

15.139 Gd₂(MoO₄)₃: A Ferroelectric Laser Host
Hans J. Borchardt and Paul E. Bierstedt (E. I. du Pont de Nemours)
Appl. Phys. Letters, Vol. 8, pp. 50–52, January 15, 1966 (3% Nd doped, ferroelectric below 159°C, 350 J threshold at −138°C, direct modulation possible)

15.140 Infrared Modulation by Means of Frustrated Total Internal Reflection
Robert W. Astheimer, Gerald Falbel, and Sheldon Minkowitz (Barnes Engineering Co.)
Appl. Optics, Vol. 5, pp. 87–91, January 1966 (variable air gap, prisms)

15.141 Optical Misalignment Due to Temperature Gradients in Electro-optic Modulator Crystals
Claire Loscoe and Herbert Mette (U. S. Army Electronics Command, Fort Monmouth)
Appl. Optics, Vol. 5, pp. 93–96, January 1966 (quartz and KDP)

15.142 Low-Voltage Light-Amplitude Modulation
J. M. Ley (Northampton College of Advanced Technology, London, England)
Electronics Letters, Vol. 2, pp. 12–13, January 1966 (y-cut ADP, ordinary and extraordinary rays, 40 V rms for 50% modulation, 0.63 μ)

15.143 Light Modulation and Beam Deflection with Potassium Tantalate-Niobate Crystals
F. S. Chen, J. E. Geusic, S. K. Kurtz, J. G. Skinner, and S. H. Wemple (B.T.L.)
J. Appl. Phys., Vol. 37, pp. 388–398, January 1966 (physical properties, modulator and deflector designs, few watts for wide band modulation)

15.144 Linear Electro-Optic Effect and Refractive Indices of Cubic ZnTe
T. R. Sliker and J. M. Jost (Clevite)
J. Opt. Soc. Am., Vol. 56, pp. 130–131, January 1966 (about 3 kV for π shift, index of about 3)

15.145 Electro-Optic Effect in Reverse-Biased GaAs p-n Junctions
W. L. Walters (RCA)
J. Appl. Phys., Vol. 37, p. 916, February 1966

15.146 Infrared and Visible Laser Modulation using Faraday Rotation in YIG
R. N. Zitter and E. G. Spencer (B.T.L.)
J. Appl. Phys., Vol. 37, pp. 1089–1090, March 1, 1966 (He–Ne laser, 13 Mc/s, spiking)

15.147 Optoelectronic Amplitude Modulators
M. I. Oduah and D. P. Howson (U. of Birmingham, England)
Electronics Letters, Vol. 2, pp. 93–94, March 1966 (GaAs emitter, PbS photoconductor modulator)

15.148 A dc-Biased Kerr Cell Light Modulator
Julian Stone, George Lynch, and Richard Pontinen (Columbia U.)
Appl. Optics, Vol. 5, pp. 653–657, April 1966 (ultra pure nitrobenzene, 5 Mc/s continuous modulation)

15.149 Low-Voltage Light-Amplitude Modulation
Comments by C. H. Clayson (U. of Birmingham, England) Rebuttal by J. M. Ley (Northampton College of Advanced Technology, England)
Electronics Letters, Vol. 2, pp. 138–139, April 1966 (comments on paper by J. M. Ley, Electronics Letters, Vol. 2, p. 12, 1966)

15.150 Low Frequency Electro-Optic and Dielectric Constants of Lithium Niobate
E. Bernal, G. D. Chen, and T. C. Lee (Honeywell)
Physics Letters, Vol. 21, pp. 259–260, May 15, 1966 (at 0.63 μ, DC and 60 Hz, modulated also at 3 GHz)

15.151 Figure of Merit for Acousto-Optical Deflection and Modulation Devices

E. I. Gordon (B.T.L.)
IEEE J. Quantum Electronics, Vol. QE-2, pp. 104–105, May

15.152 Temperature Variation of the Index of Refraction of ADP, KDP, and Deuterated KDP
Richard A. Phillips (U. of Michigan)
J. Opt. Soc. Am., Vol. 56, pp. 629–632, May 1966

15.153 Electro-Optic Coefficients in Single-Domain Ferroelectric Lithium Niobate
P. V. Lenzo, E. G. Spencer, and K. Nassau (B.T.L.)
J. Opt. Soc. Am., Vol. 56, pp. 633–635, May 1966

15.154 Optical Modulation in Bulk GaAs using the Gunn Effect
M. G. Cohen, S. Knight and J. P. Elward (B.T.L.)
Appl. Phys. Letters, Vol. 8, pp. 269–271, June 1, 1966 (He–Ne laser, 0.5 to 4 Gc/s modulation)

15.155 High-Frequency Electro-Optic Coefficients of Lithium Niobate
E. H. Turner (B.T.L.)
Appl. Phys. Letters, Vol. 8, pp. 303–304, June 1, 1966 (50 to 86 Mc/s modulation frequency, 0.633μ wavelength)

15.156 Barium Titanate Light Modulator. II
I. P. Kaminow (B.T.L.)
Appl. Phys. Letters, Vol. 8, pp. 305–307, June 1, 1966 (5 to 200 MHz)

15.157 A Simple ac Photometric Method for Measuring Complex Kerr Magnetooptic Coefficients
Carl F. Buhrer (G.T. & E. Labs.)
Appl. Optics, Vol. 5, pp. 1015–1017, June 1966

15.158 Microwave Frequency Light Modulation with Traveling-Wave Carbon Disulfide Kerr Cells
A. J. Chenoweth and O. L. Gaddy (U. of Illinois)
Proc. IEEE, Vol. 54, pp. 877–878, June 1966 (2.92 Gc/s, 300 watts for 50% modulation)

15.159 *Laser Receivers; Devices, Techniques, Systems*
Monte Ross (McDonnell Aircraft)
John Wiley & Sons, Inc., New York, 405 p., 1966 (noise, information theory, detectors, modulators, transmission, components and systems)

15.160 Longitudinal Magneto-Optical Kerr Effect in EuO and EuS
J. H. Greiner and G. J. Fan (IBM)
Appl. Phys. Letters, Vol. 9, pp. 27–29, July 1, 1966 (8 to 12°K, 350 to 900 nm band, large effect)

15.161 20 Gc/s Modulation of Light by Electro-Absorption in GaAs
H. D. Rees (Royal Radar Estab., England)
Physics Letters, Vol. 21, pp. 629–631, July 1, 1966

15.162 Internal Laser Modulation by Acoustic Lens-Like Effects
A. J. DeMaria and G. E. Danielson, Jr. (United Aircraft)
IEEE J. Quantum Electronics, Vol. QE-2, pp. 157–164, July 1966 (777 kHz cylindrical acoustic wave, increased output)

15.163 A New Microwave Light Modulator
G. K. Megla (Corning Glass Work)
Proc. IEEE, Vol. 54, p. 1006, July 1966 (electrooptical crystal between 2-wire line, light beam orthogonal to line axis, 750 MHz, 20 W modulator power)

15.164 Modulation of $10.6\text{-}\mu$ Laser Radiation by Ultrasonic Diffraction
H. R. Carleton and R. A. Soref (Sperry Rand Research Center)
Appl. Phys. Letters, Vol. 9, pp. 110–112, August 1, 1966 (CdS, GaAs, and Si crystals; 20 MHz modulation)

15.165 Effects of Piezoelectric Resonances in ADP and KDP Light-Modulator Crystals

F. Hoff and B. Stádník (IREE Acad. of Sci., Czechoslovakia)
Electronics Letters, Vol. 2, p. 293, August 1966 (introduces noise, may destroy crystals, mechanical resonances in z-cut but not x- and y-cut plates)

15.166 External Light Modulation with Low Microwave Power
H. Brand, B. Hill, E. Holtz, and G. Wencker (Tech. Hochschule Aachen, Germany)
Electronics Letters, Vol. 2, pp. 317–318, August 1966 (33 mm long resonant KDP crystal, 20 mW at 9.1 GHz)

15.167 GaAs as an Electrooptic Modulator at 10.6 Microns
A. Yariv and C. A. Mead (California Inst. of Tech.), and J. V. Parker (Electro-Optical)
IEEE J. Quantum Electronics, Vol. QE-2, pp. 243–245, August 1966

15.168 An Acoustic Light Modulator for 10.6μ
R. W. Dixon and A. N. Chester (BTL)
Appl. Phys. Letters, Vol. 9, pp. 190–192, September 1, 1966 (Bragg effect, tellurium, high efficiency)

15.169 Electric Breakdown Mechanism in Cuprous Chloride Single Crystals
A. L. Gentile (Hughes Research)
Appl. Phys. Letters, Vol. 9, pp. 237–239, September 15, 1966 (CuCl used in light modulator, method to increase resistivity)

15.170 Intra-Cavity Perturbation of a Gas Laser by Faraday Effects in Glasses
D. Chen and T. C. Lee (Honeywell)
IEEE J. Quantum Electronics, Vol. QE-2, pp. 461–463, September 1966 (internal modulation, efficient operation)

15.171 Direct Frequency Modulation of a Semiconductor Laser by Ultrasonic Waves
J. E. Ripper, G. W. Pratt, Jr., and C. G. Whitney (M.I.T.)
IEEE J. Quantum Electronics, Vol. QE-2, pp. 603–605, September 1966

15.172 Gallium-Arsenide Electro-Optic Modulators
T. E. Walsh (RCA)
RCA Rev., Vol. 27, pp. 323–335, September 1966 (for 0.9 to 16 μ, up to 1.7 GHz)

15.173 Electrooptic Light Modulators
I. P. Kaminow and E. H. Turner (BTL)
Appl. Optics, Vol. 5, pp. 1612–1628, October 1966 (review, various materials)

15.174 A Review of Acoustooptical Deflection and Modulation Devices
E. I. Gordon (BTL)
Appl. Optics, Vol. 5, pp. 1629–1639, October 1966 (momentum considerations, efficiency, bandwidth)

15.175 Carbon Disulfide Traveling-Wave Kerr Cells
A. J. Chenoweth, O. L. Gaddy, and D. F. Holshouser (U. of Illinois)
Appl. Optics, Vol. 5, pp. 1652–1656, October 1966 (44-cm long cell, 50 percent modulation, 190 W at 3 GHz)

15.176 Optical Activity and Electrooptic Effect in Bismuth Germanium Oxide ($Bi_{12}GeO_{20}$)
P. V. Lenzo, E. G. Spencer, and A. A. Ballman (BTL)
Appl. Optics, Vol. 5, pp. 1688–1689, October 1966 (modulation up to 500 MHz)

15.177 Light Modulation Using Natural Crystals of Zinc Sulphide
J. M. Ley (London City U., England)
Electronics Letters, Vol. 2, pp. 394–396, October 1966

16. DETECTION

16.20 Infrared Quantum Counter Action in Pr-Doped Fluoride Lattices
M. R. Brown (Ministry of Aviation, England), and W. A. Shand (U. of Aberdeen, Scotland)
Phys. Rev. Letters, Vol. 11, pp. 366–368, October 15, 1963

16.21 Solid-State Modulation and Direct Demodulation of Gas Laser Light at a Microwave Frequency
Kenneth M. Johnson (Texas Instruments)
Proc. IEEE, Vol. 51, pp. 1368–1369, October 1963

16.22 "Push-Pull" Optical Modulators and Demodulators
Fred Sterzer (RCA)
Appl. Optics, Vol. 2, pp. 1197–1198, November 1963

16.23 Photoconductive Mixing in CdSe Single Crystals
O. Svelto, P. D. Coleman, M. DiDomenico, Jr., and R. H. Pantell (Stanford U.)
J. Appl. Phys., Vol. 34, pp. 3182–3186, November 1963

16.24 Detection of Laser Radiation
L. Mandel (U. of London, England), and E. Wolf (U. of Rochester)
J. Opt. Soc. Am., Vol. 53, p. 1315, November 1963

16.25 Light Waves Mixing in Nonlinear Anisotropic Media
Domenico Solimini (U. di Roma, Italy)
Proc. IEEE, Vol. 51, pp. 1408–1411, November 1963

16.26 Photoemitters Having a High Quantum Efficiency
R. M. White (U. of California)
Proc. IEEE, Vol. 51, p. 1662, November 1963

16.27 Optical Heterodyning with Noncritical Angular Alignment
W. S. Read and D. L. Fried (North American Aviation)
Proc. IEEE, Vol. 51, p. 1787, December 1963

16.28 Optical Heterodyning with a CW Gaseous Laser
S. F. Jacobs and P. J. Rabinowitz (TRG)
Proc. Third Int. Conf. Quantum Electronics, Columbia U. Press, N. Y., Vol. 1, pp. 481–487, 1964

16.29 The Helium Neon Laser as a Quantum Counter at 3.39 Microns
W. E. Bell, A. Bloom, and R. C. Rempel (Spectra-Physics)
Proc. Third Int. Conf. Quantum Electronics, Columbia U. Press, N. Y., Vol. 2, pp. 1347-1352, 1964

16.30 Microwave Demodulation of Light
A. E. Siegman, S. E. Harris (Stanford U.), and B. J. McMurtry (Sylvania)
Proc. Third Int. Conf. Quantum Electronics, Columbia U. Press, N. Y., Vol. 2, pp. 1651-1658, 1964

16.31 An FM-AM Optical Converter
S. E. Harris (Stanford U.)
Proc. Third Int. Conf. Quantum Electronics, Columbia U. Press, N. Y., Vol. 2, pp. 1671-1677, 1964

16.32 Photomultiplication with Microwave Response
O. L. Gaddy and D. F. Holshouser (U. of Illinois)
Proc. Third Int. Conf. Quantum Electronics, Columbia U. Press, N. Y., Vol. 2, pp. 1717-1722, 1964

16.33 Aperçu sur les Possibilités Offertes par les Photomultiplicateurs Rapides dans la Détection de Lumiére Modulée
J. Nussli (Lab. d'Electronique et de Physique Appliquée, Paris, France)
Proc. Third Int. Conf. Quantum Electronics, Columbia U. Press, N. Y., Vol. 2, pp. 1723-1730, 1964

16.34 Detection of Coherent Light by Heterodyne Techniques Using Solid State Photodiodes
G. Lucovsky, R. B. Emmons, B. Harned, and J. K. Powers (Philco)
Proc. Third Int. Conf. Quantum Electronics, Columbia U. Press, N. Y., Vol. 2, pp. 1731-1738, 1964

16.35 Détecteur-Amplificateur et Mélangeur de Laser en Microonde par une Diode Paramétrique
S. Saito, K. Kurokawa, Y. Fujii, T. Kimura, and Y. Uno (U. of Tokyo, Japan)
Proc. Third Int. Conf. Quantum Electronics, Columbia U. Press, N. Y., Vol. 2, pp. 1739-1740, 1964

16.36 A New Principle in the Design of a Millimetric Photo-Electric Laser Mixer
A. L. Cullen and P. N. Robson (U. of Sheffield, England)
Proc. Third Int. Conf. Quantum Electronics, Columbia U. Press, N. Y., Vol. 2, pp. 1741-1750, 1964

16.37 The Theory of Optical Mixing in Semi-Conductors
R. H. Pantell, M. DiDomenico, Jr., O. Svelto, and J. N. Weaver (Stanford U.)
Proc. Third Int. Conf. Quantum Electronics, Columbia U. Press, N. Y., Vol. 2, pp. 1811-1818, 1964

16.38 Optical Frequency Mixing in Photoconductive InSb
E. N. Fuls (BTL)
Appl. Phys. Letters, Vol. 4, pp. 7–8, January 1, 1964

16.39 Irradiance Linearity Corrections for Multiplier Phototubes
D. J. Baker and C. L. Wyatt (Utah State U.)
Appl. Optics, Vol. 3, pp. 89–91, January 1964

16.40 Signal Detection with a Laser Amplifier
Herbert A. Steinberg (TRG)
Proc. IEEE, Vol. 52, pp. 28–32, January 1964

16.41 Lossless Beam Combination for Optical Heterodyning
A. E. Siegman (Stanford U.) and B. J. McMurtry (Sylvania)
Proc. IEEE, Vol. 52, p. 94, January 1964

16.42 Optical Heterodyning Using Point Contact Germanium Diodes
C. K. N. Patel and W. M. Sharpless (BTL)
Proc. IEEE, Vol. 52, pp. 107–108, January 1964

16.43 Solid-State Photodetection: A Comparison between Photodiodes and Photoconductors
M. DiDomenico, Jr. (BTL) and O. Svelto (Politecnico, Milan, Italy)
Proc. IEEE, Vol. 52, pp. 136–144, February 1964

16.44 Demodulation of Low-Level Broad-Band Optical Signals with Semiconductors: Part II—Analysis of the Photoconductive Detector
H. S. Sommers, Jr., and W. B. Teutsch (RCA)
Proc. IEEE, Vol. 52, pp. 144–153, February 1964

16.45 Noise Performance of Photo Diodes in Parametric Amplifiers
K. Garbrecht and W. Heinlein (Siemens & Halske, Germany)
Proc. IEEE, Vol. 52, pp. 192–193, February 1964

16.46 Cartridge-Type Point-Contact Photodiode
W. M. Sharpless (BTL)
Proc. IEEE, Vol. 52, pp. 207–208, February 1964

16.47 Infrared Quantum Counter Action in Er-Doped Fluoride Lattices
M. R. Brown (Ministry of Aviation, England) and W. A. Shand (U. of Aberdeen, Scotland)
Phys. Rev. Letters, Vol. 12, pp. 367–369, March 30, 1964

16.48 Optical Heterodyne Detection of Microwave-Modulated Light
R. Targ (Sylvania)
Proc. IEEE, Vol. 52, pp. 303–304, March 1964

16.49 Evaluating Light Demodulators
D. E. Caddes and B. J. McMurtry (Sylvania)
Electronics, Vol. 37, No. 13, pp. 54–61, April 6, 1964

16.50   Balanced Optical Discriminator
        I. P. Kaminow (BTL)
        Appl. Optics, Vol. 3, pp. 507–510, April 1964

16.51   Direct Demodulation and Frequency Conversion of Micro-
        wave-Modulated Light in a CdSe Bulk Photoconductor
        M. DiDomenico, Jr. (BTL)
        J. Appl. Phys., Vol. 35, pp. 1353–1354, April 1964 (3030
        Mcps experiment)

16.52   The Traveling-Wave Phototube Part I: Theoretical Analysis
        D. E. Caddes, B. J. McMurtry, and A. E. Jacquez
        (Sylvania)
        IEEE Trans. on Electron Devices, Vol. ED-11, pp. 156–
        163, April 1964

16.53   The Traveling-Wave Phototube Part II: Experimental Anal-
        ysis
        R. Targ, D. E. Caddes, and B. J. McMurtry (Sylvania)
        IEEE Trans. on Electron Devices, Vol. ED-11, pp. 164–170,
        April 1964

16.54   Effect of an Electrostatic Field in the Dynamic Crossed-Field
        Photomultiplier
        R. Hankin, E. Dallaflor (Hallicrafters), and B. Alpiner
        (Zenith)
        Proc. IEEE, Vol. 52, pp. 412–413, April 1964

16.55   Some Aspects of Electron Motion in Biased DCFEM Operation
        O. L. Gaddy and D. F. Holshouser (U. of Illinois)
        Proc. IEEE, Vol. 52, pp. 413–414, April 1964

16.56   Coherent Detection of Light Scattered from a Diffusely Re-
        flecting Surface
        G. Gould, S. F. Jacobs, J. T. LaTourrette, M. Newstein, and
        P. Rabinowitz (TRG)
        Appl. Optics, Vol. 3, pp. 648–649, May 1964

16.57   Improved Gain and Stability in the Dynamic Crossed-Field
        Photomultiplier
        O. L. Gaddy and D. F. Holshouser (U. of Illinois)
        Proc. IEEE, Vol. 52, p. 616, May 1964

16.58   Theory of Photoelectric Mixing at a Metal Surface
        R. L. Smith (United Aircraft Res. Labs.)
        Appl. Optics., Vol. 3, pp. 709–713, June 1964 (strict cophasal and
        parallel polarization not required)

16.59   Electro-Optic Frequency Modulation in Optical Resonators
        A. Yariv (Lockheed)
        Proc. IEEE, Vol. 52, pp. 719–720, June 1964 (electro-optical
        modulator in cavity, frequency stabilization)

16.60   Microwave Signal-to-Noise Performance of CdSe Bulk Photo-
        conductive Detectors
        M. DiDomenico, Jr. and L. K. Anderson (BTL)
        Proc. IEEE, Vol. 52, pp. 815–822, July 1964 (3000-Mc experi-
        ments, also high-intensity nonlinear effects)

16.61   Demodulation of Phase-Modulated Light using Birefringent Crys-
        tals
        S. E. Harris (Stanford U.)
        Proc. IEEE, Vol. 52, pp. 823-831, July 1964 (experiments be-
        tween 2 and 12 Gc)

16.62   Synchronous Detection of Microwave-Modulated Incoherent Opti-
        cal Carriers
        J. H. Ward, W. F. Davison, and E. R. Marcusen (ITT Fed.
        Labs.)
        Proc. IEEE, Vol. 52, p. 854, July 1964 (GaAs diode, 78 Mc-
        subcarrier modulation, phase-locked detector)

16.63   An Available Power-Bandwidth Product for Photodiodes
        R. B. Emmons and G. Lucovsky (Philco)
        Proc. IEEE, Vol. 52, p. 865, July 1964 (analytical)

16.64   Infra-Red Quantum Counter Action in Ho-Doped Fluoride Lat-
        tices
        M. R. Brown (England) and W. A. Shand (Scotland)
        Physics Letters, Vol. 11, pp. 219-220, August 1, 1964 (several IR
        absorption lines, outputs available at 6410, 5350 and 4850 Å)

16.65   On the Noise Performance of a Photoparametric Amplifier
        S. Saito and Y. Fujii (U. of Tokyo, Japan)
        Proc. IEEE, Vol. 52, pp. 978-979, August 1964

16.66   Erratum: On the Noise Performance of a Photoparametric Am-
        plifier
        S. Saito and Y. Fujii (U. of Tokyo, Japan)
        Proc. IEEE, Vol. 53, p. 167, February 1965 (Proc. IEEE, Vol.
        52, pp. 978–979; August 1964)

16.67   Demodulation of Phase-Modulated Optical Maser Beam by Auto-
        Correlation Technique
        S. Saito and T. Kimura (U. of Tokyo)
        Proc. IEEE, Vol. 52, p. 1048, September 1964 (experiments with
        He-Ne laser)

16.68   Gas Laser Preamplifier Performance
        W. B. Bridges and G. S. Picus (Hughes)
        Appl. Optics, Vol. 3, pp. 1189-1190, October 1964 (16-dB im-
        provement in sensitivity at 3.5 micron using xenon laser)

16.69   Spectral Characteristic, Efficiency and Speed of a $GaAs_xP_{1-x}$ GaAs
        Photodiode
        T. B. Ramachandran and W. J. Moroney (Microwave Asso-
        ciates)
        Proc. IEEE, Vol. 52, pp. 1358-1359, November 1964 (peak re-
        sponse at about 8650 Å)

16.70   Mixing and Detection of Coherent Light in a Bulk Photocon-
        ductor
        P. D. Coleman, R. C. Eden, and J. N. Weaver (Stanford U.)
        IEEE Trans. on Electron Devices, Vol. ED-11, pp. 488-497,
        November 1964 (DC or AC bias, 3Gc. experiments, noise con-
        siderations, optimization, analyses)

16.71   Detection of Weakly Modulated Light at Microwave Frequencies
        M. G. Cohen and E. I. Gordon (BTL)
        Bell Sys. Tech. J., Vol. 43, pp. 3068-3070, November 1964 (dou-
        ble system of homodyne/superheterodyne)

16.72   Theory of Optical Beating in Photoconductors (Phenomenological
        Modification of Semiclassical Treatment)
        C. M. Penchina (Syracuse U.)
        Phys. Rev., Vol. 136, pp. A911-A917, November 16, 1964

16.73   Optical Heterodyne Detection of the Forward-Stimulated Bril-
        louin Scattering
        D. A. Jennings and H. Takuma (NBS)
        Appl. Phys. Letters, Vol. 5, pp. 241-242, December 15, 1964 (123
        Mc/s beat frequency)

16.74   Laser Applications to Communication
        D. Sette (U. of Rome, Rome, Italy)
        Z. angewandte Mathematik und Physik, Vol. 16, No. 1, pp.
        156-169, January 25, 1965 (general discussion, comparison of
        modulation and detection methods)

16.75   Lateral Effects in High-Speed Photodiodes
        G. Lucovsky and R. B. Emmons (Philco)
        IEEE Trans. on Electron Devices, Vol. ED-12, pp. 5-12,
        January 1965

16.76   Microwave Photodiodes Exhibiting Microplasma-Free Carrier
        Multiplication
        L. K. Anderson, P. G. McMullin, L. A. D'Asaro, and A. Goetz-
        berger (BTL)
        Appl. Phys. Letters, Vol. 6, pp. 62-64, February 15, 1965
        (used 0.63μ HeNe laser, 1.5-10 Gc Pockels modulator)

16.77   High-Speed Photodiode Signal Enhancement at Avalanche Break-
        down Voltage
        Kenneth M. Johnson (Texas Instr.)

IEEE Trans. on Electron Devices, Vol. ED-12, pp. 55-63, February 1965 (0.63μ HeNe laser, cavity type Pockels cell, 1.45 Gc/s modulation, diode noise analyzed)

16.78 Photovoltaic Effect in Photoconductors
P. N. Keating (Tyco)
J. Appl. Phys., Vol. 36, pp. 564-570, February 1965

16.79 Avalanche Multiplication in InAs Photodiodes
G. Lucovsky and R. B. Emmons (Philco)
Proc. IEEE, Vol. 53, p. 180, February 1965 (0.63μ HeNe laser, cw and 125 Mc/s modulation)

16.80 High Speed Photomultipliers
R. C. Miller and N. C. Wittwer (BTL)
IEEE Int. Conv. Record, Vol. 13, Part 5, pp. 7-16, March 1965 (8-stages, crossed field or electrostatic, up to 4 Gc response)

16.81 Piezoelectric Detection of Laser Pulses with Cadmium Sulfide Thin Films
M. S. Bruma and M. F. Velghe (Nat. Sci. Res. Cen., France)
IEEE Int. Conv. Record, Vol. 13, Part 5, pp. 17-20, March 1965 (beam passes axially through CdS tube)

16.82 Infrared Detectors
George A. Morton (RCA)
RCA Review, Vol. 26, pp. 3-21, March 1965 (from 1.8μ to beyond 100μ, various types, detectivity)

16.83 Microwave Photomultipliers Using Transmission Dynodes
D. Blattner, H. Johnson, J. Ruedy, and F. Sterzer (RCA)
RCA Review, Vol. 26, pp. 22-41, March 1965 (1.0 Gc bandwidth at 1.5 Gc, K$_a$ band possible, helix structure, RF mixing)

16.84 Transverse Electron-Beam Wave Excitation by Photoelectric Mixing of Laser Beams
J. C. Bass (U. of Sheffield, England)
Electronics Letters, Vol. 1, pp. 38-39, April 1965 (transverse-field microwave phototube appears possible)

16.85 Performance of Laser-Pumped Quantum Counters
William F. Krupke (Aerospace)
IEEE J. Quantum Electronics, Vol. QE-1, pp. 20-28, April 1965 (proposed experiments with Er^{3+} and Sm^{2+})

16.86 A Scannable Detector of Microwave-Modulated Light
M. B. Fisher (Sylvania)
IEEE J. Quantum Electronics, Vol. QE-1, pp. 37-42, April 1965 (image dissector electron gun, 10.7 Gc/s experiments, 4 Gc/s instantaneous bandwidth)

16.87 Secondary-Emission Amplification at Microwave Frequencies
R. C. Miller and N. C. Wittwer (BTL)
IEEE J. Quantum Electronics, Vol. QE-1, pp. 49-59, April 1965 (crossed field device, also electrostatic device, responses up to 4 Gc/s, S-1 photocathode, Cu-Be secondary emitters)

16.88 Gold-Doped Germanium as an Infrared High-Frequency Photodetector
R. A. Wood (BTL)
J. Appl. Phys., Vol. 36, pp. 1490-1491, April 1965 (sensitive out to 9μ wavelength, up to 1500 Mc/s response, 77°K)

16.89 Photoparametric Amplifier
P. Penfield, Jr. (M.I.T.) and D. E. Sawyer (Sperry-Rand)
Proc. IEEE, Vol. 53, pp. 340-347, April 1965 (p-n junction photodiode, comprehensive analysis)

16.90 Detectors for Microwave-Modulated Light
L. K. Anderson (BTL)
Electro-Technology, Vol. 75, May 1965, pp. 44-48 (general review, comparisons)

16.91 Square Law Behavior of Photocathodes at High Light Intensities and High Frequencies
A. M. Johnson (BTL)
IEEE J. Quantum Electronics, Vol. QE-1, pp. 99-101, May 1965 (beats greater than ruby laser emission linewidth)

16.92 Detectors for Ultraviolet, Visible, and Infrared Radiation
R. A. Smith (M.I.T.)
Appl. Optics, Vol. 4, pp. 631-638, June 1965 (review, problems, current practice)

16.93 Extrinsic Detectors
H. Levinstein (Syracuse U.)
Appl. Optics, Vol. 4, pp. 639-647, June 1965 (Ge, Ge:Si alloys, impurity-activated InSb)

16.94 Indium Antimonide Submillimeter Photoconductive Detectors
E. H. Putley (Royal Radar Est., England)
Appl. Optics, Vol. 4, pp. 649-657, June 1965 (performance, comparisons)

16.95 Photoconductive Indium Antimonide Detectors
F. D. Morten and R. E. J. King (Assoc. Semicond. Manufacturers, Ltd., England)
Appl. Optics, Vol. 4, pp. 659-663, June 1965 (design theory, practical embodiments)

16.96 Optimum Utilization of Lead Sulfide Infrared Detectors Under Diverse Operating Conditions
J. N. Humphrey (Electronics Corp. of America)
Appl. Optics, Vol. 4, pp. 665-675, June 1965

16.97 High Speed Photodetection in Germanium and Silicon Cartridge-Type Point-Contact Photodiodes
M. DiDomenico, Jr., W. M. Sharpless, and J. J. McNicol (BTL)
Appl. Optics, Vol. 4, pp. 677-682, June 1965 (3-Gc/sec modulation experiments, 0.63-μ He-Ne laser, Ge and Si laser-formed photodiodes)

16.98 The Cooled Germanium Bolometer as a Far Infrared Detector
C. E. Jones, Jr., A. R. Hilton, J. B. Damrel, Jr. (Texas Instruments) and C. C. Helms (Perkin-Elmer)
Appl. Optics, Vol. 4, pp. 683-685, June 1965 (4.2°K, 90 to 250 μ range covered)

16.99 Photon Effects in Hg$_{1-x}$Cd$_x$Te
P. W. Kruse (Honeywell Res. Ctr.)
Appl. Optics, Vol. 4, pp. 687-692, June 1965 (1 to 14-μ spectral range, less than 10^{-7} sec response time at 77°K)

16.100 Lead Selenide Detectors for Ambient Temperature Operation
T. H. Johnson, H. T. Cozine, and B. N. McLean (Santa Barbara Res. Ctr.)
Appl. Optics, Vol. 4, pp. 693-696, June 1965 (1 to 4.8-μ region, time constant less than 3 μsec)

16.101 High Frequency Photodiodes
G. Lucovsky and R. B. Emmons (Philco)
Appl. Optics, Vol. 4, pp. 697-702, June 1965 (cutoff frequency about 20 Gc/sec)

16.102 Sensitivity of Pr^{3+}:LaCl$_3$ Infrared Quantum Counter
John F. Porter, Jr. (Johns Hopkins U.)
IEEE J. Quantum Electronics, Vol. QE-1, pp. 113-115, June 1965 (1.48-2.33 μ, 300-4.2°K, best NEP = 2 × 10^{-5} W(c/s)$^{-1/2}$ at 2.03 μ in 20-30 ms)

16.103 Directional Characteristics in Optical Heterodyne Detection Processes
V. J. Corcoran (Martin)
J. Appl. Phys., Vol. 36, pp. 1819-1825, June 1965 (single and multiple apertures)

16.104 Photomixing with Diffusely Reflected Light
G. A. Massey (Sylvania)
Appl. Optics, Vol. 4, pp. 781-784, July 1965 (0.63-μ gas laser, estimate of sensitivity)

16.105 Optimum Processing of Photoelectron Counts
J. W. Goodman (Stanford U.)
IEEE J. Quantum Electronics, Vol. QE-1, pp. 180-181, July 1965

16.106 Photomixing in a GaAs$_x$P$_{1-x}$-GaAs Heterodiode
T. B. Ramachandran, K. K. Chow, W. J. Moroney, and P. Olendzensky (Microwave Associates)
J. Appl. Phys., Vol. 36, pp. 2594–2595, August 1965 (bromine-argon laser, beats in 13–22 Gc/s region)

16.107 Determination of the Properties of Illuminated Photoconductors with Microwave Radiation
Albert P. Sheppard (U.S. Army Research Office)
Proc. IEEE, Vol. 53, p. 1151, August 1965 (9.5 Gc/s, CdS in plastic)

16.108 The Helix Loss Correction Factor of R$_{eq}$ for a Traveling-Wave Phototube
Yoichi Fujii and Hiroshi Ogawa (U. of Tokyo, Japan)
Proc. IEEE, Vol. 53, p. 1239, September 1965

16.109 Measurement of the Transient Parameters of a High-Power Laser with a Photodiode
R. V. Ambartsumyan, N. G. Basov, P. G. Eliseyev, V. S. Zuyev P. G. Kryukov, and Yu. Yu. Stoylov
Radio Eng. Electronic Physics, Vol. 9, pp. 1487–1488, September 1965 (GaAs photodiode, Q-sw laser)

16.110 Two-Step Excitation in Erbium-Doped Cadmium Fluoride
Leon Esterowitz and Jon Noonan (U.S. Army ERDL)
Appl. Phys. Letters, Vol. 7, pp. 281–283, November 15, 1965 (infrared quantum counter)

16.111 Laser-Quenching of Photoconductivity and Recombination Processes in Sensitive Photoconductors
Jose Saura and Richard H. Bube (Stanford U.)
J. Appl. Phys., Vol. 36, pp. 3660–3662, November 1965

16.112 A High Gain Silicon Photodetector
S. W. Ing., Jr., and G. C. Gerhard (GE and Xerox)
Proc. IEEE, Vol. 53, pp. 1714–1722, November 1965

16.113 Information Capacity of a Photoelectric Detector
B. E. Goodwin and L. P. Bolgiano, Jr. (U. of Delaware)
Proc. IEEE, Vol. 53, pp. 1745–1746, November 1965

16.114 Optimum Detection Thresholds in Optical Communications
T. F. Curran and M. Ross (Hallicrafters)
Proc. IEEE, Vol. 53, pp. 1770–1771, November 1965

16.115 Response Times of Ge:Cu Infrared Detectors
James T. Yardley and C. Bradley Moore (U. of California)
Appl. Phys. Letters, Vol. 7, pp. 311–312, December 1, 1965 (p-type Ge:Cu, 3.39 μ radiation, response time less than 2.2 ns)

16.116 Tracking Heterodyne Detection
W. S. Read and R. G. Turner (North American Aviation)
Appl. Optics, Vol. 4, pp. 1570–1573, December 1965 (angular alignment for mixing, gas laser experiment)

16.117 Rejection of Coherent Interference in Optical Modulation-Demodulation Experiments
J. Richard Kerr (Sylvania)
IEEE Trans. on Instrumentation and Measurement, Vol. IM-14, pp. 209–214, December 1965 (homodyne and phase sensitive detection)

16.118 A Proposal for a New Sensitivity Measure for Optical and IR Detectors
Monte Ross (McDonnel Aircraft)
Proc. IEEE, Vol. 53, pp. 2160–2161, December 1965 (noise and thermal equivalent powers, minimum discernible signal)

16.119 Saturation of Electroluminescent Image-Retaining Panels by Laser Beams
J. C. Bass (U. of Sheffield, England)
Appl. Optics, Vol. 5, pp. 169–170, January 1966 (Thorn Image Retaining Panel MK. II, paralysis due to heating effects)

16.120 Observation of Microwave Beats in a Parallel-Plate Transmission-Line Photomixer
J. C. Bass and J. A. Jones (U. of Sheffield, Sheffield, England)
Electronics Letters, Vol. 2, pp. 25–26, January 1966 (ruby laser input, 10–40 Gc/s output)

16.121 Power Loss in Propagation through a Turbulent Medium for an Optical-Heterodyne System with Angle Tracking
David M. Chase (TRG)
J. Opt. Soc. Am., Vol. 56, pp. 33–44, January 1966

16.122 Response of a Spectrometer to a Modulated Plane Wave
Casper W. Barnes (Stanford Research Institute)
J. Opt. Soc. Am., Vol. 56, pp. 53–58, January 1966

16.123 Variation of Time-of-Flight of Electrons through a Photomultiplier
James D. Rees, Jr. and M. Parker Givens (U. of Rochester)
J. Opt. Soc. Am., Vol. 56, pp. 93–95, January 1966 (uses gas laser with several axial modes, variations of up to 5 ns measured)

16.124 Application of CCl$_4$ and CCl$_2$:CCl$_2$ Ultrasonic Modulators to Infrared Optical Heterodyne Experiments
F. E. Goodwin and M. E. Pedinoff (Hughes)
Appl. Phys. Letters, Vol. 8, pp. 60–61, February 1, 1966

16.125 A Transverse Wave Phototube for Detection of Microwave-Frequency-Modulated Light
J. Richard Kerr (Sylvania)
IEEE J. Quantum Electronics, Vol. QE-2, pp. 21–29, February 1966

16.126 A Balanced Mixer for Optical Heterodyning: The ANN Detector
T. Waite and R. A. Gudmundsen (Autonetics)
Proc. IEEE, Vol. 54, pp. 297–299, February 1966 (interference fringes move at difference frequency)

16.127 A Balanced Mixer for Optical Heterodyning: The Magic T Optical Mixer
Tom Waite (Autonetics)
Proc. IEEE, Vol. 54, pp. 334–335, February 1966

16.128 Detection of Weak Microwave Modulation of Light
H. Pursey, Patricia A. Merran and B. Trevelyan (National Physical Lab., England)
Nature, Vol. 210, pp. 511–512, April 30, 1966 (5 W X-band source for KDP modulator, crossed polarizer and analyzer)

16.129 Sensitive S-Band Travelling-Wave Phototube
J. R. Mansell and J. L. Phillips (Mullard Res. Labs., England)
Electronics Letters, Vol. 2, pp. 155–156, April 1966 (S-20 photocathode, 2–4 Gc/s)

16.130 Detection of Coherent Optical Radiation
Marvin E. Lasser (Philco)
IEEE Spectrum, Vol. 3, No. 4, pp. 73–78, April 1966 (review)

16.131 Theory of Far Infrared Detection Using Nonlinear Optical Mixing
Masamoto Takatsuji (Hitachi, Japan)
Japanese J. Appl. Phys., Vol. 5, pp. 389–400, May 1966 (longer than 50 μ, laser noise dominant)

16.132 Two-Step Excitation in Trivalent Thulium
L. Esterowitz, J. Noonan, and A. Schnitzler (Fort Belvoir)
Appl. Phys. Letters, Vol. 8, pp. 271–273, June 1, 1966 (quantum counter, room temperature)

16.133 Laser Receivers; Devices, Techniques, Systems
Monte Ross (McDonnell Aircraft)
John Wiley & Sons, Inc., New York, 405 p., 1966 (noise, information theory, detectors, modulators, transmission, components, and systems)

16.134 Measurement of Low Light Intensities by Synchronous Single Photon Counting

F. T. Arecchi, E. Gatti, and A. Sona (Lab. CISE, Italy)
Rev. Sci Instr., Vol. 37, pp. 942–948, July 1966 (chopped light, synchronous detection)

16.135 Refrigeration for Photomultipliers
A. L. Broadfoot (Kitt Peak Nat'l Observatory)
Appl. Optics, Vol. 5, pp. 1259–1263, August 1966 (closed cycle, +5° to −55°C)

16.136 Accurate Method for Determining Photometric Linearity
H. E. Bennett (Michelson Lab.)
Appl. Optics, Vol. 5, pp. 1265–1270, August 1966 (rotate middle polarizer of three in series)

16.137 Infrared Quantum Counter Action in Rare Earth Doped Fluoride Lattices
M. R. Brown (Signals R & D Estab., England), and W. A. Shand (U. of Aberdeen, Scotland)
IEEE J. Quantum Electronics, Vol. QE-2, pp. 251–253, August 1966

16.138 Infrared Detection by Parametric Frequency Up-Conversion
F. M. Johnson and J. A. Duardo (Electro-Optical)
IEEE J. Quantum Electronics, Vol. QE-2, p. 296, August 1966 (abstract, IR of 0.9 to 2.2 μ, KDP and LiNbO$_3$, into visible region)

16.139 A Traveling-Wave Photomultiplier
M. B. Fisher and R. T. McKenzie (Sylvania)
IEEE J. Quantum Electronics, Vol. QE-2, pp. 322–327, August 1966 (four-stage electron multiplier, 1 to 3 GHz frequency range)

16.140 New Method of Detecting Weak Light Signals
Yoh-Han Pao, R. N. Zitter, and J. E. Griffiths (BTL)
J. Opt. Soc. Am., Vol. 56, pp. 1133–1135, August 1966 (detect larger "shot" noise which increases with light flux, 0.63-μ He-Ne laser, detect 459 cm^{-1} Raman line of CCl$_4$)

16.141 Polarization Selectivity of Optical Superheterodyne Receiving Apparatus
A. I. Filatov
Optics and Spectroscopy, Vol. 21, pp. 145–146, August 1966

16.142 Optimization of Preamplifiers for Detection of Short Pulses of Light with Photodiodes
Byron N. Edwards (Aeronutronic)
Appl. Optics, Vol. 5, pp. 1423–1425, September 1966

16.143 Closed Cycle Cryogenic Refrigerators as Integrated Cold Sources for Infrared Detectors
A. Daniels and F. K. du Pré (Philips, Netherlands)
Appl. Optics, Vol. 5, 1457–1460, September 1966 (Stirling cycle)

16.144 Avalanche Multiplication Photodiodes
L. A. D'Asaro and L. K. Anderson
Bell Lab. Record, Vol. 44, pp. 277–280, September 1966 (very little noise, combined light detector and microwave amplifier, Ge, 3 GHz modulation, 30 dB gain)

16.145 Semiconductor Lasers and Fast Detectors in the Infrared (3 to 15 Microns)
M. Rodot, C. Verie, and Y. Marfaing (CNRS, France), and J. Besson and H. Leblock (SAT, France)
IEEE J. Quantum Electronics, Vol. QE-2, pp. 586–593, September 1966

16.146 Heterodyne Detection of a Weak Light Beam
L. Mandel (U. of Rochester)
J. Opt. Soc. Am., Vol. 56, pp. 1200–1206, September 1966

16.147 Photovoltaic Effect in Pb$_x$Sn$_{1-x}$Te Diodes
I. Melngailis and A. R. Calawa (M.I.T. Lincoln Lab.)
Appl. Phys. Letters, Vol. 9, pp. 304–306, October 15, 1966 (response to 9.6 μ at 77°K and 12 μ at 12°K)

16.148 High-Speed Photodetectors
L. K. Anderson (BTL), and B. J. McMurtry (Sylvania)
Appl. Optics, Vol. 5, pp. 1573–1587, October 1966 (status report, vacuum and solid-state detectors, noise, comparison)

16.149 The Antenna Properties of Optical Heterodyne Receivers
A. E. Siegman (Stanford U.)
Appl. Optics, Vol. 5, pp. 1588–1594, October 1966 (effects of noise)

16.150 Laser Saturation of Photoconductivity and Determination of Imperfection Parameters in Sensitive Photoconductors
Richard H. Bube and Ching-Tao Ho (Stanford U.)
J. Appl. Phys., Vol. 37, pp. 4132–4138, October 1966

16.151 High-Speed Photodetectors
L. K. Anderson (BTL), and B. J. McMurtry (Sylvania)
Proc. IEEE, Vol. 54, pp. 1335–1349, October 1966 (status report, vacuum and solid-state detectors, noise, comparison)

16.152 The Antenna Properties of Optical Heterodyne Receivers
A. E. Siegman (Stanford U.)
Proc. IEEE, Vol. 54, pp. 1350–1356, October 1966 (effects of noise)

16.153 Electrooptical Effect in GaAs
V. S. Bagaev, Yu. N. Berozashvili, and L. V. Keldysh (Lebedev Phys. Inst., Acad. of Sci., USSR)
JETP Letters, Vol. 4, pp. 246–249, November 1, 1966 (change in refractive index, band edge shift)

16.154 Optimum Heterodyne Detection at 10.6 μm in Photoconductive Ge:Cu
M. C. Teich, R. J. Keyes, and R. H. Kingston (M.I.T. Lincoln Lab.)
Appl. Phys. Letters, Vol. 9, pp. 357–360, November 15, 1966 (1.3×10^{-19} W in 1-Hz bandwidth, within a factor of 10 of perfect photon counter)

16.155 Optimization of Optical Systems
Edwin B. Champagne (Wright-Patterson AFB)
Appl. Optics, Vol. 5, pp. 1843–1845, November 1966 (detection, SNR)

16.156 Coherent Homodyne Detection at 10.6 μm with an Extrinsic Photoconductor
R. A. Soref (Sperry Rand)
Electronics Letters, Vol. 2, pp. 410–411, November 1966 (Si: Al detector, system detectivity at 15 kHz of 3 x 10^{11} cm W^{-1}Hz$^{\frac{1}{2}}$)

16.157 Heterodyne Detection and Linewidth Measurement with High Power CO$_2$ Lasers
R. A. Brandewie, W. T. Haswell, III, and R. H. Harada (Autonetics)
IEEE J. Quantum Electronics, Vol. QE-2, pp. 756–757, November 1966 (Cu doped Ge mixer, near 4°K, linewidth less than 710 Hz)

16.158 Demodulation of Low-Level Broad-Band Optical Signals with Semiconductors. Part III: Experimental Study of the Photoconductive Detector
H. S. Sommers, Jr., and E. K. Gatchell (RCA)
Proc. IEEE, Vol. 54, pp. 1553–1568, November 1966 (Ge, InAs, InSb, to 5 μm, hundreds of MHz)

17. BEAM QUALITY, CONTROL, AND DEFLECTION

17.1   Far-Field Fabry–Perot Diffraction Patterns of a Neodymium in Glass Laser
        C. Martin Stickley and Rudolph A. Bradbury (AFCRL)
        Appl. Optics, Vol. 2, pp. 867–869, August 1963

17.2   Electroacoustic Deflection of a Coherent Light Beam
        A. J. Giarola and T. R. Billeter (Boeing)
        Proc. IEEE, Vol. 51, pp. 1150–1151, August 1963

17.3   Change of the Refractive Index of Ruby During Optical Pumping
        U. J. Schmidt (Ramo-Wooldridge)
        J. Appl. Phys., Vol. 35, pp. 259–260, January 1964

17.4   Electro-Optic Light Beam Deflector
        V. J. Fowler, C. F. Buhrer, and L. R. Bloom (G. T. & E. Labs.)
        Proc. IEEE, Vol. 52, pp. 193–194, February 1964

17.5   Correction to "Electro-Optic Light Beam Deflector"
        V. J. Fowler, C. F. Buhrer, and L. R. Bloom (Gen. Tel. & Electronics Lab.)
        Proc. IEEE, Vol. 52, p. 862, July 1964 (Proc. IEEE, Vol. 52, pp. 193–194; February 1964)

17.6   Thermal Effects in Laser Amplifiers and Oscillators
        A. E. Blume and K. F. Tittel (G.E.)
        Appl. Optics, Vol. 3, pp. 527–530, April 1964

17.7   Digital Light Deflection
        T. J. Nelson
        Bell Sys. Tech. J., Vol. 43, pp. 821–845, May 1964 (employs n optical modulators and n uniaxial crystals to provide 2^n beam positions, some experiments)

17.8   The Use of Wollaston Prisms for a High-Capacity Digital Light Deflector
        W. J. Tabor
        Bell Sys. Tech. J., Vol. 43, pp. 1153–1154, May 1964

17.9   The Quadrant Multiplier Phototube, a New Star-Tracker Sensor
        M. Rome, H. G. Fleck, and D. C. Hines (Electro-Mechanical Research)
        Appl. Optics, Vol. 3, pp. 691–695, June 1964 (high-precision nulling-type tracker for single image)

17.10   Non-Mechanical Scanning of Light Using Acoustic Waves
        R. Lipnick, A. Reich, and G. A. Schoen (Lockheed Electronics)
        Proc. IEEE, Vol. 52, pp. 853–854, July 1964 (Osram 100-watt lamp, ⅓ degree 100-Kc scan)

17.11   Light Beam Deflection Using the Kerr Effect in Single Crystal Prisms of $BaTiO_3$
        W. Haas, R. Johannes, and P. Cholet (Philco)
        Appl. Optics, Vol. 3, pp. 988–989, August 1964 (small deflections, temperature sensitive)

17.12   Control of Laser Radiation with Birefringent Crystals: the Microwave Circuit Viewpoint
        M. R. Wohlers and K. G. Leib (Grumman)
        J. Appl. Phys., Vol. 35, pp. 2311–2312, August 1964

17.13   The Use of Perovskite Paraelectrics in Beam Deflectors and Light Modulators
        F. S. Chen, J. E. Geusic, S. K. Kurtz, J. G. Skinner, and S. H. Wemple (BTL)
        Proc. IEEE, Vol. 52, pp. 1258–1259, October 1964 (KTN prism beam deflector, 17 volts for π shift in modulator)

17.14   A High Speed Digital Light Beam Deflector
        U. J. Schmidt (Space Tech. Labs.)
        Physics Letters, Vol. 12, pp. 205–206, October 1, 1964

17.15   Electro-Optic [$KTa_xNb_{1-x}O_3$ (KTN)] Gratings for Light Beam Modulation and Deflection
        M. G. Cohen and E. I. Gordon (BTL)
        Appl. Phys. Letters, Vol. 5, pp. 181–182, November 1, 1964 (interaction of light and acoustic waves, photoelastic effect, ridged waveguide transducer, 9.5 Gc, many resolvable spots)

17.16   A Variable Beamwidth Millimetric Wave Antenna
        M. A. Kott (Johns Hopkins U.)
        IEEE Trans. on Antennas and Propagation, Vol. AP-12, pp. 662–667, November 1964 (design technique based on and applicable to optical zoom lens, beamwidth variable by a factor of 7.7 to 1)

17.17   Comment on Light Beam Deflectors
        J. G. Skinner (BTL)
        Appl. Optics, Vol. 3, p. 1504, December 1964 (figure of merit is maximum number of resolvable spots)

17.18   On the Possibility of Using Conical Refraction Phenomena for Laser Beam Steering
        R. P. Burns (RADC)
        Appl. Optics, Vol. 3, pp. 1505–1506, December 1964 (emergent beam parallel with input, circular scan of output)

17.19   Bidirectional Electrically Switched Laser
        R. V. Pole, R. A. Myers, and J. Nuñez (IBM)
        Appl. Optics, Vol. 4, pp. 119–121, January 1965 (17.5° between laser beam axes in double rooftop ruby, alternately switched)

17.20   Laser Deflection with the Conjugate Plano-Concentric Resonator
        R. A. Myers, R. V. Pole, and J. Nuñez (IBM)
        Appl. Optics, Vol. 4, pp. 140–141, January 1965 (6° between two beams)

17.21   Nonmechanical Scanning of Light in One and Two Dimensions
        R. Lipnick, A. Reich, and G. A. Schoen (Lockheed)
        Proc. IEEE, Vol. 53, p. 321, March 1965 (acoustic scanner, 0.63μ He–Ne laser)

17.22   An Ultrasonic Light Deflection System
        A. Korpel, R. Adler, P. Desmares, and T. M. Smith (Zenith)
        IEEE J. Quantum Electronics, Vol. QE-1, pp. 60–61, April 1965 (Bragg reflection in water, 70 resolvable points, 40–45 Mc/s sound frequency)

17.23   Scanning, Coherent Light Beams—Plane-Wave Expansions and Interactions with Dispersive Systems
        Casper W. Barnes (S.R.I.)
        J. Opt. Soc. Am., Vol. 55, pp. 382–391, April 1965 (interaction of beam with diffraction grating, dispersive dielectric, prism)

17.24   Optical Beam Deflection by Pulsed Temperature Gradients in Bulk GaAs
        S. G. Liu and W. L. Walters (RCA)
        Proc. IEEE, Vol. 53, pp. 522–523, May 1965 (deflection of 0.15 degree)

17.25   Laser Deflection Modulation in a CdS Prism
        R. Kalibjian, T. Huen, C. Maninger, and J. Yee (U. of California)
        Proc. IEEE, Vol. 53, p. 539, May 1965 (ohmic contacts on prism, 2° deflection with 2 amps, index change)

17.26   Correction to "Laser Deflection Modulation in a CdS Prism"
        R. Kalibjian, T. Huen, C. Maninger, and J. Yee (Lawrence Rad. Lab., U. of California)
        Proc. IEEE, Vol. 53, p. 1225, September 1965 (Proc. IEEE, Vol. 53, p. 539, May 1965)

17.27   Device for Accurately Changing the Direction of a Light Beam through Small Angles
        J. G. Hirschberg (Princeton U.)
        Appl. Optics, Vol. 4, p. 759, June 1965 (variable-angle prism, flexible bellows between two silica windows filled with liquid)

Appl. Phys. Letters, Vol. 8, pp. 198–199, April 15, 1966 (electro-optical crystal in interferometer cavity)

17.51   Dynamic Optical Path Distortions in Laser Rods
S. D. Sims, A. Stein, and C. Roth (TRG)
Appl. Optics, Vol. 5, pp. 621–626, April 1966 (Mach-Zehnder interferometer, significant diverging effect during pump pulse)

17.52   Thermal Distortion of Diffraction-Limited Optical Elements
F. W. Quelle, Jr. (Office of Naval Research, Boston)
Appl. Optics, Vol. 5, pp. 633–637, April 1966

17.53   Dynamic Measurements of Phase Shifts in Laser Amplifiers
A. Y. Cabezas, L. G. Komai, and R. P. Treat (Hughes)
Appl. Optics, Vol. 5, pp. 647–651, April 1966 (1.5 and 2.2λ change per °C for 6.3-cm long rods of Nd glass and ruby, respectively)

17.54   Inhomogeneities in Optical Crystals Resulting from Nonplanar Solid-Melt Interface during Growth
James F. Nester (Perkin-Elmer)
J. Appl. Phys., Vol. 37, pp. 2002–2004, April 1966 ($CaWO_4:Nd^{3+}$ crystals, interferogram, radially asymmetric temperature gradient, laser beam 10 times diffraction limit)

17.55   High-Resolution Digital Light Deflector
W. Kulcke, K. Kosanke, E. Max, H. Fleisher and T. J. Harris (IBM)
Appl. Phys. Letters, Vol. 8, pp. 266–268, May 15, 1966 (16 stages)

17.56   Figure of Merit for Acousto-Optical Deflection and Modulation Devices
E. I. Gordon (B.T.L.)
IEEE J. Quantum Electronics, Vol. QE-2, pp. 104–105, May 1966

17.57   Spatial and Temporal Variation of the Optical Path Length in Flash-Pumped Laser Rods
Herbert Welling and Charles J. Bickart (U. S. Army Electronics Command)
J. Opt. Soc. Am., Vol. 56, pp. 611–618, May 1966

17.58   Electro-Optic Deflection with $BaTiO_3$ Prisms
E. P. Ippen (U. of California)
IEEE J. Quantum Electronics, Vol. QE-2, p. 152, June 1966

17.59   Electron Beam Scanlaser
R. V. Pole and R. A. Myers (IBM Watson)
IEEE J. Quantum Electronics, Vol. QE-2, pp. 182–184, July 1966 (electron beam induces birefringence in KDP at one mirror, quartz plate, Hg^+ laser emits at 6150Å)

17.60   Total Internal Reflection Light Deflector
M. A. Habegger, T. J. Harris, and J. Lipp (IBM)
Appl. Optics, Vol. 5, pp. 1403–1405, September 1966 (digital, electrooptic)

17.61   An Internally Scanned Laser
E. S. Kohn and V. J. Fowler (GT & E Labs.)
IEEE J. Quantum Electronics, Vol. QE-2, pp. 464–466, September 1966 (enlarged laser cavity, pulsed optical delay line with shear pulses, high deflection rates)

17.62   Laser Brightness Gain and Mode Control by Compensation for Thermal Distortion
C. Martin Stickley (AF Cambridge Research Lab.)
IEEE J. Quantum Electronics, Vol. QE-2, pp. 511–518, September 1966 (brightness gain of 100)

17.63   Multiple Acoustic Diffraction Techniques for Frequency Shifting of Laser Sources
C. S. Tsai (Lockheed Research Lab.), and B. A. Auld (Stanford U.)
Proc. IEEE, Vol. 54, pp. 1217–1218, September 1966

17.64   A Review of Acoustooptical Deflection and Modulation Devices
E. I. Gordon (BTL)
Appl. Optics, Vol. 5, pp. 1629–1639, October 1966 (momentum considerations, efficiency, bandwidth)

17.65   Digital Light Deflectors
W. Kulcke, K. Kosanke, and E. Max (IBM, Germany), M. A. Habegger, T. J. Harris, and H. Fleisher (IBM, New York)
Appl. Optics, Vol. 5, pp. 1657–1667, October 1966 (several types, experiments)

17.66   A Survey of Laser Beam Deflection Techniques
V. J. Fowler and J. Schlafer (GT & E Labs.)
Appl. Optics, Vol. 5, pp. 1675–1682, October 1966

17.67   Design of an Electro-Optic Polarization Switch for a High-Capacity High-Speed Digital Light Deflection System
S. K. Kurtz
Bell Sys. Tech. J., Vol. 45, pp. 1209–1246, October 1966 (KTN, ZnTe, 10^6 addresses per second, 2.6 W, temp fluctuations less than 0.01°C)

17.68   A Review of Acoustooptical Deflection and Modulation Devices
E. I. Gordon (BTL)
Proc. IEEE, Vol. 54, pp. 1391–1401, October 1966 (momentum considerations, efficiency, bandwidth)

17.69   Digital Light Deflectors
W. Kulcke, K. Kosanke, and E. Max (IBM, Germany), and M. A. Habegger, T. J. Harris, and H. Fleisher (IBM Poughkeepsie)
Proc. IEEE, Vol. 54, pp. 1419–1429, October 1966 (several types, experiments)

17.70   A Survey of Laser Beam Deflection Techniques
V. J. Fowler and J. Schlafer (GT & E Labs.)
Proc. IEEE, Vol. 54, pp. 1437–1444, October 1966

17.71   The Far Infrared Optical Properties of $LiNbO_3$
D. R. Bosomworth (RCA)
Appl. Phys. Letters, Vol. 9, pp. 330–331, November 1, 1966 (300 and 80°K, 100- to 1000-μ region, large birefringence, appreciable dispersion)

17.72   Biased Resonance Circuits for Electrooptic Digital Deflectors
C. F. Haugh (IBM Poughkeepsie)
Appl. Optics, Vol. 5, pp. 1777–1781, November 1966 (low power dissipation)

17.73   Electronic Beam Scanning of Injection Lasers
G. E. Fenner (GE)
J. Appl. Phys., Vol. 37, pp. 4991–4994, December 1966 (GaAs at 77°K, several sections, nonuniform current distribution)

18.   COHERENCE, INTERFERENCE, QUANTUM NOISE, NARROW-LINEWIDTH LASERS, AND FREQUENCY CONTROL

18.1   Splitting of Fabry–Perot Rings by Microwave Modulation of Light
I. P. Kaminow (BTL)
Appl. Phys. Letters, Vol. 2, pp. 41–42, January 15, 1963

18.2   The Quantum Theory of Optical Coherence

Roy J. Glauber (Harvard U.)
Phys. Rev., Vol. 130, pp. 2529–2539, June 15, 1963

18.3   Interference Rings in Ruby Maser Beams
B. P. Stoicheff and A. Szabo (National Res. Council, Can.)
Appl. Optics, Vol. 2, pp. 811–815, August 1963

Academy of Sciences, U.S.S.R.)
Soviet Physics JETP, Vol. 21, pp. 509–510, August 1965 (ring laser, 3.39-μ He–Ne laser, amplitude correlation data)

18.66   Proposed Frequency Stabilization of the FM Laser
S. E. Harris (Stanford U.), M. Kenneth Oshman, B. J. McMurtry, and E. O. Ammann (Sylvania)
Appl. Phys. Letters, Vol. 7, pp. 185–187, October 1, 1965 (detect ratio of first- and second-beat amplitudes)

18.67   Amplitude and Phase of an Anharmonic Oscillator
M. W. Muller (Varian)
IEEE J. Quantum Electronics, Vol. QE-1, p. 321, October 1965

18.68   A Two-Channel Laser Frequency Control System
A. D. White (BTL)
IEEE J. Quantum Electronics, Vol. QE-1, pp. 322–323, October 1965 (accuracy of ± 1 Mc/s)

18.69   Coherent Division of Quanta
L. I. Gudzenko and G. M. Guro (Lebedev Physics Inst., Academy of Sciences, U.S.S.R.)
Soviet Physics JETP, Vol. 21, pp. 756–760, October 1965

18.70   Oscillations in a System Comprising Two-Level Molecules and a Radiation Field
V. F. Chel'tsov
Soviet Physics JETP, Vol. 21, pp. 761–764, October 1965

18.71   Stimulated Raman Scattering in the Anti-Stokes Region
V. N. Lugovoï (Lebedev Physics Inst., Academy of Sciences, U.S.S.R.)
Soviet Physics JETP, Vol. 21, pp. 811–813, October 1965 (generation mechanism and properties)

18.72   Frequency Selective Coupling to the FM Laser
S. E. Harris (Stanford U.) and B. J. McMurtry (Sylvania)
Appl. Phys. Letters, Vol. 7, pp. 265–267, November 15, 1965

18.73   Measurement of the 633-nm Wavelength of Helium-Neon Lasers
K. D. Mielenz, K. F. Nefflen, K. E. Gillilland, and R. B. Stephens (NBS) and R. B. Zipin (Sheffield Corp.)
Appl. Phys. Letters, Vol. 7, pp. 277–279, November 15, 1965 (632.99145 and 632.99147 nm in vacuum)

18.74   Stimulated Emission of Radiation in a Single Mode
F. W. Cummings (U. of California)
Phys. Rev., Vol. 140, pp. A1051–A1056, November 15, 1965

18.75   A Study of Partial Coherence and Its Application to the Collimation of Pulsed Multimode Laser Radiation
W. S. C. Chang and N. R. Kilcoyne (Ohio State U.)
Appl. Optics, Vol. 4, pp. 1404–1411, November 1965 (long-time-averaged measurements, beam divergence, output energy)

18.76   Mach-Zehnder Interferometer with Adjustable Compensation
A. Stein and T. Shultz (Control Data Corp.)
Appl. Optics, Vol. 4, pp. 1510–1511, November 1965 (dual prism adjustable compensator)

18.77   Stabilized, Single-Frequency Output from a Long Laser Cavity
P. W. Smith (BTL)
IEEE J. Quantum Electronics, Vol. QE-1, pp. 343–348, November 1965 (external and internal mode selectors, 1.5-m-long laser cavity, 15 mW, 0.63 μ from He–Ne laser)

18.78   Frequency Stabilization of Gas Lasers
A. D. White (BTL)
IEEE J. Quantum Electronics, Vol. QE-1, pp. 349–357, November 1965 (survey, causes of instability, system considerations)

18.79   Investigation of the Correlation of Phase Fluctuations of an Optical Field by a Photoelectric Predetection Interferometer

Emmanuel J. Vourgourakis (Hughes)
J. Opt. Soc. Am., Vol. 55, pp. 1455–1456, November 1965 (two-element interferometer)

18.80   Coherence Function for Waves in Random Media
David M. Chase (TRG)
J. Opt. Soc. Am., Vol. 55, pp. 1559–1560, November 1965

18.81   On the Statistics of Kastler's Multimode Radiation Ensemble
G. J. Troup (Monash U., Australia)
Proc. IEEE, Vol. 53, p. 1732, November 1965

18.82   Simple Improvement of Amplitude Stability in Helium-Neon Gas-Lasers
Viktor Met (Electro Optics Associates)
Proc. IEEE, Vol. 53, pp. 1780–1781, November 1965 (oscillating laser cavity mirror)

18.83   Doppler Shift of Ruby Laser Light by Means of a Kerr Cell Traveling Wave Line
F. P. Küpper and E. Fünfer (Institut für Plasmaphysik, Germany)
Physics Letters, Vol. 19, pp. 486–487, December 1, 1965 (60-cm-long cell, theoretical shift of 2.3×10^{10} c/s)

18.84   Measurement of the Statistical Distribution of Gaussian and Laser Sources
F. T. Arecchi (Laboratori Centro Informazioni Studi Esperienze and Istituto di Fisica dell'Università, Milan, Italy)
Phys. Rev. Letters, Vol. 15, pp. 912–916, December 13, 1965

18.85   Photoelectron Statistics Produced by a Laser Operating below the Threshold of Oscillation
Charles Freed (Lincoln Lab., M.I.T.) and Hermann A. Haus (M.I.T.)
Phys. Rev., Letters, Vol. 15, pp. 943–946, December 20, 1965

18.86   Automatic Frequency Control of a Laser Local Oscillator for Heterodyne Detection of Microwave-Modulated Light
Russell Targ and W. D. Bush (Sylvania)
Appl. Optics, Vol. 4, pp. 1523–1527, December 1965 (maintains 2.5 Mc/s offset frequency, 0.63-μ He–Ne laser)

18.87   Noise Properties of Pulsed Ruby Laser Amplifiers
I. J. D'Haenens and C. R. Giuliano (Hughes)
IEEE J. Quantum Electronics, Vol. QE-1, pp. 393–397, December 1965 (spontaneous fluorescence has broader beam and wider bandwidth than laser beam)

18.88   A Technique for Determining the Photoelectric Counting Distributions for Some Ideal Systems
G. J. Troup (Monash U., Victoria, Australia)
IEEE J. Quantum Electronics, Vol. QE-1, p. 398, December 1965

18.89   Photon-Counting Statistics
F. A. Johnson, T. P. McLean, and E. R. Pike (Royal Radar Establishment, England)
Proc. Physics of Quantum Electronics Conf., McGraw-Hill Book Co., New York, pp 706–714, 1966

18.90   Amplitude Noise in Gas Lasers Below and Above the Threshold of Oscillation
Charles Freed (Lincoln Lab., M.I.T.) and Hermann A. Haus (M.I.T.)
Proc. Physics of Quantum Electronics Conf., McGraw-Hill Book Co., New York, pp. 715–724, 1966

18.91   Quantum Noise V: Phase Noise in a Homogeneously Broadened Maser
Melvin Lax (B.T.L.)
Proc. Physics of Quantum Electronics Conf., McGraw-Hill Book Co., New York, pp. 735–747, 1966

Appl. Optics, Vol. 5, pp. 823–826, May 1966 (two spherical mirrors, wavelength-sensitive discriminator)

18.120 A Method of Producing an Unmodulated Laser Output at a Controlled Frequency
D. C. Wilson and W. R. C. Rowley (Nat'l Phys. Lab., England)
J. Sci Instrum., Vol. 43, pp. 314–316, May 1966 (control 0.63-μ He–Ne laser frequency to within 10^6 Hz for long periods, second coupled and modulated laser as reference)

18.121 On the Theory of a Two-Photon Quantum Amplifier
P. Lambropoulos (Bendix)
Physics Letters, Vol. 21, pp. 418–419, June 1, 1966

18.122 Laser Photon Counting Distributions Near Threshold
Archibald W. Smith and J. A. Armstrong (IBM)
Phys. Rev. Letters, Vol. 16, pp. 1169–1172, June 20, 1966 (0.63-μ He–Ne laser, twice threshold)

18.123 Stimulated Effects in N_2 and CH_4 Gases
T. A. Wiggins, R. V. Wick, and D. H. Rank (Penn. State U.)
Appl. Optics, Vol. 5, pp. 1069–1072, June 1966 (Brillouin scattering, Q-sw ruby laser)

18.124 Absolute Frequency Stabilization of a Gas Laser Using Optical Resonance Amplification Techniques
H. P. Brändli, R. Dänkliker and K. P. Meyer (U. of Berne, Switzerland)
IEEE J. Quantum Electronics, Vol. QE-2, pp. 152–153, June 1966 (proposal, Zeeman splitting, He–Ne laser)

18.125 Classical Behavior of Systems of Quantum Oscillators
R. J. Glauber (Harvard U.)
Physics Letters, Vol. 21, pp. 650–652, July 1, 1966

18.126 Quantum Theory of a Laser Model
Charles R. Willis (Boston U.)
Phys. Rev., Vol. 147, pp. 406–414, July 8, 1966

18.127 Dynamic Laser Wavelength Selection
M. A. Habegger, T. J. Harris, and E. Max (IBM)
IBM J. Res. & Dev., Vol. 10, pp. 346–350, July 1966 (electrooptic Q-spoiler, five wavelengths of argon ion laser)

18.128 Gas Discharge Modulation Noise in He–Ne Lasers
L. J. Prescott and A. van der Ziel (U. of Minnesota)
IEEE J. Quantum Electronics, Vol. QE-2, pp. 173–177, July 1966 (0.63 μ, strong correlation between laser light noise and discharge current noise)

18.129 Frequency Stabilization of a Gas Laser
T. G. Polanyi and M. L. Skolnick (Laser, Inc.), and I. Tobias (Rutgers U.)
IEEE J. Quantum Electronics, Vol. QE-2, pp. 178–179, July 1966 (0.63 μ laser, axial magnetic field, circular polarization, within 50 kHz of line center)

18.130 Properties of Partially Coherent Light
J. G. Meadors and W. S. C. Chang (Ohio State U.)
J. Opt. Soc. Am., Vol. 56, pp. 865–868, July 1966

18.131 Measurement of the Time Evolution of a Radiation Field by Joint Photocount Distributions
F. T. Arecchi, A. Berné, and A. Sona (Lab. CISE, Italy)
Phys. Rev. Letters, Vol. 17, pp. 260–263, August 1, 1966

18.132 Photoelectron Statistics Produced by a Laser Operating Below and Above the Threshold of Oscillation
C. Freed and H. A. Haus (M.I.T.)
IEEE J. Quantum Electronics, Vol. QE-2, pp. 190–195, August 1966

18.133 Intensity Fluctuations in the Output of Laser Oscillators
D. E. McCumber (BTL)
IEEE J. Quantum Electronics, Vol. QE-2, pp. 219–221, August 1966

18.134 Build-Up of Laser Oscillations from Quantum Noise
M. Scully, W. E. Lamb, Jr., and M. Sargent, III (Yale U.)
IEEE J. Quantum Electronics, Vol. QE-2, p. 294, August 1966 (abstract)

18.135 Large Photoelastic Anisotropy of Sapphire
D. E. Caddes and C. D. W. Wilkinson (Stanford U.)
IEEE J. Quantum Electronics, Vol. QE-2, pp. 330–331, August 1966 (for longitudinal acoustic waves along c-axis, light frequency shifted by 730 MHz)

18.136 Effect of Plasma Fluctuations on Gas Laser Noise
Uichi Kubo, Kazuo Kawabe, and Yoshio Inuishi (Osaka U., Japan)
Japanese J. Appl. Phys., Vol. 5, p. 731, August 1966 (He–Ne laser, dc discharge current fluctuations, laser output fluctuations)

18.137 Theory of Interferometric Analysis of Laser Phase Noise
J. A. Armstrong (IBM)
J. Opt. Soc. Am., Vol. 56, pp. 1024–1031, August 1966 (discusses experiments on random phase fluctuation process)

18.138 Statistics of Laser Radiation
S. L. Boersma (Delft, Holland)
Proc. IEEE, Vol. 54, p. 1070, August 1966

18.139 A Proposed Method of Shifting the Frequency of Light Waves
C. Yeh (U. of Southern California)
Appl. Phys. Letters, Vol. 9, pp. 184–185, September 1, 1966 (moving mirror)

18.140 Measurement of Photon Time-of-Arrival Distribution in Partially Coherent Light
D. B. Scarl (Cornell U.)
Phys. Rev. Letters, Vol. 17, pp. 663–666, September 19, 1966

18.141 Excess Photon Noise in Multimode Lasers
H. Hodara (Tetra Tech.), and N. George (California Inst. of Tech.)
IEEE J. Quantum Electronics, Vol. QE-2, pp. 337–340, September 1966 (phase locked and uncoupled modes, up to 20 dB excess noise)

18.142 Photocount Distributions and Field Statistics
F. T. Arecchi, A. Berné, and A. Sona (Lab. CISE, Italy), and P. Burlamacchi (Centro Microonde, Italy)
IEEE J. Quantum Electronics, Vol. QE-2, pp. 341–350, September 1966 (nonlinear and linear counting distributions, joint photocount distributions)

18.143 The Measurement of the Frequency Fluctuations of a Laser Field
Benedetto Daino (Stanford U.)
IEEE J. Quantum Electronics, Vol. QE-2, pp. 351–354, September 1966 (proposed double heterodyning experiment)

18.144 Coherence Effects in Multiphoton Absorption Processes
John G. Meadors (Ohio U.)
IEEE J. Quantum Electronics, Vol. QE-2, pp. 638–644, September 1966 (to determine coherence properties of light beams)

18.145 On the Stabilization of a High-Power Single-Frequency Laser
P. W. Smith (BTL)
IEEE J. Quantum Electronics, Vol. QE-2, pp. 666–668, September 1966 (0.63-μ He–Ne laser, 50 mW output, within ±1 MHz, longitudinal mode selector, feedback control)

19. RAMAN AND BRILLOUIN SCATTERING

Appl. Phys. Letters, Vol. 5, pp. 58–60, August 1, 1964 (cylindrical lens, nitrobenzene, Stokes shift Fabry-Perot)

19.15 Stimulated Raman Scattering and Induced Optical Absorption in Liquids
R. G. Brewer (IBM, San Jose, California)
Physics Letters, Vol. 11, pp. 294–295, August 15, 1964 (power losses primarily into Stokes shift)

19.16 Stimulated Brillouin Scattering in Liquids
E. Garmire and C. H. Townes (MIT)
Appl. Phys. Letters, Vol. 5, pp. 84–86, August 15, 1964 (many liquids, lowest threshold with CS_2)

19.17 Raman Maser Study of Optical Difference Frequency Production
L. W. Davis, S. L. McCall, and A. P. Rodgers (Philco WDL)
J. Appl. Phys., Vol. 35, pp. 2289–2290, August 1964 (beat frequency in ADP between 2nd H of ruby and Raman shift from benzene)

19.18 Raman Laser Action in Mixed Liquids
J. A. Calviello and Z. H. Heller (AIL)
Appl. Phys. Letters, Vol. 5, pp. 112–113, September 1, 1964 (benzene-toluene, benzene-fluorobenzene, and benzene-acetone mixtures, independent shifts)

19.19 Stimulated Brillouin Scattering in Liquids
R. G. Brewer and K. E. Rieckhoff (IBM)
Phys. Rev. Letters, Vol. 13, pp. 334–336, September 14, 1964 (Q-switched ruby, water and benzene, breakdown and cavitation in water)

19.20 The Ruby Laser as a Brillouin Light Amplifier
R. G. Brewer (IBM)
Appl. Phys. Letters, Vol. 5, pp. 127–128, October 1, 1964 (Stokes and anti-Stokes shift of Brillouin scattered light from CCl_4, Q-switched ruby laser, focused beam, scattered light recollimated by lens, amplified twice by ruby and then detected)

19.21 The Raman Spectrum of Gallium Phosphide
M. V. Hobden and J. P. Russell (Royal Radar Est., England)
Physics Letters, Vol. 13, pp. 39–41, November 1, 1964 (used 0.63-μ gas laser)

19.22 Inverse Raman Spectra: Induced Absorption at Optical Frequencies
W. J. Jones and B. P. Stoicheff (Natl. Res. Council, Canada)
Phys. Rev. Letters, Vol. 13, pp. 657–659, November 30, 1964

19.23 Characteristics of an Ideal Raman Oscillator-Amplifier
W. H. Culver and E. J. Seppi (I. D. A.)
J. Appl. Phys., Vol. 35, pp. 3421-3422, November 1964 (propose Raman resonant cavity)

19.24 Multimode Effects in Stimulated Raman Emission
N. Bloembergen and Y. R. Shen (U. of California)
Phys. Rev. Letters, Vol. 13, pp. 720–724, December 14, 1964 (complicating factor)

19.25 Stimulated Brillouin Scattering in the Off-Axis Resonator
H. Takuma and D. A. Jennings (NBS)
Appl. Phys. Letters, Vol. 5, pp. 239–241, December 15, 1964 (effect using CS_2, tunable wavelength depends on angle)

19.26 Optical Heterodyne Detection of the Forward-Stimulated Brillouin Scattering
D. A. Jennings and H. Takuma (NBS)
Appl. Phys. Letters, Vol. 5, pp. 241–242, December 15, 1964 (123 Mc/s beat frequency)

19.27 Angular Dependence of the Raman Scattering from Benzene Excited by the He–Ne CW Laser
T. C. Damen, R. C. C. Leite, and S. P. S. Porto (BTL)
Phys. Rev. Letters, Vol. 14, pp. 9–11, January 4, 1965 (Raman cross section about 5×10^{-29} cm²)

19.28 Vibrational Interaction in Mixed Liquids During Stimulated Raman Action
W. Kaiser and M. Maier (Technische Hochschule Munchen, Germany) and J. A. Giordmaine (BTL)
Appl. Phys. Letters, Vol. 6, pp. 25–26, January 15, 1965

19.29 Giant Pulse Lasers and Stimulated Raman Scattering
R. W. Hellwarth, F. J. McClung, W. G. Wagner, and D. Weiner (Hughes)
Z. angewandte Mathematik und Physik, Vol. 16, No. 1, pp. 27–32, January 25, 1965

19.30 Laser-Stimulated Raman Effect in Alkali Halides
E. L. Cook (Auburn U.)
Bull. Am. Phys. Soc., Series II, Vol. 10, No. 2, p. 257–J10, February 1965 (theoretical)

19.31 Characteristics of a Raman Laser Excited by an Ordinary Ruby Laser
H. Takuma and D. A. Jennings (NBS)
Proc. IEEE, Vol. 53, pp. 146–149, February 1965 (normal pulse mode, benzene, 9.5 kw threshold, about 5×10^{-29} cm² cross section, high efficiency)

19.32 Theory of Stimulated Brillouin and Raman Scattering
Y. R. Shen and N. Bloembergen (Harvard U.)
Phys. Rev., Vol. 137, pp. A1787–A1805, March 15, 1965

19.33 Multiple Quantum Transition and Raman Action in Three-Level Systems
J. S. Margolis and G. Birnbaum (North American Aviation)
J. Appl. Phys., Vol. 36, pp. 726–731, Part 1, March 1965 (appropriate from microwave to optical frequencies)

19.34 Stimulated Brillouin Scatterings in Anisotropic Media and Observation of Phonons
H. Hsu and W. Kavage (Ohio State U.)
Physics Letters, Vol. 15, pp. 207–208, April 1, 1965 (three possible types of scattering, 75 Mc/s phonons in quartz generated by ruby laser)

19.35 Generation of Anti-Stokes Radiation in the Higher Order Coherent Raman Processes
C. L. Tang and T. F. Deutsch (Raytheon)
Phys. Rev., Vol. 138, p. A1–A8, April 5, 1965 (experiments, volume amplification effect observed)

19.36 Rayleigh and Brillouin Scattering in Benzene: Depolarization Factors
H. Z. Cummins and R. W. Gammon (Johns Hopkins U.)
Appl. Phys. Letters, Vol. 6, pp. 171–173, April 15, 1965

19.37 Intensive Stimulated Brillouin Scattering in a Parallel Laser Beam
E. Burlefinger and H. Puell (Technischen Hochschule München, Germany)
Physics Letters, Vol. 15, pp. 313–314, April 15, 1965 (cryptocyanine inside ruby laser cavity)

19.38 Multimode Effects in the Gain of Raman Amplifiers and Oscillators I. Oscillators
P. Lallemand and N. Bloembergen (Harvard U.)
Appl. Phys. Letters, Vol. 6, pp. 210–212, May 15, 1965

19.39 Multimode Effects in the Gain of Raman Amplifiers and Oscillators II. Amplifiers
P. Lallemand and N. Bloembergen (Harvard U.)
Appl. Phys. Letters, Vol. 6, pp. 212–213, May 15, 1965

19.40 Amplification of Coherent Radiation Using Stimulated Raman Scattering
B. A. Akanaev, S. A. Akhmanov, and R. V. Khokhlov (Moscow State U., U.S.S.R.)
JETP Letters, Vol. 1, pp. 104–107, May 15, 1965

19.41 Stimulated Raman Emission Using a Nd^{3+}/Glass Laser
M. D. Martin and E. L. Thomas (Ministry of Aviation, Signals Res. and Dev. Estab., England)
Physics Letters, Vol. 16, p. 132, May 15, 1965 (benzene within Q-sw laser cavity)

19.42 Generation of Stokes and Anti-Stokes Radiation in Raman Media
H. A. Haus, P. L. Kelley, and H. J. Zeiger (MIT)
Phys. Rev., Vol. 138, pp. A960–A971, May 17, 1965 (3-dimensional analysis, cylindrical laser beam)

19.43 Raman Scattering in Silicon
J. P. Russell (Royal Radar Est., England)

Appl. Phys. Letters, Vol. 6, pp. 223–224, June 1, 1965 (opaque silicon, 523 cm⁻¹ shift, focused 0.63-μ HeNe laser)

19.44 Theory of Stimulated Raman Effect. II
Y. R. Shen (U. of California)
Phys. Rev., Vol. 138, pp. A1741–A1746, June 14, 1965 (extension of coupled-wave approach)

19.45 A Tandem Spectrometer to Detect Laser-Excited Raman Radiation
D. Landon (Spex Inc.) and S. P. S. Porto (BTL)
Appl. Optics, Vol. 4, pp. 762–763, June 1965 (measures weak Raman lines in the presence of strong Rayleigh radiation)

19.46 Stimulated Brillouin Scattering as a Parametric Interaction
D. A. Sealer and H. Hsu (Ohio State U.)
IEEE J. Quantum Electronics, Vol. QE-1, pp. 116–124, June 1965 (interaction between optical and hypersonic waves, Q-sw ruby laser pump, K-band waveguide output)

19.47 The Angular Distribution of Stimulated Raman Emission in Liquids
E. Garmire (M.I.T.)
Physics Letters, Vol. 17, pp. 251–252, July 15, 1965 (anomalous radiation pattern)

19.48 Investigations of Induced Raman Scattering in Mixtures
V. A. Zubov, M. M. Sushchinskiĭ, and I. K. Shuvalov (Lebedev Physics Institute, Academy of Sciences, U.S.S.R.)
Soviet Physics JETP, Vol. 21, pp. 249–250, July 1965 (CS_2 and C_6H_6)

19.49 Precise Measurements of Wave Vectors for Stimulated Raman Emission
V. G. Cooper and A. D. May (U. of Toronto, Canada)
Appl. Phys. Letters, Vol. 7, pp. 74–76, August 1, 1965 (interferometry, in 3 liquids, near 7000 Å)

19.50 Three-Photon Molecular Scattering of Light
S. A. Akhmanov and D. N. Klyshko (Moscow State U., U.S.S.R.)
JETP Letters, Vol. 2, pp. 108–111, August 15, 1965 (estimated cross sections of Rayleigh and Raman scattering)

19.51 Double-Photon Excitation of Excimer Fluorescence of Pyrene Solutions
J. B. Birks and H. G. Seifert (U. of Manchester, England)
Physics Letters, Vol. 18, pp. 127–128, August 15, 1965

19.52 Intensity Correlations in Raman Scattering
Alexander L. Fetter (U. of California)
Phys. Rev., Vol. 139, pp. A1616–A1623, August 30, 1965 (scattering by crystals, lifetime and spatial coherence of vibrational states, about 50 μs using gas laser)

19.53 Comparison of Observed and Predicted Stimulated Raman Scattering Conversion Efficiencies
D. Weiner, S. E. Schwarz, and F. J. McClung (Hughes)
J. Appl. Phys., Vol. 36, pp. 2395–2399, August 1965 (in nitrobenzene, more efficient than prediction)

19.54 Brillouin Spectra of Liquids Using He–Ne Laser
D. H. Rank, Edward M. Kiess, Uwe Fink, and T. A. Wiggins (Pennsylvania State U.)
J. Opt. Soc. Am., Vol. 55, pp. 925–927, August 1965 (sonic velocity and Brillouin component measurements in six liquids)

19.55 He–Ne Laser as a Light Source for High-Resolution Raman Spectroscopy
Alfons Weber (Fordham U.) and S. P. S. Porto (BTL)
J. Opt. Soc. Am., Vol. 55, pp. 1033–1034, August 1965 (Raman spectrum of methylacetylene)

19.56 Stimulated Brillouin Scattering in Liquids
T. A. Wiggins, R. V. Wick, and D. H. Rank (Pennsylvania State U.), and A. H. Guenther (Kirtland AFB)

Appl. Optics, Vol. 4, pp. 1203–1205, September 1965 (two back-scattered and reamplified pulses)

19.57 Optical Mixing in Raman-Active Medium
Kanji Kubota (Osaka U., Japan)
J. Phys. Soc. Japan, Vol. 20, pp. 1738–1739, September 1965 (third-order electric polarization in benzene, Q-sw ruby)

19.58 Analysis of Stimulated Raman Scattering in Waveguides
William G. Wagner and Shaul Yatsiv (Hughes)
IEEE J. Quantum Electronics, Vol. QE-1, pp. 287–294, October 1965 (forward-backward ratio depends on laser pulse)

19.59 Interaction of Light and Microwave Sound
C. F. Quate, C. D. W. Wilkinson, and D. K. Winslow (Stanford U.)
Proc. IEEE, Vol. 53, pp. 1604–1623, October 1965 (comprehensive, parametric equations, diffraction patterns, experiments)

19.60 Brillouin Scattering in Liquids
G. Benedek and T. Greytak (M.I.T.)
Proc. IEEE, Vol. 53, pp. 1623–1629, October 1965 (velocity and decay rate, water and toluene)

19.61 The Threshold and Intensity of Stimulated Raman Radiation Lines in Liquids
S. A. Akhmanov, A. I. Kovrigin, N. K. Kulakova, A. K. Romanyuk, M. M. Strukov and R. V. Khokhlov (Moscow State U., U.S.S.R.)
Soviet Physics JETP, Vol. 21, pp. 801–803, October 1965 (benzene and cyclohexane, 2nd H of Q-sw Nd glass, lower threshold than ruby)

19.62 Stimulated Brillouin and Raman Scattering in Gases
E. E. Hagenlocker and W. G. Rado (Ford)
Appl. Phys. Letters, Vol. 7, pp. 236–238, November 1, 1965 (nitrogen and hydrogen)

19.63 Stimulated Raman Scattering and Parametric Processes
V. T. Platonenko and R. V. Khokhlov (Moscow State U., U.S.S.R.)
JETP Letters, Vol. 2, pp. 269–270, November 1, 1965 (pump by stimulated Raman scattering, prediction)

19.64 Growth of Optical Plane Waves in Stimulated Brillouin Scattering
Richard G. Brewer (IBM)
Phys. Rev., Vol. 140, pp. A800–A805, November 1, 1965 (liquid cell length variable up to 92 cm, acoustic bursts, material fracture, stream of bubbles)

19.65 Time-Resolved Interferometry in Stimulated Brillouin Scattering
Concetto R. Giuliano (Hughes)
Appl. Phys. Letters, Vol. 7, pp. 279–281, November 15, 1965 (successive Stokes orders occur in an iterative process)

19.66 Gain in a Diffusely Pumped Raman Amplifier
D. P. Bortfeld and W. R. Sooy (Hughes)
Appl. Phys. Letters, Vol. 7, pp. 283–285, November 15, 1965 (about 8 percent gain)

19.67 Induced Mandel'shtam-Brillouin Scattering in Gases
D. I. Mash, V. V. Morozov, V. S. Starunov, and I. L. Fabelinskii (Lebedev Physics Institute, Academy of Sciences, U.S.S.R.)
JETP Letters, Vol. 2, pp. 349–351, December 15, 1965

19.68 Beam Deterioration and Stimulated Raman Effect
Y. R. Shen and Y. J. Shaham (U. of California)
Phys. Rev. Letters, Vol. 15, pp. 1008–1010, December 27, 1965 (filaments from homogeneous beam to explain large effect)

19.69 Self-Focusing of Laser Beams and Stimulated Raman Gain in Liquids

P. Lallemand and N. Bloembergen (Harvard U.)
Phys. Rev. Letters, Vol. 15, pp. 1010–1012, December 27, 1965
(self-focusing largest in liquids with large quadratic Ker effect, nitrobenzene experiments)

19.70 Basic Equations and Conservation Theorems for the Electro-striction Phonon Maser
H. A. Haus and Paul Penfield, Jr. (M.I.T.)
J. Appl. Phys., Vol. 36, pp. 3735–3739, December 1965

19.71 Nonlinear Interactions of Radiation in Plasmas
Gordon Baym and R. W. Hellwarth (U. of Illinois)
Proc. Physics of Quantum Electronics Conf., McGraw-Hill Book Co., New York, pp. 105–110, 1966

19.72 Coupling of Light with Phonons, Magnons, and Plasmons
N. Bloembergen and Y. R. Shen (U. of California)
Proc. Physics of Quantum Electronics Conf., McGraw-Hill Book Co., New York, pp. 119–128, 1966

19.73 Light Waves with Exponential Gain
N. Bloembergen (U. of California), and P. Lallemand (Harvard U.)
Proc. Physics of Quantum Electronics Conf., McGraw-Hill Book Co., New York, pp. 137–154, 1966

19.74 Anomalies in Stimulated Raman Scattering Conversion Efficiencies
F. J. McClung, W. G. Wagner, and D. Weiner (Hughes)
Proc. Physics of Quantum Electronics Conf., McGraw-Hill Book Co., New York, pp. 155–158, 1966

19.75 Effect of Laser Pump Modulation on Stokes Radiation in Stimulated Raman Scattering
William G. Wagner, Shaul Yatsiv, and Robert W. Hellwarth (Hughes)
Proc. Physics of Quantum Electronics Conf., McGraw-Hill Book Co., New York, pp. 159–166, 1966

19.76 Stimulated Raman Emission in Liquids
E. Garmire (M.I.T.)
Proc. Physics of Quantum Electronics Conf., McGraw-Hill Book Co., New York, pp. 167–179, 1966

19.77 Forward Emission of Raman Radiation in Various Liquids
Georges Bret and Guy Mayer (C.S.F., France)
Proc. Physics of Quantum Electronics Conf., McGraw-Hill Book Co., New York, pp. 180–191, 1966

19.78 Induced Absorption Spectra at Optical Frequencies
A. K. MacQuillan and B. P. Stoicheff (U. of Toronto, Canada)
Proc. Physics of Quantum Electronics Conf., McGraw-Hill Book Co., New York, pp. 192–199, 1966

19.79 Interference Between Stimulated Brillouin and Raman Scattering
Toshio Ito and Hiroshi Takuma (U. of Tokyo, Japan)
Proc. Physics of Quantum Electronics Conf., McGraw-Hill Book Co., New York, pp. 200–206, 1966

19.80 Stimulated Brillouin Scattering and Stimulated Phonon Generation in Anisotropic Media
H. Hsu and W. Kavage (Ohio State U.)
Proc. Physics of Quantum Electronics Conf., McGraw-Hill Book Co., New York, pp. 207–215, 1966

19.81 Multiple Stimulated Brillouin Scattering
Richard G. Brewer and Donald C. Shapero (IBM)
Proc. Physics of Quantum Electronics Conf., McGraw-Hill Book Co., New York, pp. 216–222, 1966

19.82 Multiple Stimulated Brillouin Scattering in Solids
P. E. Tannenwald (Lincoln Lab., M.I.T.)
Proc. Physics of Quantum Electronics Conf., McGraw-Hill Book Co., New York, pp. 223–230, 1966

19.83 Brillouin Scattering and the Dispersion of Hypersonic Waves
R. Y. Chiao and P. A. Fleury (M.I.T.)
Proc. Physics of Quantum Electronics Conf., McGraw-Hill Book Co., New York, pp. 241–252, 1966

19.84 Coherent Raman Effect Produced by Second-Harmonic Frequency of Ruby Laser Light in Benzene
Shogo Yoshikawa and Yoshiyasu Matsumura (Nippon Electric, Japan) and Humio Inaba (Tohoku U., Sendai, Japan)
Appl. Phys. Letters, Vol. 8, pp. 27–28, January 1, 1966 (Stokes and anti-Stokes near 3472 Å)

19.85 Concerning Induced Mandel'shtam-Brillouin Scattering
A. A. Chaban (Acoustics Institute, Moscow, U.S.S.R.)
JETP Letters, Vol. 3, pp. 45–47, January 15, 1966

19.86 Complex Intensity-Dependent Index of Refraction, Frequency Broadening of Stimulated Raman Lines, and Stimulated Rayleigh Scattering
N. Bloembergen and P. Lallemand (Harvard U.)
Phys. Rev. Letters, Vol. 16, pp. 81–84, January 17, 1966 (several liquids)

19.87 Stimulated Raman Effect in Some Tetrahedral Molecules
D. H. Rank, R. V. Wick, and T. A. Wiggins (Pennsylvania State U.)
Appl. Optics, Vol. 5, pp. 131–133, January 1966 ($SnCl_4$, methane, Q-sw ruby laser, also Brillouin effect)

19.88 Selection of Raman Laser Materials
Gisela Eckhardt (Hughes)
IEEE J. Quantum Electronics, Vol. QE-2, pp. 1–8, January 1966

19.89 Angular Distribution of Stimulated Raman Radiation
Koichi Shimoda (U. of Tokyo, Japan)
Japanese J. Appl. Phys., Vol. 5, pp. 86–92, January 1966

19.90 Stimulated Raman Emission Frequencies in 21 Organic Liquids
Joseph J. Barrett and Marvin C. Tobin (Perkin-Elmer)
J. Opt. Soc. Am., Vol. 56, pp. 129–130, January 1966 (used Q-sw ruby)

19.91 Theory of the Stimulated Raman Effect in Plasmas
G. G. Comisar (Aerospace)
Phys. Rev., Vol. 141, pp. 200–203, January 1966 (weak effects predicted, technical limitations)

19.92 Optical Nonlinearities of a Plasma
N. Bloembergen and Y. R. Shen (U. of California)
Phys. Rev., Vol. 141, pp. 298–305, January 1966 (predicts weak stimulated Raman and second harmonic generation effects)

19.93 Tunability of the Raman Laser
H. E. Puthoff, R. H. Pantell, and B. G. Huth (Stanford U.)
J. Appl. Phys., Vol. 37, pp. 860–864, February 1966

19.94 Brillouin Spectra of Viscous Liquids
D. H. Rank, Edward M. Kiess, and Uwe Fink (Pennsylvania State U.)
J. Opt. Soc. Am., Vol. 56, pp. 163–166, February 1966 (glycerine, ethylene glycol and normal octyl alcohol)

19.95 Stimulated Brillouin Effect in High-Pressure Gases
D. H. Rank, T. A. Wiggins, R. V. Wick, and D. P. Eastman (Pennsylvania State U.), and A. H. Guenther (Air Force Weapons Lab.)
J. Opt. Soc. Am., Vol. 56, pp. 174–176, February 1966 (nitrogen, methane, and carbon dioxide, 19 to 385 amagat densities)

19.96 Raman Effect in Zinc Oxide
T. C. Damen, S. P. S. Porto, and B. Tell (B.T.L.)
Phys. Rev., Vol. 142, pp. 570–574, February 1966 (He–Ne and argon lasers)

19.97   Threshold Intensities of Stimulated Raman Scattering in Mixed Liquids
M. Maier (Technischen Hochschule München, Germany)
Physics Letters, Vol. 20, pp. 388–389, March 1, 1966

19.98   Near-Forward Raman Scattering in Zinc Oxide
S. P. S. Porto, B. Tell, and T. C. Damen (B.T.L.)
Phys. Rev. Letters, Vol. 16, pp. 450–452, March 14, 1966 (up to 3.4° off axis)

Stimulated Raman Effect in Acetone and Acetone-Carbon-Disulfide Mixtures
Georges G. Bret (CEPCA Compagnie Générale de Télégraphie Sans Fil, France) and Marguerite M. Denariez (Laboratoire de Chimie Physique Faculté des Sciences, France)
Appl. Phys. Letters, Vol. 8, pp. 151–154, March 15, 1966

19.100 Length Dependence of Stimulated Raman Effect in Benzene
Charles C. Wang (Philco)
J. Appl. Phys., Vol. 37, pp. 1943–1945, March 15, 1966 (lower threshold with longer cell length)

19.101 Threshold for Stimulated Raman Spectra
R. V. Wick, T. A. Wiggins, and D. H. Rank (Penn. State U.)
Appl. Optics, Vol. 5, pp. 473–474, March 1966 (Q-sw ruby, CH_4 and H_2, varied pressure, lens focal length, cell length)

19.102 Calculation of the Polar Diagram and Power Output of Stimulated Raman Radiation
P. N. Butcher and N. R. Ogg (Royal Radar Estab., England)
British J. Appl. Phys., Vol. 17, pp. 387–389, March 1966

19.103 Theoretical Discussion of the Inverse Faraday Effect, Raman Scattering, and Related Phenomena
P. S. Pershan, J. P. van der Ziel, and L. D. Malmstrom (Harvard U.)
Phys. Rev., Vol. 143, pp. 574–583, March 1966

19.104 Polarization of the Second Order Raman Spectrum of Gallium Phosphide
P. G. Marlow, J. P. Russell, and C. T. Sennett (Royal Radar Estab., England)
Physics Letters, Vol. 20, pp. 610–612, April 1, 1966

19.105 Light Scattering by Coherently Driven Lattice Vibrations
J. A. Giordmaine and W. Kaiser (B.T.L.)
Phys. Rev., Vol. 144, pp. 676–688, April 15, 1966 (stimulated Raman effect, low intensity probe light)

19.106 Raman Effect in Cadmium Sulfide
B. Tell, T. C. Damen, and S. P. S. Porto (B.T.L.)
Phys. Rev., Vol. 144, pp. 771–774, April 15, 1966 (continuous He–Ne and argon ion laser sources)

19.107 The First-Order Raman Spectrum of Magnesium Fluoride
R. S. Krishnan (Indian Inst. of Sciences, India) and J. P. Russell (Royal Radar Estab., England)
British J. Appl. Physics, Vol. 17, pp. 501–503, April 1966 (0.63-μ gas laser)

19.108 Secondary Stimulated Raman Emission in Liquids
M. K. Dheer, D. Madhavan and T. S. Jaseja (Indian Inst. of Tech., India)
Appl. Phys. Letters, Vol. 8, pp. 225–227, May 1, 1966 (Q-sw ruby, primary Raman from cyclohexane, secondary Raman from CS_2)

19.109 Induced Mandel'shtam-Brillouin Scattering in Single-Crystal Quartz at Temperatures 2.1–300°K
S. V. Krivokhizha, D. I. Mash, V. V. Morozov, V. S. Starunov, and I. L. Fabelinskii (Lebedev Physics Inst., Academy of Sciences, U.S.S.R.)

JETP Letters, Vol. 3, pp. 245–248, May 1, 1966 (focused Q-sw ruby laser beam)

19.110 A New Method for Observing the Inverse Raman (Absorption) Spectra
J. A. Duardo and F. M. Johnson (Electro-Optical Systems) and M. A. El-Sayed (U. of California)
Physics Letters, Vol. 21, pp. 168–169, May 1, 1966 (acetonitrile, second harmonic of Nd-glass laser)

19.111 Spontaneous Brillouin Scattering in Gases
D. P. Eastman, T. A. Wiggins, and D. H. Rank (Pennsylvania State U.)
Appl. Optics, Vol. 5, pp. 879–880, May 1966 (CO_2 and N_2)

19.112 Brillouin Scattering in Liquids at 4880 Å
S. L. Shapiro, M. McClintock, D. A. Jennings, and R. L. Barger (N.B.S., Boulder)
IEEE J. Quantum Electronics, Vol. QE-2, pp. 89–93, May 1966 (many liquids, Brillouin linewidths, depolarization, Rayleigh lines and wings)

19.113 Stimulated Raman and Brillouin Scattering in Selective Resonators
E. B. Aleksandrov, A. M. Bonch-Bruevich, N. N. Kostin, and V. A. Khodovoĭ
Soviet Physics JETP, Vol. 22, pp. 986–992, May 1966 (cell containing liquid within ruby laser cavity, used transverse selective resonator)

19.114 Temperature Variation of the Width of Stimulated Raman Lines in Liquids
P. Lallemand (Harvard U.)
Appl. Phys. Letters, Vol. 8, pp. 276–277, June 1, 1966 (benzaldehyde, width decreases with decreasing temperature)

19.115 Frequency Shifts in the Raman Lines Due to the High-Frequency Stark Effect
M. K. Dheer and T. S. Jaseja (Indian Inst. of Technology, Kanpur, India)
Physics Letters, Vol. 21, pp. 415–417, June 1, 1966 (cyclohexane)

19.116 Threshold of Stimuated Raman Scattering in Liquids and Self-Focusing of Laser Beams
M. Maier and W. Kaiser (Tech. Hochschule München, Germany)
Physics Letters, Vol. 21, pp. 529–530, June 15, 1966 (pure and mixed liquids)

19.117 Intensity and Width of the Fine Structure Components of Scattered Light in Liquids and the Attenuation of Hypersound
D. I. Mash, V. S. Starunov, E. V. Tiganov, and I. L. Fabelinskii (Lebedev Phys. Inst., Acad. of Sci., USSR)
Soviet Phys. JETP, Vol. 22, pp. 1205–1211, June 1966 (0.63-μ He–Ne laser light, five liquids)

19.118 Stimulated Brillouin Scattering in Ferroelectric TGS and Rochelle Salt. Effect of the Critical Phonon Relaxation in Rochelle Salt Near the Upper λ-Point
F. DeMartini (M.I.T.)
Appl. Phys. Letters, Vol. 9, pp. 31–33, July 1, 1966 (triglycine sulphate, giant pulse ruby laser)

19.119 Raman Lasers Using Secondary Raman Lines
S. E. Schwarz (U. of California), and A. Pine (Harvard U.)
Appl. Phys. Letters, Vol. 9, pp. 49–51 July 1, 1966 (cyclohexane, selective feedback, saturation mechanisms)

19.120 Stimulated Brillouin Shifts by Optical Beats
Richard G. Brewer (IBM)
Appl. Phys. Letters, Vol. 9, pp. 51–53, July 1, 1966 (sensitive optical heterodyne, temperature dependence for toluene)

19.121 Enhancement of Raman Cross Section in CdS Due to Resonant Absorption

R. C. C. Leite and S. P. S. Porto (BTL)
Phys. Rev. Letters, Vol. 17, pp. 10–12, July 4, 1966

19.122 Depolarization of Raman Scattering in Calcite
S. P. S. Porto, J. A. Giordmaine, and T. C. Damen (BTL)
Phys. Rev., Vol. 147, pp. 608–611, July 15, 1966

19.123 New Width Data of the A_{1g} Raman Line in Calcite
K. Park (BTL)
Physics Letters, Vol. 22, pp. 39–41, July 15, 1966 (1.1 cm^{-1} at 300°K, narrower with decreasing temperature)

19.124 Stimulated Raman Scattering in Hydrogen: A Measurement of the Vibrational Lifetime
F. DeMartini and J. Ducuing (M.I.T.)
Phys. Rev. Letters, Vol. 17, pp. 117–119, July 18, 1966

19.125 Brillouin-Scattering Dispersion in Ferroelectric Triglycine Sulfate
Robert W. Gammon and Herman Z. Cummins (Johns Hopkins U.)
Phys. Rev. Letters, Vol. 17, pp. 193–195, July 25, 1966

19.126 Gain, Frequency Shift, and Angular Distribution of Stimulated Raman Radiations under Multimode Excitation
Koichi Shimoda (U. of Tokyo, Japan)
Japanese J. Appl. Phys., Vol. 5, pp. 615–623, July 1966

19.127 Saturation and Spectral Characteristics of the Stokes Emission in the Stimulated Brillouin Process
C. L. Tang (Cornell U.)
J. Appl. Phys., Vol. 37, pp. 2945–2955, July 1966 (also photoelastic amplification of noise at Stokes frequency)

19.128 Mode Coupling in an External Raman Resonator
R. H. Pantell, B. G. Huth, H. E. Puthoff, and R. L. Kohn (Stanford U.)
Appl. Phys. Letters, Vol. 9, pp. 104–106, August 1, 1966 (benzene, modulator in Raman resonator)

19.129 Stimulated Pure Rotational Raman Scattering in Deuterium
R. W. Minck, E. E. Hagenlocker, and W. G. Rado (Ford)
Phys. Rev. Letters, Vol. 17, pp. 229–231, August 1, 1966

19.130 Brillouin Scattering in Liquid Helium II
Michael A. Woolf, P. M. Platzman, and M. G. Cohen (BTL)
Phys. Rev. Letters, Vol. 17, pp. 294–297, August 8, 1966 (propagation of high-frequency phonons in liquid helium)

19.131 Identification of a New Spectral Component in Brillouin Scattering of Liquids
W. S. Gornall, G. I. A. Stègeman, B. P. Stoicheff, R. H. Stolen, and V. Volterra (U. of Toronto, Canada)
Phys. Rev. Letters, Vol. 17, pp. 297–299, August 8, 1966 (density fluctuation effects in CCl₄)

19.132 Brillouin Scattering Spectra of Crystalline Quartz, Fused Quartz, and Glass
Stephen M. Shapiro, Robert W. Gammon, and Herman Z. Cummins (Johns Hopkins U.)
Appl. Phys. Letters, Vol. 9, pp. 157–159, August 15, 1966

19.133 Spatial Coherence of Light Scattered from Liquids
F. Gori and D. Sette (U. of Rome, Italy)
Phys. Rev. Letters, Vol. 17, pp. 361–363, August 15, 1966

19.134 Infrared Difference Frequency Generation
M. D. Martin and E. L. Thomas (Ministry of Aviation, England)
IEEE J. Quantum Electronics, Vol. QE-2, pp. 196–201, August 1966 (mixing in CdS, between laser and Raman shifted frequency, list of stimulated Raman emission materials)

19.135 The Influence of Self-Focusing on the Stimulated Brillouin, Raman, and Rayleigh Effects

N. Bloembergen, P. Lallemand, and A. Pine (Harvard U.)
IEEE J. Quantum Electronics, Vol. QE-2, pp. 246–248, August 1966 (anomalously high gain, wider Raman linewidth with increasing temperature of benzaldehyde)

19.136 Threshold of Stimulated Raman Scattering in Pure and Mixed Liquids and Self-Focusing of Laser Beams
M. Maier and W. Kaiser (Tech. Hochschule München, Germany)
IEEE J. Quantum Electronics, Vol. QE-2, pp. 296–297, August 1966 (abstract)

19.137 New Method of Detecting Weak Light Signals
Yoh-Han Pao, R. N. Zitter, and J. E. Griffiths (BTL)
J. Opt. Soc. Am., Vol. 56, pp. 1133–1135, August 1966 (detect larger "shot" noise which increases with light flux, 0.63-μ He-Ne laser, detect 459 cm^{-1} Raman line of CCl₄)

19.138 Stimulated Brillouin Scattering in Liquids
Alan S. Pine (Harvard U.)
Phys. Rev., Vol. 149, pp. 113–117, September 9, 1966 (transverse resonator and also backward-wave oscillator, self-focusing effects, hypersonic properties)

19.139 Brillouin Scattering in Cubic Crystals
G. B. Benedek and K. Fritsch (M.I.T.)
Phys. Rev., Vol. 149, pp. 647–662, September 16, 1966 (He-Ne laser source, crystals at room temperature, theory of scattering from thermally excited sound waves)

19.140 Theory of Enhanced Raman Scattering and Virtual Quasiparticles in Crystals
Joseph L. Birman and Achintya K. Ganguly (New York U.)
Phys. Rev. Letters, Vol. 17, pp. 647–649, September 19, 1966 (in CdS)

19.141 On the Theory of Stimulated Brillouin Scattering with Stokes Feedback
P. Lambropoulos, S. Kern, and R. K. Mueller (Bendix)
IEEE J. Quantum Electronics, Vol. QE-2, pp. 649–658, September 1966

19.142 Theory of a Tunable Raman Laser
P. A. Wolff (BTL)
IEEE J. Quantum Electronics, Vol. QE-2, pp. 659–665, September 1966 (electron cyclotron resonance in InSb, wavelengths beyond 60 μ possible using 10.6-μ CO₂ laser for pumping)

19.143 Optical Heterodyne Detection of Stimulated Brillouin Scattering in Quartz
Alan S. Pine (Harvard U.)
IEEE J. Quantum Electronics, Vol. QE-2, pp. 673–674, September 1966 (increased precision and sensitivity)

19.144 Emission of Stokes and Anti-Stokes Components of Arbitrary Orders in Stimulated Raman Scattering
V. N. Lugovoi
Optics and Spectroscopy, Vol. 21, pp. 171–175, September 1966

19.145 Spectrum of Light Scattered by Density and Anisotropy Fluctuations in Liquid Nitrobenzene
V. S. Starunov, E. V. Tiganov, and I. L. Fabelinskii (Lebedev Phys. Inst., Acad. of Sci., USSR)
JETP Letters, Vol. 4, pp. 176–179, October 1, 1966 (narrow diffuse wing and fine-structure lines, Mandel'shtam-Brillouin components)

19.146 Raman Scattering from Mixed Crystals (Ca$_x$Sr$_{1-x}$)F₂ and (Sr$_x$Ba$_{1-x}$)F₂
R. K. Chang, Brad Lacina, and P. S. Pershan (Harvard U.)
Phys. Rev. Letters, Vol. 17, pp. 755–758, October 3, 1966 (Raman

frequency varies linearly with concentration, maximum line-width near 50 percent mixture)

19.147   Erratum: Raman Scattering from Mixed Crystals $(Ca_xSr_{1-x})F_2$ and $(Sr_xBa_{1-x})F_2$
R. K. Chang, Brad Lacina, and P. S. Pershan
Phys. Rev. Letters, Vol. 17, p. 945, October 24, 1966
(Phys. Rev. Letters, Vol. 17, p. 755, 1966)

19.148   Transmission and Intense Reflection of Laser Light in CS_2
M. Maier, W. Rother, and W. Kaiser (Tech. Hochschule München, Germany)
Physics Letters, Vol. 23, pp. 83–85, October 3, 1966 (high Brillouin reflection)

19.149   Stimulated Brillouin Scattering in Shock-Compressed Fluids
R. N. Keeler, G. H. Bloom, and A. C. Mitchell (Lawrence Rad. Lab., U. of California)
Phys. Rev. Letters, Vol. 17, pp. 852–854, October 17, 1966 (liquid acetone)

19.150   Intensity of the Stimulated Raman Emission in CS_2-Nitrobenzene Mixtures
F. Barocchi (Centro Microonde, Italy), and M. Mancini (U. of Firenze, Italy)
Physics Letters, Vol. 23, pp. 230–231, October 17, 1966

19.151   Time Variations of Stimulated Raman Process and Effect of Relaxation
Tadao Shimizu (Inst. Phys. and Chem. Res., Tokyo, Japan), and Fujio Shimizu (U. of Tokyo, Japan)
Japanese J. Appl. Phys., Vol. 5, pp. 948–956, October 1966

19.152   Normal Brillouin Scattering in Compressed Gases
E. G. Rawson, E. H. Hara, A. D. May, and H. L. Welsh (U. of Toronto, Canada)
J. Opt. Soc. Am., Vol. 56, pp. 1403–1405, October 1966

19.153   Laser-Excited Electronic Raman Spectrum of Europium Yttrium Gallium Garnet (Eu^{3+}YGaG)
J. A. Koningstein (Carleton U., Canada)
J. Opt. Soc. Am., Vol. 56, pp. 1405–1406, October 1966 (excited by 0.63-μ He–Ne laser)

19.154   Nonlinear Effects Produced by Raman Maser Radiations
D. Madhavan, M. K. Dheer, and T. S. Jaseja (Indian Inst. of Tech., India)
Appl. Optics, Vol. 5, pp. 1823–1828, November 1966 (second harmonic generation, mixing of Raman lines, several liquids)

19.155   Theory of the Infrared Generation by Coherently Driven Molecular Vibrations
Francesco De Martini (M.I.T.)
J. Appl. Phys., Vol. 37, pp. 4503–4507, November 1966 (proposal, 10-μ wavelength from calcite, Q-sw ruby pump, Raman effect, 10-kW output)

19.156   Angular Dependence and Depolarization Ratio of the Raman Effect
S. P. S. Porto (BTL)
J. Opt. Soc. Am., Vol. 56, pp. 1585–1589, November 1966.

19.157   Raman Spectra of YPO_4 and $YbPO_4$
I. Richman (Aerospace)
J. Opt. Soc. Am., Vol. 56, pp. 1589–1590, November 1966 (with 0.63-μ He–Ne laser)

19.158   Comparison of the He–Ne Laser with the Hg 4358 and 5461 Å Lines as a Raman Excitation Source
M. V. Evans, T. M. Hard, and W. F. Murphy (U. of Wisconsin)
J. Opt. Soc. Am., Vol. 56, pp. 1638–1639, November 1966 (roughly comparable)

19.159   Analysis of Raman Scattering by F Centers
Charles H. Henry (BTL)
Phys. Rev., Vol. 152, pp. 699–704, December 9, 1966 (cross section of 5×10^{-24} cm²)

19.160   Raman Scattering by Lithium Niobate
R. F. Schaufele (Esso), and M. J. Weber (Raytheon)
Phys. Rev., Vol. 152, pp. 705–708, December 9, 1966

19.161   Raman Scattering by Local Modes in Germanium-Rich Silicon-Germanium Alloys
D. W. Feldman, M. Ashkin, and James H. Parker, Jr. (Westinghouse)
Phys. Rev. Letters, Vol. 17, pp. 1209–1212, December 12, 1966 (4880-Å argon laser, 300°K)

19.162   Brillouin and Critical Light Scattering in $SrTiO_3$-Crystals
W. Kaiser and R. Zurek (Tech. Hochschule München, Germany)
Physics Letters, Vol. 23, pp. 668–670, December 12, 1966 (80 to 300°K)

19.163   Pressure-Induced Line Shift and Collisional Narrowing in Hydrogen Gas Determined by Stimulated Raman Emission
P. Lallemand, P. Simova, and G. Bret (Harvard U.)
Phys. Rev. Letters, Vol. 17, pp. 1239–1241, December 19, 1966

19.164   Intense Light Bursts in the Stimulated Raman Effect
M. Maier, W. Kaiser, and J. A. Giordmaine (Tech. Hochschule München, Germany)
Phys. Rev. Letters, Vol. 17, pp. 1275–1277, December 26, 1966 (backward scattering, Stokes shift, higher peak power than incident, CS_2)

19.165   Stimulated Raman Spectrum of Cyclohexane
T. S. Jaseja, M. K. Dheer, and D. Madhavan (Indian Inst. of Tech., India)
J. Appl. Phys., Vol. 37, p. 4996, December 1966

20. HARMONIC GENERATION, MIXING, AND OTHER PARAMETRIC INTERACTIONS

20.1   Observation of Saturation Effects in Optical Harmonic Generation
R. W. Terhune, P. D. Maker, and C. M. Savage (Ford)
Appl. Phys. Letters, Vol. 2, pp. 54–55, February 1, 1963

20.2   Generation of Laser Axial Mode Difference Frequencies in a Nonlinear Dielectric
Kenneth E. Niebuhr (IBM)
Appl. Phys. Letters, Vol. 2, pp. 136–137, April 1, 1963

20.3   Proposal for Beating Two Optical Masers
G. D. Mahan and J. J. Hopfield (U. of California)
J. Appl. Phys., Vol. 34, pp. 1531–1534, May 1963

20.4   Observation of Reflected Light Harmonics at the Boundary of Piezoelectric Crystals
J. Ducuing and N. Bloembergen (Harvard U.)
Phys. Rev. Letters, Vol. 10, pp. 474–476, June 1, 1963

20.34 Electromagnetic Mode Mixing in Nonlinear Media
R. H. Kingston and A. L. McWhorter (M.I.T.)
Proc. IEEE, Vol. 53, pp. 4–12, January 1965

20.35 Nonlinear Optical Interactions in LiNbO₃ without Double Refraction
R. C. Miller, G. D. Boyd, and A. Savage (BTL)
Appl. Phys. Letters, Vol. 6, pp. 77–79, February 15, 1965 (1.15μ HeNe and 1.06μ Nd lasers, temperature and angle effects)

20.36 Second-Harmonic Generation of Light with Double Refraction
G. D. Boyd, A. Ashkin, J. M. Dziedzic, and D. A. Kleinman (BTL)
Phys. Rev., Vol. 137, pp. A1305–A1320, February 15, 1965 (long crystals, double refraction and absorption limitations, HeNe laser, 1.15μ)

20.37 Second Harmonic Radiation from Inhomogeneities in an Intense Light Field
G. A. Askar'yan (Lebedev Institute, Academy of Sciences, U.S.S.R.)
Soviet Physics JETP, Vol. 20, pp. 522–523, February 1965 (inhomogeneities small compared to wavelength, theory)

20.38 The Mixing of Raman and Laser Frequencies
M. D. Martin, E. L. Thomas, and J. K. Wright (Ministry of Aviation, Signals Res. and Dev. Estab., England)
Physics Letters, Vol. 15, pp. 136–137, March 15, 1965 (Q-sw ruby laser, mixing in ADP)

20.39 On Non-Linear Resonant Light Mixing in Plasmas
A. Salat (Institut für Plasmaphysik GmbH, Germany)
Physics Letters, Vol. 15, pp. 139–141, March 15, 1965 (two and three laser systems)

20.40 The Effect of Birefringence on Second Harmonic Generation in Thick Crystals
D. D. Bhawalkar, W. A. Gambling, R. C. Smith, and L. S. Watkins (U. of Southampton, England)
Physics Letters, Vol. 15, pp. 220–221, April 1, 1965

20.41 Intrinsic and Extrinsic Photomixing in Semi-Insulating GaAs
R. J. Strain and C. C. Tooke (Standard Telecommunication Labs., England)
Appl. Phys. Letters, Vol. 6, pp. 157–158, April 15, 1965 (mixing axial modes of either 0.63 or 1.15μ gas lasers)

20.42 Harmonic Generation of Hypersonic Waves in Liquids
R. G. Brewer (IBM)
Appl. Phys. Letters, Vol. 6, pp. 165–166, April 15, 1965 (Q-sw ruby laser, Brillouin scattering)

20.43 Measurement of Parametric Gain Accompanying Optical Difference Frequency Generation
C. C. Wang and G. W. Racette (Philco)
Appl. Phys. Letters, Vol. 6, pp. 169–171, April 15, 1965 (ADP crystal, Q-sw 2nd H of ruby for pump, 1 db gain of 0.63 μ signal)

20.44 Quantum Theory for Parametric Interactions of Light and Hypersound
A. Yariv (Calif. Inst. of Tech.)
IEEE J. Quantum Electronics, Vol. QE-1, pp. 28–36, April 1965 (parametric amplification, frequency conversion, traveling wave interaction)

20.45 Parametric Principles in Optics
I. P. Kaminow (BTL)
IEEE Spectrum, Vol. 2, No. 4, pp. 35–43, April 1965 (nonlinear optics, interaction, mixing, modulation, Raman effect)

20.46 Improved Geometry for Quantitative Measurements of Optical Frequency Doubling
A. Savage (BTL)
J. Appl. Phys., Vol. 36, pp. 1496–1497, April 1965 (wedge-shaped sample of crystal)

20.47 Phase-Matched Second-Harmonic Generation in KDP without Double Refraction
A. Ashkin, G. D. Boyd, and D. A. Kleinman (BTL)
Appl. Phys. Letters, Vol. 6, pp. 179–180, May 1, 1965

20.48 Coupled-Modes Analysis of the Phonon-Photon Parametric Backward-Wave Oscillator
D. L. Bobroff (Raytheon)
J. Appl. Phys., Vol. 36, pp. 1760–1769, May 1965 (electrostriction effects couple intense laser beam with an acoustic wave)

20.49 The Effect of Polarization Selectivity on Optical Mixing in Photoelectric Surfaces
A. J. Bahr (U. Sheffield, England)
Proc. IEEE, Vol. 53, p. 513, May 1965

20.50 Nonlinear Optical Properties of Liquids
J. A. Giordmaine (BTL)
Phys. Rev., Vol. 138, pp. A1599–A1606, June 14, 1965 (optical mixing but not second harmonic)

20.51 Tunable Coherent Parametric Oscillation in LiNbO₃ at Optical Frequencies
J. A. Giordmaine and R. C. Miller (BTL)
Phys. Rev. Letters, Vol. 14, pp. 973–976, June 14, 1965 (Q-sw Nd laser, second harmonic is pump, oscillates between 0.97 and 1.15 μ)

20.52 Nonlinear Optical Sum Generation in ADP using only Incoherent Light
D. H. McMahon and A. R. Franklin (Sperry Rand)
J. Appl. Phys., Vol. 36, pp. 2073–2074, June 1965 (infrared from xenon arc lamp)

20.53 Effect of Gaussian Beam Spread on Phase Velocity Matching in cw Optical Second-Harmonic Generation
G. E. Francois and A. E. Siegman (Stanford U.)
Phys. Rev., Vol. 139, pp. A4–A9, July 5, 1965 (angular dependence, ADP, 0.63-μ He–Ne gas laser)

20.54 Relative Phase Measurement between Fundamental and Second-Harmonic Light
R. K. Chang (Harvard U.), J. Ducuing (M.I.T.), and N. Bloembergen (U. of California)
Phys. Rev. Letters, Vol. 15, pp. 6–8, July 5, 1965 (second harmonic produced by reflection from nonlinear mirror of GaAs, InAs, and ZnTe)

20.55 Ferroelectric Behavior of Lithium Niobate
K. Nassau and H. J. Levinstein (BTL)
Appl. Phys. Letters, Vol. 7, pp. 69–70, August 1, 1965 (Curie temperature 1210°C, preparation techniques)

20.56 Parametric Phenomena Observed on Ultrasonic Waves in Water
A. Korpel and Robert Adler (Zenith)
Appl. Phys. Letters, Vol. 7, pp. 106–108, August 15, 1965 (4.68 Mc/s ultrasonic, diffraction pattern)

20.57 The Effects of Phase Matching Method and of Uniaxial Crystal Symmetry on the Polar Distribution of Second-Order Non-Linear Optical Polarization
J. E. Midwinter and J. Warner (Royal Radar Estab., England)
British J. Appl. Phys., Vol. 16, pp. 1135–1142, August 1965 (three-wave interactions)

20.58 Photomixing in a GaAsₓP₁₋ₓ-GaAs Heterodiode
T. B. Ramachandran, K. K. Chow, W. J. Moroney, and P. Olendzensky (Microwave Associates)
J. Appl. Phys., Vol. 36, pp. 2594–2595, August 1965 (bromine-argon laser, beats in 13–22 Gc/s region)

20.59 Optical Harmonic Generation in Semiconductors and in Dielectrics Close to an Absorption Band Edge
A. M. Afanas'ev and É. A. Manykin (Moscow Engineering Physics Inst.)
Soviet Physics JETP, Vol. 21, pp. 323–325, August 1965 (second harmonic of Nd laser in CdS, also gas, third harmonic)

20.60 Generation of Ultraviolet Radiation by Using Cascade Frequency Conversion

S. A. Akhmanov, A. I. Kovrigin, A. S. Piskarskas, and R. V. Khokhlov (Moscow State U., U.S.S.R.)
JETP Letters, Vol. 2, pp. 141–143, September 1, 1965 (3rd and 4th H of Q-sw Nd laser, in KDP, 8 and 3 MW/cm² respectively)

20.61 Detection of Nonlinear Optical Sum Spectra in ADP using Incoherent Light
D. H. McMahon and A. R. Franklin (Sperry Rand)
J. Appl. Phys., Vol. 36, pp. 2807–2810, September 1965 (xenon arc source, sum range from 5000 to 2700 Å)

20.62 Optical Mixing in Raman-Active Medium
Kanji Kubota (Osaka U., Japan)
J. Phys. Soc. Japan, Vol. 20, pp. 1738–1739, September 1965 (third-order electric polarization in benzene, Q-sw ruby)

20.63 Resonance Phenomena in Nonlinear Optics
E. A. Manykin and A. M. Afanas'ev
Soviet Physics JETP, Vol. 21, pp. 619–623, September 1965 (second and third harmonics)

20.64 Observation of Parametric Amplification in the Optical Range
S. A. Akhmanov, A. I. Kovrigin, A. S. Piskarskas, V. V. Fadeev, and R. V. Khokhlov (Moscow State U., U.S.S.R.)
JETP Letters, Vol. 2, pp. 191–193, October 1, 1965 (Q-sw Nd glass laser, 0.53 μ pump using 2nd H, amplification of 1.06 μ signal in KDP)

20.65 Nonlinear Optics of Many-Particle Systems
Gordon Baym (U. of Illinois) and R. W. Hellwarth (Hughes)
IEEE J. Quantum Electronics, Vol. QE-1, pp. 309–320, October 1965

20.66 Saturation Effects in Second-Harmonic Generation of Light Using Unfocused Laser Beams
Charles C. Wang and George W. Racette (Philco)
J. Appl. Phys., Vol. 36, pp. 3281–3284, October 1965 (ADP and KDP crystals, Q-sw ruby, angular dependence of 2nd harmonic)

20.67 Stimulated Raman Scattering and Parametric Processes
V. T. Platonenko and R. V. Khokhlov (Moscow State U., U.S.S.R.)
JETP Letters, Vol. 2, pp. 269–270, November 1, 1965 (pump by stimulated Raman scattering, prediction)

20.68 Generation of the Difference Frequency by Non-Collinear Light Beams in K.D.P. Crystal
N. Van Tran, J. Spalter, J. Hanus, J. Ernest, and D. Kehl (C.G.E., France)
Physics Letters, Vol. 19, pp. 285–287, November 1, 1965 (pump with 2nd H of ruby laser, Nd glass laser signal, idler at 5150 Å)

20.69 Efficient Continuous Optical Second-Harmonic Generation
R. G. Smith, K. Nassau, and M. F. Galvin (BTL)
Appl. Phys. Letters, Vol. 7, pp. 256–258, November 15, 1965 (lithium niobate crystal inside cavity of 1.06-μ Nd:YAG laser, 0.3 mW second, 20 mW fundamental)

20.70 Observation of Two-Dimensional Parametric Interaction of Light Waves
S. A. Akhmanov, A. G. Ershov, V. V. Fadeev, R. V. Khokhlov, O. N. Chunaev, and E. M. Shvom (Moscow State U.)
JETP Letters, Vol. 2, pp. 285–288, November 15, 1965 (Q-sw ruby)

20.71 The Effects of Phase Matching Method and of Crystal Symmetry on the Polar Dependence of Third-Order Non-Linear Optical Polarization
J. E. Midwinter and J. Warner (Royal Radar Estab., England)
British J. Appl. Phys., Vol. 16, pp. 1667–1674, November 1965 (four wave interactions)

20.72 Theory of Optical Harmonic Generation at a Metal Surface
Sudhanshu S. Jha (Stanford U.)
Phys. Rev., Vol. 140, pp. A2020–A2030, December 13, 1965

20.73 Generation of Far Infrared as a Difference Frequency
Frits Zernike, Jr., and Paul R. Berman (Perkin-Elmer)
Phys. Rev. Letters, Vol. 15, pp. 999–1001, December 27, 1965 (in quartz, Nd glass laser, 100 cm⁻¹)

20.74 Efficient Phase-Matched Harmonic Generation in Tellurium with a CO_2 Laser at 10.6 μ
C. K. N. Patel (BTL)
Phys. Rev. Letters, Vol. 15, pp. 1027–1030, December 27, 1965 (4000 times greater nonlinear coefficient than from KDP)

20.75 Influence of the Finite Aperture of a Light Beam on Nonlinear Effects in an Anisotropic Medium
S. A. Akhmanov, A. I. Kovrigin, and N. K. Kulakova (Moscow State U., U.S.S.R.)
Soviet Physics JETP, Vol. 21, pp. 1034–1040, December 1965 (spatial structure of second harmonic radiation, electric field on KDP)

20.76 Second-Harmonic Generation and Parametric Amplification Using Intense Unfocused Laser Beams
Charles C. Wang and George W. Racette (Philco)
Proc. Physics of Quantum Electronics Conf., McGraw-Hill Book Co., New York, pp. 20–30, 1966

20.77 Optical Parametric Oscillation in LiNbO₃
J. A. Giordmaine and Robert C. Miller (B.T.L.)
Proc. Physics of Quantum Electronics Conf., McGraw-Hill Book Co., New York, pp. 31–42, 1966

20.78 Nonlinear Effects at Multiples of Laser Frequencies
S. A. Akhmanov, V. G. Dmitriev, R. V. Khokhlov, and A. I. Kovrygin (Moscow U., U.S.S.R.)
Proc. Physics of Quantum Electronics Conf., McGraw-Hill Book Co., New York, pp. 43–48, 1966

20.79 Sum and Difference Frequency Generation in Gases and Liquids
P. N. Butcher (Royal Radar Estab., England), W. H. Kleiner, P. L. Kelley, and H. J. Zeiger (Lincoln Lab., M.I.T.)
Proc. Physics of Quantum Electronics Conf., McGraw-Hill Book Co., New York, pp. 49–59, 1966

20.80 Dispersion of the Optical Nonlinearity in Semiconductors
N. Bloembergen (U. of California), R. K. Chang (Harvard U.), and J. Ducuing (M.I.T.)
Proc. Physics of Quantum Electronics Conf., McGraw-Hill Book Co., New York, pp. 67–79, 1966

20.81 Second-Harmonic Generation of Light from Surface Layers of Media with Inversion Symmetry
N. Bloembergen (U. of California), and R. K. Chang (Harvard U.
Proc. Physics of Quantum Electronics Conf., McGraw-Hill Book Co., New York, pp. 80–85, 1966

20.82 Nonlinear Optical Effects in Plasmas
Norman M. Kroll (U. of California)
Proc. Physics of Quantum Electronics Conf., McGraw-Hill Book Co., New York, pp. 86–104, 1966

20.83 Nonlinear Interactions of Radiation in Plasmas
Gordon Baym and R. W. Hellwarth (U. of Illinois)
Proc. Physics of Quantum Electronics Conf., McGraw-Hill Book Co., New York, pp. 105–110, 1966

20.84 Coupling of Light with Phonons, Magnons, and Plasmons
N. Bloembergen and Y. R. Shen (U. of California)
Proc. Physics of Quantum Electronics Conf., McGraw-Hill Book Co., New York, pp. 119–128, 1966

20.85 The Generation of Molecular Vibrational Frequencies by Optical Mixing
M. D. Martin and E. L. Thomas (Ministry of Aviation, S.R.D.E., Christchurch, Hants)
Physics Letters, Vol. 19, pp. 651–652, January 1, 1966

20.86 Observation of Microwave Beats in a Parallel-Plate Transmission-Line Photomixer
J. C. Bass and J. A. Jones (U. of Sheffield, Sheffield, England)
Electronics Letters, Vol. 2, pp. 25–26, January 1966 (ruby laser input, 10–40 Gc/s output)

20.87 Optical Nonlinearities of a Plasma
N. Bloembergen and Y. R. Shen (U. of California)
Phys. Rev., Vol. 141, pp. 298–305, January 1966 (predicts weak stimulated Raman and second harmonic generation effects)

20.88 The Temperature Dependence of Optical Birefringence in Lithium Niobate
J. Warner, D. S. Robertson and K. F. Hulme (Royal Radar Estab., England)
Physics Letters, Vol. 20, pp. 163–164, February 1, 1966 (for optical mixing, 0.63 and 1.15 μ, birefringence decreases with increasing temperature)

20.89 Optical Second-Harmonic Generation Using a Focused Gaussian Laser Beam
J. E. Bjorkholm (Stanford U.)
Phys. Rev., Vol. 142, pp. 126–136, February 4, 1966 (0.63-μ He–Ne laser, ADP crystal, limiting effects)

20.90 Parametric Interactions of Optical Modes
Amnon Yariv (California Inst. of Technology)
IEEE J. Quantum Electronics, Vol. QE-2, pp. 30–37, February 1966

20.91 Magnetic-Dipole Contribution to Optical Harmonics in Silver
Fielding Brown and Robert E. Parks (Williams College)
Phys. Rev. Letters, Vol. 16, pp. 507–509, March 21, 1966 (second harmonic)

20.92 cw Measurement of the Optical Nonlinearity of Ammonium Dihydrogen Phosphate
G. E. Francois (Stanford U.)
Phys. Rev., Vol. 143, pp. 597–600, March 1966 (0.63-μ He–Ne laser)

20.93 Optical Harmonic Generation in the Infrared Using a CO_2 Laser
C. K. N. Patel (B.T.L.)
Phys. Rev. Letters, Vol. 16, pp. 613–616, April 4, 1966 (several crystals, Q-sw CO_2 laser, 10 kW peak, 300 ns)

20.94 Experimental Verification of the Laws for the Reflected Intensity of Second-Harmonic Light
R. K. Chang and N. Bloembergen (Harvard U.)
Phys. Rev., Vol. 144, pp. 775–780, April 15, 1966 (GaAs reflector, Q-sw ruby and Nd glass lasers)

20.95 Second-Harmonic Enhancement with an Internally-Modulated Ruby Laser
Roger L. Kohn and Richard H. Pantell (Stanford U.)
Appl. Phys. Letters, Vol. 8, pp. 231–233, May 1, 1966 (loss modulator to couple axial modes, 7 ns spike period)

20.96 Tunable Parametric Light Generator with KDP Crystal
S. A. Akhmanov, A. I. Kovrigin, V. A. Kolosov, A. S. Piskarskas, V. V. Fadeev, and R. V. Khokhlov (Moscow State U., U.S.S.R.)
JETP Letters, Vol. 3, pp. 241–245, May 1, 1966 (tunable from 9575 to 11,775 Å, rotated KDP within optical resonator)

20.97 Coherent Optical Mixing in Optically Active Liquids
P. M. Rentzepis, J. A. Giordmaine, and K. W. Wecht (B.T.L.)

Phys. Rev. Letters, Vol. 16, pp. 792–794, May 2, 1966 (quadratic polarizability, arabinose solutions, Q-sw ruby plus 2nd harmonic)

20.98 Second-Harmonic Generation of Light by Focused Laser Beams
D. A. Kleinman, A. Ashkin, and G. D. Boyd (B.T.L.)
Phys. Rev., Vol. 145, pp. 338–379, May 6, 1966

20.99 Second-Harmonic Generation of Light in Reflection from Media with Inversion Symmetry
N. Bloembergen, R. K. Chang, and C. H. Lee (Harvard U.)
Phys. Rev. Letters, Vol. 16, pp. 986–989, May 30, 1966 (core electrons contribute significantly in silver)

20.100 Obtaining the Equations of Motion for Parametrically Coupled Oscillators or Waves
A. E. Siegman (Stanford U.)
Proc. IEEE, Vol. 54, pp. 756–762, May 1966

20.101 Theory of Parametric Oscillator Threshold with Single-Mode Optical Masers and Observation of Amplification in $LiNbO_3$
G. D. Boyd and A. Ashkin (B.T.L.)
Phys. Rev., Vol. 146, pp. 187–198, June 3, 1966

20.102 Enhancement of Optical Second-Harmonic Generation (SHG) by Reflection Phase Matching in ZnS and GaAs
G. D. Boyd and C. K. N. Patel (B.T.L.)
Appl. Phys. Letters, Vol. 8, pp. 313–315, June 15, 1966 (TIR within optical material, Q-sw CO_2 laser)

20.103 Resonant Optical Second Harmonic Generation and Mixing
A. Ashkin, G. D. Boyd, and J. M. Dziedzic (B.T.L.)
IEEE J. Quantum Electronics, Vol. QE-2, pp. 109–124, June 1966

20.104 Generation of Coherent Radiation in the Infrared Band by Nonlinear-Optics Methods
S. A. Akhmanov, G. V. Venkin, B. V. Zubov, and R. V. Khokhlov (Moscow State U., USSR)
JETP Letters, Vol. 4, pp. 14–17, July 1, 1966 (2–5 μ range, mixing of Q-sw ruby beam with first Stokes component of Raman scattering, $LiNbO_3$ crystal)

20.105 Properties of a Phase-Matchable Nonlinear Optical Crystal: Potassium Dithionate
M. V. Hobden, D. S. Robertson, P. H. Davies, K. F. Hulme, J. Warner, and J. Midwinter (Royal Radar Estab., England)
Physics Letters, Vol. 22, pp. 65–66, July 15, 1966 (smaller nonlinear coefficient than for KDP)

20.106 Proposed Backward Wave Oscillation in the Infrared
S. E. Harris (Stanford U.)
Appl. Phys. Letters, Vol. 9, pp. 114–116, August 1, 1966

20.107 Dependence of Second-Harmonic Generation on the Position of the Focus
D. A. Kleinman and R. C. Miller (BTL)
Phys. Rev., Vol. 148, pp. 302–312, August 5, 1966 (coherence length effects, He–Ne laser, $LiNbO_3$)

20.108 Temperature Dependence of the Optical Properties of Ferroelectric $LiNbO_3$ and $LiTaO_3$
R.C. Miller and A. Savage (BTL)
Appl. Phys. Letters, Vol. 9, pp. 169–171, August 15, 1966 (also second harmonic generation with 1.06 μ Nd laser)

20.109 The Temperature Dependence of the Refractive Indices of Pure Lithium Niobate
M. V. Hobden and J. Warner (Royal Radar Estab., England)
Physics Letters, Vol. 22, pp. 243–244, August 15, 1966 (up to 500°C, 0.4 to 4.0 μ)

20.110 Stimulated Optical Frequency Mixing in Liquids and Solids
R. V. Wick, D. H. Rank, and T. A. Wiggins (Penn. State U.)
Phys. Rev. Letters, Vol. 17, pp. 466–467, August 29, 1966

21. THOMSON SCATTERING, COMPTON SCATTERING, AND PLASMA DIAGNOSTICS

21.1 Measurement of Plasma Density Using a Gas Laser as an Infrared Interferometer
D. E. T. F. Ashby and D. F. Jephcott (A.E.R.E., England)
Appl. Phys Letters, Vol. 3, pp. 13–16, July 1, 1963

21.2 High-Intensity Limit of Thomson Scattering
P. Stehle (Space Tech. Labs.)
J. Opt. Soc. Am., Vol. 53, p. 1003, August 1963

21.3 Plasma Refractive Index by a Laser Phase Measurement
J. B. Gerardo and J. T. Verdeyen (U. of Illinois)
Appl. Phys. Letters, Vol. 3, pp. 121–123, October 1, 1963

21.4 Scattering of Optical Pulses from a Nonequilibrium Plasma
S. E. Schwarz (Calif. Inst. of Tech.)
Proc. IEEE, Vol. 51, p. 1362, October 1963

21.5 Thomson Scattering of Intense Light Beams
L. Mandel (Imperial College, England)
J. Opt. Soc. Am., Vol. 54, p. 265, February 1964

21.6 Plasma Diagnostics Using the Raman Effect
R. P. Urtz, Jr. (RADC)
IEEE Int. Conv. Record, Vol. 12, Pt. 2, pp. 289-295, 1964 (atmospheric plasma, pulsed ruby laser, Stokes and anti-Stokes)

21.7 The Laser Interferometer: Application to Plasma Diagnostics
J. B. Gerardo and J. T. Verdeyen (U. of Illinois)
Proc. IEEE, Vol. 52, pp. 690-697, June 1964 (He-Ne gas laser as both source and detector, helium plasma afterglow study)

21.8 Optical Mixing as a Plasma Density Probe
N. M. Kroll, A. Ron, and N. Rostoker (U. of California)
Phys. Rev. Letters, Vol. 13, pp. 83–86, July 20, 1964 (analytical, two crossed laser beams)

21.9 Spatial Density Measurements in Fast Theta-Pinch Plasma by Laser Excitation of Coupled Infrared Resonators
R. F. Gribble, J. P. Craig, and A. A. Dougal (U. of Texas)
Appl. Phys. Letters, Vol. 5, pp. 60-62, August 1, 1964 (3.39-μ He-Ne laser)

21.10 Time-Space Resolved Experimental Diagnostics of Theta-Pinch Plasma by Faraday Rotation of Infrared He-Ne Maser Radiation
A. A. Dougal, J. P. Craig, and R. F. Gribble (U. of Texas)
Phys. Rev. Letters, Vol. 13, pp. 156-158, August 3, 1964

21.11 Plasma Experiments with a 570-kJ Theta-Pinch
F. C. Jahoda, E. M. Little, W. E. Quinn, F. L. Ribe and G. A. Sawyer (Los Alamos)
J. Appl. Phys., Vol. 35, pp. 2351-2363, August 1964 (Q-switched ruby laser used as interferometer source)

21.12 Electron Density and Temperature Measurements in a 26 kj Θ-Pinch by Light Scattering
H. J. Kunze, A. Eberhagen, and E. Fünfer (Inst. f. Plasmaphysik, Germany)
Physics Letters, Vol. 13, pp. 38-39, November 1, 1964 (Q-switched ruby laser, 16×10^{16} e/cm^3, 43 eV)

21.13 A Zebra-Stripe Display for an Optical Interferometer, and Its Use to measure Plasma Density in the Presence of Vibration
A. Gibson and G. W. Reid (Atomic Energy Res. Est., England)
Appl. Phys. Letters, Vol. 5, pp. 195-197, November 15, 1964 (He-Ne laser)

21.14 Interferometric Measurements of Rapid Phase Changes in the Visible and Near Infrared using a Laser Light Source
R. G. Buser and J. J. Kainz (USAERDL)
Appl. Optics, Vol. 3, pp. 1495-1499, December 1964 (on plasma, He-Ne laser, Michelson interferometer, fringe shifts measured by photomultiplier)

21.15 Measurement of the Faraday Effect in a θ-Pinch Plasma Using a He-Ne Laser
I. S. Falconer, R. Benesch, and S. A. Ramsden (National Res. Council, Canada)
Physics Letters, Vol. 14, pp. 38–40, January 1, 1965 (0.63-μ laser, 12 kilogauss, 6×10^{16} e/cm^3)

21.16 A Moving Mirror Laser Interferometer for Plasma Diagnostics
W. A. Kricker and W. I. B. Smith (U. of Sydney, Australia)
Physics Letters, Vol. 14, pp. 102–103, January 15, 1965 (50 c/sec mirror, distinguish between rising and falling density, 0.63-μ He-Ne laser, 2.7×10^{15} e/cm^3)

21.17 Plasma Diagnostics by Non-Linear Resonant-Angle Scattering
A. Salat and A. Schlüter (Institut für Plasmaphysik GmbH, Germany)
Physics Letters, Vol. 14, pp. 106–107, January 15, 1965

21.18 Laser Applications in the Field of Plasma Physics
B. Kronast (Institut fur Plasmaphysik, Munich, Germany)
Z. angewandte Mathematik und Physik, Vol. 16, No. 1, pp. 120–121, January 25, 1965 (25 references)

21.19 Performance of the He-Ne Gas Laser as an Interferometer for Measuring Plasma Density
D. E. T. F. Ashby, D. F. Jephcott, A. Malein, and F. A. Raynor (UK AEA, England)
J. Appl. Phys., Vol. 36, pp. 29–34, January 1965 (independent of electron temperature, particularly useful with very dense plasmas)

21.20 Extension of the CW Laser Interferometry Technique Measuring Plasma Density
D. A. Baker, J. E. Hammel, and F. C. Jahoda (Los Alamos)
Bull. Am. Phys. Soc., Series II, Vol. 10, No. 2, p. 218-M6, February 1965 (7 mc/s modulation of 3.39μ He-Ne using external mirror)

21.21 Multiple-Pass System for the Measurement of the Faraday Rotation Produced by a Plasma
K. Halbach (Lawrence Rad. Lab.)
Bull. Am. Phys. Soc., Series II, Vol. 10, No. 2, p. 219-M7, February 1965 (plasma within Fabry-Perot cavity)

21.22 Observation of the Faraday Effect in a θ Pinch Plasma
I. S. Falconer and S. A. Ramsden (National Res. Council, Canada)
Bull. Am. Phys. Soc., Series II, Vol. 10, No. 2, p. 219-M8, February 1965 (0.63μ He-Ne laser, dual polarization photodiodes)

21.23 Electron-Density Measurements in a θ Pinch by an Infrared-Maser Technique
R. F. Gribble, J. P. Craig, and A. A. Dougal (U. of Texas)
Bull. Am. Phys. Soc., Series II, Vol. 10, No. 2, p. 224-P6, February 1965 (time and space resolved electron density, 6×10^{16} e/cm^3)

21.24 Thomson Scattering of CW Laser Light from Electrons in a Plasma
L. A. Farrow and S. J. Buchsbaum (BTL)
Bull. Am. Phys. Soc., Series II, Vol. 10, No. 2, p. 226-Q1, February 1965 (He-Ne 0.63μ, He 8 Torr, laser chopped at 33 cps)

21.25 Thomson Scattering from a Hollow-Cathode Arc Plasma
E. Gerry (MIT)
Bull. Am. Phys. Soc., Series II, Vol. 10, No. 2, p. 226-Q3, February 1965 (ruby laser, 5×10^{13} e/cm^3, 45° scattering)

21.26 Thomson Scattering of Laser Light from a Plasma
D. R. Sigman, J. F. Holt, and M. L. Pool (Ohio State U.)
Bull. Am. Phys. Soc., Series II, Vol. 10, No. 2, p. 226-Q4, February 1965 (20j ruby laser, minimum of 10^{13} e/cm^3, 90° scattering)

21.27 Thomson-Scattering Observations from MHD Shock-Heated Plasmas
    R. M. Patrick and E. T. Gerry (Avco)
    Bull. Am. Phys. Soc., Series II, Vol. 10, No. 2, p. 226–Q5, February 1965 (Q-switched laser, 90° scattering, 4–10 × 10^{15} ions/cm³)

21.28 Magnetic-Field Measurements in a θ Pinch by Faraday Rotation of Infrared Maser Radiation
    A. A. Dougal, J. P. Craig and R. F. Gribble (U. of Texas)
    Bull. Am. Phys. Soc., Series II, Vol. 10, No. 2, p. 226–Q6, February 1965 (3.39μ HeNe laser, 6 × 10^{16} e/cm³)

21.29 Scattering of a Laser Beam from a θ-Pinch Plasma
    S. A. Ramsden and W. E. R. Davies (National Res. Council, Canada)
    Bull. Am. Phys. Soc., Series II, Vol. 10, No. 2, p. 227–Q7, February 1965 (ruby laser, 90° scattering)

21.30 Observation of Thomson and Cooperative Scattering of Ruby-Laser Light by a Plasma
    A. W. DeSilva, D. E. Evans, and M. J. Forrest (UKAEA, England)
    Bull. Am. Phys. Soc., Series II, Vol. 10, No. 2, p. 227–Q8, February 1965 (2.2 eV from Doppler broadened spectrum of 70 Å)

21.31 A Ruby Laser Diagnostic Plasma Probe
    D. O. Kingsland (G.E.)
    Proc. IEEE, Vol. 53, p. 196, February 1965 (Q-sw ruby, 2 Torr He discharge, Doppler profile, 2 × 10^{15} e/cm³, 30,000°K electrons)

21.32 Extension of Plasma Interferometry Technique with a He–Ne Laser
    D. A. Baker, J. E. Hammel, and F. C. Johoda (Los Alamos Sci. Lab.)
    Rev. Sci. Instr., Vol. 36, pp. 395–396, March 1965 (simultaneous 0.63 and 3.39μ operation, some feedback into optical cavity)

21.33 Correlated Interferometric Measurements of Plasma Electron Densities at Optical and Microwave Frequencies
    J. B. Gerardo and J. T. Verdeyen (U. of Illinois)
    Appl. Phys. Letters, Vol. 6, pp. 185–187, May 1, 1965 (triple mirror, plasma within HeNe laser cavity, also 70 Gc/s)

21.34 Temperature Determination of Plasmas Produced by Giant Laser Pulses
    H. Opower and E. Burlefinger (Technischen Hochschule München, Germany)
    Physics Letters, Vol. 16, pp. 37–38, May 1, 1965 (about 10^6 °K carbon plasma using 30 MW Q-sw ruby laser)

21.35 Frequency Shift in High-Intensity Compton Scattering
    T. W. B. Kibble (Imperial College, London, England)
    Phys. Rev., Vol. 138, pp. B740–B753, May 10, 1965

21.36 Nonlinear Scattering of Radiation from Plasmas
    D. F. Dubois (Hughes)
    Phys. Rev. Letters, Vol. 14, pp. 818–822, May 17, 1965

21.37 Observation of Stimulated Compton Scattering of Electrons by Laser Beam
    L. S. Bartell, H. Bradford Thompson, and R. R. Roskos (Iowa State U.)
    Phys. Rev. Letters, Vol. 14, pp. 851–852, May 24, 1965 (1.65-kV electron beam at right angle inside 30-J ruby laser cavity)

21.38 Electron-Density Measurements with a Laser Interferometer
    E. B. Hooper, Jr. and G. Bekefi (M.I.T.)
    Bull. Am. Phys. Soc., Series II, Vol. 10, No. 5, p. 605–ED11, June 1965 (discharge within 0.63-μ laser cavity)

21.39 Plasma Diagnosis by Means of Optical Scattering
    S. E. Schwarz (Calif. Inst. of Tech.)
    J. Appl. Phys., Vol. 36, pp. 1836–1841, June 1965 (electron density and velocity distribution, experiments near 10^{13} e/cm³)

21.40 Gas Laser Measurements of the Electron Density of a Plasma Produced by a Very Fast Theta-Pinch Preheater
    A. C. C. Warnock, W. M. Deuchars, J. Irving, and D. E. Kidd (U. of Strathclyde, Glasgow, Scotland)
    Appl. Phys. Letters, Vol. 7, pp. 29–30, July 15, 1965 (interferometry, simultaneous 0.63 and 3.39 μ wavelengths, hydrogen gas)

21.41 Time-Dependent Electron Density Measurements in a Fast Theta-Pinch Discharge
    W. M. Deuchars, D. E. Kidd, J. Irving, and A. C. C. Warnock (U. of Strathclyde, Glasgow, Scotland)
    Appl. Phys. Letters, Vol. 7, pp. 30–32, July 15, 1965 (0.63 μ gas laser, interferometer, up to 10^{16} e/cm³)

21.42 High-Frequency Laser Interferometry in Plasma Diagnostics
    J. B. Gerardo, J. T. Verdeyen, and M. A. Gusinow (U. of Illinois)
    J. Appl. Phys., Vol. 36, pp. 2146–2151, July 1965 (plasma within 3- and 4-mirror Fabry Perot cavities, 0.63-μ gas laser, detect 50 resonances per μs)

21.43 Corrections to Thompson Scattering for Intense Laser Beams
    J. J. Sanderson (English Electric Co., England)
    Physics Letters, Vol. 18, pp. 114–115, August 15, 1965 (includes radiation damping term)

21.44 Laser-Induced Perturbation in a Plasma
    S. F. Paik and O. Ben-Dov (Northwestern U.)
    Proc. IEEE, Vol. 53, pp. 1145–1146, August 1965 (plasma affected, detected by Langmuir probes)

21.45 A Laser Interferometer for Repetitively Pulsed Plasmas
    E. B. Hooper, Jr., and G. Bekefi (M.I.T.)
    Appl. Phys. Letters, Vol. 7, pp. 133–135, September 1, 1965 (triple-mirror interferometer, 10^{13}–10^{14} e/cm³)

21.46 Infrared Laser Interferometry Utlizing Quantum-Electronic Cross Modulation
    R. J. Freiberg, L. A. Weaver, and J. T. Verdeyen (U. of Illinois)
    J. Appl. Phys., Vol. 36, p. 3352, October 1965 (detection of cross-modulation sidelight at visible wavelength)

21.47 Measurement of Low Electron Density in Plasmas Using Laser Interferometry
    Herbert Malamud (Sperry Gyroscope)
    Rev. Sci. Instr., Vol. 36, pp. 1507–1508, October 1965 (two-mode laser interferometer, different polarizations, one reference beam, expected minimum detectable plasma is 1.2 × 10^{12} electrons cm⁻²)

21.48 Note on Radiative Corrections to Thomson Scattering in Intense Laser Beams
    J. H. Eberly (U. of Rochester)
    Physics Letters, Vol. 19, pp. 284–285, November 1, 1965

21.49 Determination of Plasma Density by Laser Interferometric and Continuum Radiation Intensity Measurements
    A. Boornard, L. J. Nicastro, and James Vollmer (RCA)
    Appl. Phys. Letters, Vol. 7, pp. 258–260, November 15, 1965

21.50 Spatially and Temporally Resolved Electron and Atom Concentrations in an Afterglow Gas Discharge
    J. B. Gerardo, J. T. Verdeyen, and M. A. Gusinow (U. of Illinois)
    J. Appl. Phys., Vol. 36, pp. 3526–3534, November 1965 (laser interferometer, detect 2 × 10^{13} e/cm³, also 4 and 8 mm microwaves)

21.51 Incorporation of a Laser into the Arm of an Interferometer for Measurement of Transient Phase Changes

Edward Thornton (IIT Research Inst.)
J. Appl. Phys., Vol. 36, pp. 3539–3541, November 1965 (0.63-μ laser)

21.52 Nonlinear Interactions of Radiation in Plasmas
Gordon Baym and R. W. Hellwarth (U. of Illinois)
Proc. Physics of Quantum Electronics Conf., McGraw-Hill Book Co., New York, pp. 105–110, 1966

21.53 Plasmon Scattering of Light and Stimulated Emission of Plasmons in Solids
A. L. McWhorter (Lincoln Lab., M.I.T.)
Proc. Physics of Quantum Electronics Conf., McGraw-Hill Book Co., New York, pp. 111–118, 1966

21.54 Stimulated Compton Scattering of Electrons by Laser Beam
L. S. Bartell and H. Bradford Thompson (Iowa State U.)
Proc. Physics of Quantum Electronics Conf., McGraw-Hill Book Co., New York, pp. 129–133, 1966

21.55 Collective Scattering of Laser Light by a Plasma
P. W. Chan and R. A. Nodwell (U. of British Columbia, Canada)
Phys. Rev. Letters, Vol. 16, pp. 122–124, January 24, 1966 (focused Q-sw ruby, strong intensities near 6910 and 6985 Å, 6.3×10^{16} e/cm³)

21.56 The Laser Interferometer as a Diagnostic Tool in Shock-Tube Experiments
Robert W. Deuel, Leonard P. Kirchner, and Edward Thornton (IIT Research Institute)
Appl. Phys. Letters, Vol. 8, pp. 59–60, February 1, 1966

21.57 Quantum Theory of Photon Interaction in a Plasma
H. Cheng (Harvard U. and B.T.L.) and Y. C. Lee (B.T.L.)
Phys. Rev., Vol. 142, pp. 104–114, February 4, 1966 (light-light scattering, optical mixing)

21.58 Observation of Cooperative Effects in the Scattering of a Laser Beam from a Plasma
S. A. Ramsden and W. E. R. Davies (National Res. Council, Canada)
Phys. Rev. Letters, Vol. 16, pp. 303–306, February 21, 1966 (Q-sw ruby, ±8 Å wavelength shift)

21.59 Radiative Corrections to Thomson Scattering from Laser Beams
T. W. B. Kibble (Imperial College, England)
Physics Letters, Vol. 20, pp. 627–628, April 1, 1966

21.60 Two Wavelength Interferometry of a Laser-Induced Spark in Air
A. J. Alcock and S. A. Ramsden (National Research Council, Canada)
Appl. Phys. Letters, Vol. 8, pp. 187–188, April 15, 1966 (two wavelength probe, 10^{19} e/cm³)

21.61 Laser and Microwave Interferometric Study of Recombination in a Highly Ionised Helium Plasma
A. A. Newton and J. M. P. Quinn (UKAEA, England) and M. C. Sexton (U. College, Ireland)
Electronics Letters, Vol. 2, pp. 157–158, April 1966 (3.3-μ He–Ne laser, 2.26 and 8.6 mm microwaves, 8×10^{11} to 1.8×10^{15} e/cm³)

21.62 On the Scattering of Light by a Plasma
Nguyen-Quang-Dong (Association Euratom-CEA sur la Fusion, France)
Physics Letters, Vol. 21, pp. 159–160, May 1, 1966

21.63 Measurements of the Ion and Electron Temperature in a Theta-Pinch Plasma by Forward Scattering
B. Kronast, H. Röhr, E. Glock, H. Zwicker, and E. Fünfer (Institut für Plasmaphysik, Germany)
Phys. Rev. Letters, Vol. 16, pp. 1082–1085, June 13, 1966 (ruby laser)

21.64 Plasma Diagnostics using Lasers: Relations between Scattered Spectrum and Electron-Velocity Distribution
Terrence S. Brown and D. J. Rose (M.I.T.)
J. Appl. Phys., Vol. 37, pp. 2709–2714, June 1966

21.65 Plasma Diagnostics by Thomson Scattering of a Laser Beam
Edward T. Gerry and D. J. Rose (M.I.T.)
J. Appl. Phys., Vol. 37, pp. 2715–2724, June 1966

21.66 Intensity Effects in Thomson Scattering
N. D. Sen Gupta (Tata Inst. of Fundamental Research, India)
Physics Letters, Vol. 21, pp. 642–643, July 1, 1966

21.67 Co-Operative Scattering of Laser Light by a Thetatron Plasma
D. E. Evans, M. J. Forrest, and J. Katzenstein (Atomic Energy Research Estab. Culham Lab., England)
Nature, Vol. 211, pp. 23–24, July 2, 1966 (ruby laser, 5° off-axis forward scattering, 30 Å spectra shift, 6×10^{15} e/cm³, 103 eV electron temperature)

21.68 The Use of the First and Second Harmonic of Ruby Laser Light in the Study of a Fast Theta-Pinch
S. Martellucci and E. Mazzucato (Lab. Gas Ionizzati, Italy)
Nuovo Cimento, Vol. 44B, pp. 107–118, July 11, 1966

21.69 Time and Spatially Resolved Interferometry on Pulsed-Laser-Induced Plasmas
C. W. Bruce, J. Deacon, and D. F. Vonderhaar (AF Weapons Lab.)
Appl. Phys. Letters. Vol. 9, pp. 164–166, August 15, 1966

21.70 Observation of Cooperative Effects and Determination of the Electron and Ion Temperatures in a Plasma from the Scattering of a Ruby Laser Beam
S. A. Ramsden, R. Benesch, W. E. R. Davies, and P. K. John (Nat'l Research Council, Canada)
IEEE J. Quantum Electronics, Vol. QE-2, pp. 267–270, August 1966 (hydrogen plasma)

21.71 Theoretical and Experimental Investigation of Regenerative Laser Amplifiers and Their Applications
H. Boersch and G. Herziger (Tech. Univ. Berlin, Germany)
IEEE J. Quantum Electronics, Vol. QE-2, pp. 549–552, September 1966 (oscillator and resonator in series, measure electron densities of 10^{13} to 10^{15} cm⁻³)

21.72 Laser Interferometer for Repetitively Pulsed Plasmas
E. B. Hooper, Jr., and G. Bekefi (M.I.T.)
J. Appl. Phys., Vol. 37, pp. 4083–4094, October 1966 (three mirrors, few microseconds time resolution)

21.73 Spatially Resolved Laser Heterodyne Measurements of Plasma Densities in Weakly Ionized Gases
J. T. Verdeyen, B. E. Cherrington, and M. E. Fein (U. of Illinois)
Appl. Phys. Letters, Vol. 9, pp. 360–362, November 15, 1966 (0.63 and 1.15 μ He–Ne laser, 24-cm path length, 10^{11} to 4×10^{12} e/cm³)

22. GAS BREAKDOWN

22.1 Gas Breakdown at Optical Frequencies
R. G. Meyerand, Jr., and A. F. Haught (United Aircraft)
Phys. Rev. Letters, Vol. 11, pp. 401–403, November 1, 1963

22.2 Optical Frequency Electrical Discharges in Gases
R. W. Minck (Ford)
J. Appl. Phys., Vol. 35, pp. 252–254, January 1964

22.3 Gas Breakdown Criterion for Pulsed Optical Radiation
R. G. Tomlinson (Ohio State U.)
Proc. IEEE, Vol. 52, pp. 721–722, June 1964 (calculated breakdown density is 4×10^{20} electrons/m^3)

22.4 Optical-Energy Absorption and High-Density Plasma Production
R. G. Meyerand, Jr. and A. F. Haught (United Aircraft)
Phys. Rev. Letters, Vol. 13, pp. 7–9, July 6, 1964 (20-ns ruby laser, ½ J absorbed, gas temperature increased)

22.5 Theory of the Electrical Breakdown of Gases by Intense Pulses of Light
J. K. Wright (Central Electricity Res. Lab., Leatherhead, Surrey, England)
Proc. Phys. Soc., Vol. 84, pp. 41–46, July 1964 (inverse bremsstrahlung absorption)

22.6 A Radiative Detonation Model for the Development of a Laser-Induced Spark in Air
S. A. Ramsden and P. Savic (Natl. Res. Council, Canada)
Nature, Vol. 203, pp. 1217–1219, September 19, 1964 (radiation supports shock wave towards lens)

22.7 Experimental and Theoretical Studies of Air Breakdown by Intense Pulse of Light
P. Nelson, P. Veyrie, M. Berry, and Y: Durand (AEC, France)
Physics Letters, Vol. 13, pp. 226–228, December 1, 1964 (multiphoton inverse bremsstrahlung effect)

22.8 Theory of Multiphoton Ionization
A. Gold and H. B. Bebb (U. of Rochester)
Phys. Rev. Letters, Vol. 14, pp. 60–63, January 18, 1965 (optical breakdown in gas)

22.9 Optical Frequency Electrical Discharge in Air
P. P. Pashinin, S. L. Mandelstam, A. M. Prokhorov, and N. K. Suhodrev (P. N. Lebedev Physical Inst., Academy of Sciences, Moscow, USSR)
Z. angewandte Mathematik und Physik, Vol. 16, No. 1, pp. 125–126, January 25, 1965

22.10 Recombination Mechanism for Laser-Produced Discharges in Argon
D. F. Edwards and M. M. Litvak (M.I.T.)
Bull. Am. Phys. Soc., Series II, Vol. 10, No. 1, p. 73–FE6, January 1965 (several ms afterglow in argon, 2.5×10^{15}e/cm^3, 2.1×10^4 °K electron temperature)

22.11 Gas Breakdown by Optical-Frequency Radiation
A. F. Haught and R. G. Meyerand, Jr. (United Aircraft)
Bull. Am. Phys. Soc., Series II, Vol. 10, No. 2, p. 189–I8, February 1965 (produced 10^{19} e/cm^3, 5–9 ns delay)

22.12 Shock-Wave Implosions Produced by Laser Breakdown
J. W. Daiber, C. W. Wittliff and A. Hertzberg (Cornell Aeronautical Lab.)
Bull. Am. Phys. Soc., Series II, Vol. 10, No. 2, p. 225–P7, February 1965 (several simultaneously fired lasers to focus shock wave)

22.13 Interpretation of the Development of a Laser-Induced Spark in Air in Terms of a Radiation-Supported Shock Wave
P. Savic and S. A. Ramsden (National Res. Council, Canada)
Bull. Am. Phys. Soc., Series II, Vol. 10, No. 2, p. 225–P8, February 1965 (wave moves toward source)

22.14 Stability of Plasma Produced by Laser
W. I. Linlor (Hughes)
Bull. Am. Phys. Soc., Series II, Vol. 10, No. 2, p. 227–Q11, February 1965 (not unstable as in fusions)

22.15 Theory of the Ionization of Gases by Laser Beams
B. A. Tozer (C.E.R.L., England)
Phys. Rev., Vol. 137, pp. A1665–A1667, March 15, 1965 (multiple photon absorption initially)

22.16 Light Spark in a Magnetic Field
G. A. Askar'yan, M. S. Rabinovich, M. M. Savchenko, and A. D. Smirnova (Lebedev Physics Inst., Academy of Sciences, U.S.S.R.)

JETP Letters, Vol. 1, pp. 5–8, April 1, 1965 (initiated by Q-sw laser)

22.17 Frequency Dependence of the Threshold of Optical Breakdown in Air
S. A. Akhmanov, A. I. Kovrigin, M. M. Strukov, and R. V. Khokhlov (Moscow State U., U.S.S.R.)
JETP Letters, Vol. 1, pp. 25–29, April 1, 1965 (Q-sw Nd glass laser, lower threshold at fundamental than second harmonic)

22.18 Ionization of the Xenon Atom by the Electric Field of Ruby Laser Emission
G. S. Voronov and N. B. Delone (Lebedev Physics Inst., Academy of Sciences, U.S.S.R.)
JETP Letters, Vol. 1, pp. 66–68, April 15, 1965 (10^{-2} torr of xenon)

22.19 Spectra of Laser-Produced Breakdown in Gases
J. W. Daiber (Cornell Aeronautical Lab.)
Bull. Am. Phys. Soc., Series II, Vol. 10, No. 4, p. 477–DH11, April 1965

22.20 Observation of a Fast Photoionization Aureole and of a Concentrated Long Lived Ionization Cloud due to a Shock Wave from a Spark in a Laser Beam
G. A. Askar'yan, M. S. Rabinovich, M. M. Savchenko, and A. D. Smirnova (Lebedev Physics Inst., Academy of Sciences, U.S.S.R.)
JETP Letters, Vol. 1, pp. 162–164, June 15, 1965 (plasma lifetime of hundreds of μs, interaction between 8-mm microwaves and optical spark)

22.21 Multiphoton Ionization of the Hydrogen Molecule in the Strong Electric Field of Ruby Laser Emission
G. S. Voronov, G. A. Delone, N. B. Delone, and O. V. Kudrevatova (Lebedev Physics Inst., Academy of Sciences, U.S.S.R.)
JETP Letters, Vol. 2, pp. 237–239, October 15, 1965 (10^{-4} torr, produced H_2^+ and H^+)

22.22 Breakdown Minima due to Electron-Impact Ionization in Super-High-Pressure Gases Irradiated by a Focused Giant-Pulse Laser
Dennis H. Gill and Arwin A. Dougal (U. of Texas)
Phys. Rev. Letters, Vol. 15, pp. 845–847, November 29, 1965 (up to 30 000 psi, minima at about 1000 to 10 000 psi, A, N_2, He)

22.23 Frequency Dependence of Optically Induced Gas Breakdown
Harold T. Buscher, Richard G. Tomlinson, and Edward K. Damon (Ohio State U.)
Phys. Rev. Letters, Vol. 15, pp. 847–849, November 29, 1965 (Nd and ruby lasers, also 2nd harmonic, thresholds do not increase monotonically with decreasing wavelengths)

22.24 Heating of a Gas by a Powerful Light Pulse
Yu. P. Raĭzer
Soviet Physics JETP, Vol. 21, pp. 1009–1017, November 1965 (light absorption, energy balance, detonation, breakdown wave, gas temperature)

22.25 Polarization of the Ionization Aureole of a Light Spark in a Constant Electric Field
G. A. Askar'yan, M. S. Rabinovich, A. D. Smirnova, and V. B. Studenov (Lebedev Physics Institute, Academy of Sciences, U.S.S.R.)
JETP Letters, Vol. 2, pp. 314–316, December 1, 1965 (Q-sw laser produces spark, 10 to 1000 V/cm field)

22.26 Multiphoton Ionization of Rare Gas and Hydrogen Atoms
H. Barry Bebb and Albert Gold (U. of Rochester)
Proc. Physics of Quantum Electronics Conf., McGraw-Hill Book Co., New York, pp. 489–498, 1966

22.27 Electrical Breakdown of Gases by Optical Frequency Radiation
Alan F. Haught, Russell G. Meyerand, Jr., and David C. Smith

(United Aircraft Research Labs.)
Proc. Physics of Quantum Electronics Conf., McGraw-Hill Book Co., New York, pp. 509–519, 1966

22.28 The Breakdown of Noble and Atmospheric Gases by Ruby and Neodymium Laser Pulses
Richard G. Tomlinson, Edward K. Damon, and Harold T. Buscher (Ohio State U.)
Proc. Physics of Quantum Electronics Conf., McGraw-Hill Book Co., New York, pp. 520–526, 1966

22.29 Investigation of Optical Frequency Breakdown Phenomena
R. W. Minck and W. G. Rado (Ford Motor Co.)
Proc. Physics of Quantum Electronics Conf., McGraw-Hill Book Co., New York, pp. 527–537, 1966

22.30 Theory of Growth of Ionization During Laser Breakdown
A. V. Phelps (Westinghouse)
Proc. Physics of Quantum Electronics Conf., McGraw-Hill Book Co., New York, pp. 538–547, 1966

22.31 Optical Frequency Electrical Discharge in Air
S. L. Mandelstam, P. P. Pashinin, A. M. Prokhorov, and N. K. Sukhodrev (Lebedev Institute of Physics, Academy of Sciences of the USSR, Moscow, USSR)
Proc. Physics of Quantum Electronics Conf., McGraw-Hill Book Co., New York, pp. 548–553, 1966

22.32 Laser-Induced Breakdown in Oxygen Gas at High Pressure
T. H. Wiggins, R. V. Wick, D. H. Rank (Pennsylvania State U.), and A. H. Guenther (Air Force Weapons Lab., Kirtland Air Force Base)
Appl. Optics, Vol. 5, pp. 166–167, January 1966 (395 atmospheres, damaged cell, 0.9 J, 10 ns pulse)

22.33 Optical Frequency Plasma Resonance in Gases
R. W. Minck and W. G. Rado (Ford Motor Co.)
J. Appl. Phys., Vol. 37, pp. 355–358, January 1966 (10% reflection from spark, postulate high electron density for plasma resonance)

22.34 Investigation of the Spark Discharge Produced in Air by Focusing Laser Radiation. II
S. L. Mandel'shtam, P. P. Pashinin, A. M. Prokhorov, Yu. P. Raĭzer, and N. K. Sukhodrev (Lebedev Physics Institute, Academy of Sciences, U.S.S.R.)
Soviet Physics JETP, Vol. 22, pp. 91–96, January 1966 ($T_e = 50$–60 eV, 10^7 cm/s ionization front velocity towards Q-sw ruby laser source, hydrodynamic breakdown mechanism)

22.35 Electromagnetic Breakdown of Air
P. A. Clavier and L. Webb (National Engineering Science Company)
J. Appl. Phys., Vol. 37, pp. 742–744, February 1966

22.36 Mesures sur un Plasma Produit par Laser
E. Fabre, P. Vasseur, and G. Bevernage (Ecole Polytechnique)
Physics Letters, Vol. 20, pp. 381–382, March 1, 1966

22.37 Multiphoton Ionization of Hydrogen and Rare-Gas Atoms
H. Barry Bebb and Albert Gold (U. of Rochester)
Phys. Rev., Vol. 143, pp. 1–24, March 4, 1966 (direct multiple-photon ionization provides free electrons to initiate breakdown, up to 14 photons, Q-sw laser)

22.38 Use of a Laser Operating in the Spike Mode to Obtain a High-Temperature Plasma
M. P. Vanyukov, V. I. Isaenko, V. V. Lyubimov, V. A. Serebryakov, and O. A. Shorokhov
JETP Letters, Vol. 3, p. 205, April 15, 1966 (60 cm long, 4.5 cm diameter Nd glass rod, 1400 J output, 1.2 ms pulse, focused density of 3 GW/cm², photo of air plasma)

22.39 Experimental Evidence of Inverse Bremsstrahlung and Electron-Impact Ionization in Low-Pressure Argon Ionized by a Giant-Pulse Laser
Che Jen Chen (Calif. Inst. of Tech.)
Phys. Rev. Letters, Vol. 16, pp. 833–835, May 9, 1966

22.40 Laser-Induced Prebreakdown and Breakdown Phenomena Observed in Cloud Chamber
C. S. Naiman, M. Y. DeWolf, I. Goldblatt, and J. Schwartz (Mithras)
Phys. Rev., Vol. 146, pp. 133–135, June 3, 1966

22.41 Energy-Loss Processes in Optical-Frequency Gas Breakdown
David C. Smith and Alan F. Haught (United Aircraft)
Phys. Rev. Letters, Vol. 16, pp. 1085–1088, June 13, 1966

22.42 Multiphoton Ionization of Krypton and Argon by Ruby Laser Radiation
G. S. Voronov, G. A. Delone, and N. B. Delone (Lebedev Phys. Inst., Acad. of Sci., USSR)
JETP Letters, Vol. 3, pp. 313–315, June 15, 1966

22.43 Many-Photon Ionization of the Xenon Atom by Ruby Laser Radiation
G. S. Voronov and N. B. Delone (Lebedev Phys. Inst., Acad. of Sci., USSR)
Soviet Phys. JETP, Vol. 23, pp. 54–58, July 1966 (six-photon process)

22.44 Investigation of the Magnetic Field of a Spark Produced by Focusing Laser Radiation
V. V. Korobkin and R. V. Serov (Lebedev Physics Inst., Acad. of Sci., USSR)
JETP Letters, Vol. 4, pp. 70–72, August 1, 1966 (produced magnetic moment of about 4×10^{-2} Oe/cm³)

22.45 Holographic Investigation of a Laser Spark
G. V. Ostrovskaya and Yu. I. Ostrovskii (Ioffe Phys. Tech. Inst., USSR)
JETP Letters, Vol. 4, pp. 83–85, August 15, 1966 (three holograms during one spark, structure of spark)

22.46 Breakdown at Optical Frequencies in the Presence of Diffusion Losses
V. E. Mitsuk, V. I. Savoskin, and V. A. Chernikov (Moscow State U., USSR)
JETP Letters, Vol. 4, pp. 88–90, August 15, 1966 (at low pressures, Kr and Xe, increased field intensities)

22.47 Statistical Models for Laser-Induced Ionisation of Gases
J. W. Gardner (English Electric, England)
Electronics Letters, Vol. 2, pp. 297–298, August 1966 (Poisson distribution predicts too low breakdown threshold, Polya distribution more accurate)

22.48 A Study of the Soft X-Ray Emission and Dynamics of a Laser Spark
A. J. Alcock, P. P. Pashinin, and S. A. Ramsden (Nat'l Research Council, Canada)
IEEE J. Quantum Electronics, Vol. QE-2, p. 296, August 1966 (abstract)

22.49 Study of Ionization, Heating, and Plasma Generation in Gases by an Intense Laser Beam
J. L. Champetier, J. C. Fecan, F. Floux, P. Nelson, and P. Veyrie (CEA, France)
IEEE J. Quantum Electronics, Vol. QE-2, p. 297, August 1966 (abstract)

22.50 The Mechanism of the Optical Breakdown in a Gas
V. A. Barynin and R. V. Khoklov (Moscow State U., USSR)
Soviet Phys. JETP, Vol. 23, pp. 314–315, August 1966 (contribution by photoionization)

23.   OTHER NONLINEAR, THERMAL, AND INTERACTION EFFECTS WITH MATTER

Proc. Third Int. Conf. Quantum Electronics, Columbia U. Press, N. Y., Vol. 1, pp. 691–698, 1964

23.17 Usinage Photonique Avec Generateur Laser
M. S. Bruma (Lab. de Chimie Physique, Faculté des Sciences de Paris, France)
Proc. Third Int. Conf. Quantum Electronics, Columbia U. Press, N. Y., Vol. 2, pp. 1333–1337, 1964

23.18 The Conditions of Plasma Heating by the Optical Generator Radiation
N. G. Basov and O. N. Krokhin (Lebedev Institute, Academy of Sciences, U.S.S.R.)
Proc. Third Int. Conf. Quantum Electronics, Columbia U. Press, N. Y., Vol. 2, pp. 1373–1377, 1964

23.19 Survey of Non-Linear Optical Effects.
P. A. Franken (U. of Michigan)
Proc. Third Int. Conf. Quantum Electronics, Columbia U. Press, N. Y., Vol. 2, pp. 1495–1500, 1964

23.20 Optique Non-Linéaire
N. Bloembergen (Harvard U.)
Proc. Third Int. Conf. Quantum Electronics, Columbia U. Press, N. Y., Vol. 2, pp. 1501–1512, 1964

23.21 Nonlinear Optical Properties of Solids
P. S. Pershan (Harvard U.)
Proc. Third Int. Conf. Quantum Electronics, Columbia U. Press, N. Y., Vol. 2, pp. 1513–1520, 1964

23.22 Nonlinear Effects in Solid State Plasmas
B. Lax, A. L. McWhorter, and J. G. Mavroides (Natl. Magnet Lab., and Lincoln Lab., MIT)
Proc. Third Int. Conf. Quantum Electronics, Columbia U. Press, N. Y., Vol. 2, pp. 1521–1526, 1964

23.23 Recent Experiments in Non-Linear Optics
J. A. Giordmaine (BTL)
Proc. Third Int. Conf. Quantum Electronics, Columbia U. Press, N. Y., Vol. 2, pp. 1549–1558, 1964

23.24 Orientation Optique de Molecules
A. Piekara and S. Kielich (Lab. d'Etude des Dielectriques, Institut de Physique de l'Academie Polonaise des Sciences, Poznan, Poland)
Proc. Third Int. Conf. Quantum Electronics, Columbia U. Press, N. Y., Vol. 2, pp. 1603–1607, 1964

23.25 Scattering of Light by Light in a Non-Linear Medium
H. R. Robl (U. S. Army Res. Office, Durham, N. C.)
Proc. Third Int. Conf. Quantum Electronics, Columbia U. Press, N. Y., Vol. 2, pp. 1635–1642, 1964

23.26 Influence de la Largeur Spectrale sur l'Interaction d'Ondes Planes au Sein d'un Diélectrique Non-Linéaire
J. Ducuing and J. A. Armstrong (Harvard U.)
Proc. Third Int. Conf. Quantum Electronics, Columbia U. Press, N. Y., Vol. 2, pp. 1643–1650, 1964

23.27 Experiments on Photo-Field-Emission
A. J. Alcock, H. Motz, and D. Walsh (Oxford U., England)
Proc. Third Int. Conf. Quantum Electronics, Columbia U. Press, N. Y., Vol. 2, pp. 1687–1693, 1964

23.28 Interaction Between Two Nd^{3+} Glass Lasers
C. J. Koester, R. F. Woodcock, E. Snitzer (American Optical), and H. M. Teager (MIT)
Proc. Third Int. Conf. Quantum Electronics, Columbia U. Press, N. Y., Vol. 2, pp. 1703–1710, 1964

23.29 Quantum-Theoretical Comparison of Nonlinear Susceptibilities in Parametric Media, Lasers, and Raman Lasers
N. Bloembergen and Y. R. Shen (Harvard U.)
Phys. Rev., Vol. 133, pp. A37–A49, January 6, 1964

23.30 Structure of Nonlinear Optical Phenomena in Potassium Dihydrogen Phosphate

J. F. Ward and P. A. Franken (U. of Michigan)
Phys. Rev., Vol. 133, pp. A183–A190, January 6, 1964

23.31 Optically-Induced Ultrasonic Waves in Transparent Dielectrics
A. J. DeMaria (United Aircraft)
Proc. IEEE, Vol. 52, pp. 96–97, January 1964

23.32 Interaction of Intense Laser Beams with Electrons
Lowell S. Brown and T. W. B. Kibble (Imperial College, England)
Phys. Rev., Vol. 133, pp. A705–A719, February 3, 1964

23.33 Laser-Heated Cathode
G. C. Dalman and T. S. Wen (Chiao Tung U., Taiwan)
Proc. IEEE, Vol. 52, pp. 200–201, February 1964

23.34 Generation of Acoustic Signals in Liquids by Ruby Laser-Induced Thermal Stress Transients
E. F. Carome, N. A. Clark, and C. E. Moeller (John Carroll U.)
Appl. Phys. Letters, Vol. 4, pp. 95–97, March 15, 1964

23.35 Some Properties of Plasma Produced by Laser Giant Pulse
William I. Linlor (Hughes Res.)
Phys. Rev. Letters, Vol. 12, pp. 383–385, April 6, 1964

23.36 Metal Ion Emission Velocity Dependence on Laser Giant Pulse Height
N. R. Isenor (U. of Waterloo, Canada)
Appl. Phys. Letters, Vol. 4, pp. 152–153, April 15, 1964

23.37 Momentum Transfer and Cratering Effects Produced by Giant Laser Pulses
Frank Neuman (NASA, Ames Res. Center)
Appl. Phys. Letters, Vol. 4, pp. 167–169, May 1, 1964
(vacuum environment)

23.38 Nonlinear Optical Theory in Solids
H. Cheng and P. B. Miller (IBM)
Phys. Rev., Vol. 134, pp. A683–A687, May 4, 1964

23.39 Nonlinear Optical Frequency Polarization in a Dielectric
E. Adler (Columbia U.)
Phys. Rev., Vol. 134, pp. A728–A733, May 4, 1964

23.40 Intensity-Dependent Changes in the Refractive Index of Liquids
P. D. Maker, R. W. Terhune, and C. M. Savage (Ford)
Phys. Rev. Letters, Vol. 12, pp. 507–509, May 4, 1964 (Q-switched ruby, water, chloroform, carbon disulfide, etc., elliptically polarized output)

23.41 Competition between Internal and External Modes of a Ruby Laser
A. Korpel and J. Free (Zenith Radio)
Proc. IEEE, Vol. 52, pp. 619–620, May 1964 (0.6° between internal and external cavity axes, spikes from two cavities generally coincident)

23.42 Three-Photon Absorption in Napthalene Crystals by Laser Excitation
S. Singh and L. T. Bradley (Natl. Res. Council, Ottawa, Canada)
Phys. Rev. Letters, Vol. 12, pp. 612–614, June 1, 1964 (Q-switched ruby, 10^{-10} efficiency, 83-ns fluorescence lifetime)

23.43 Gases from Flash and Laser Irradiation of Coal
A. G. Sharkey, Jr., J. L. Shultz, and R. A. Friedel (U. S. Bur. of Mines)
Nature, Vol. 202, pp. 988–989, June 6, 1964 (different gases compared to high temperature carbonization)

23.44 Presence of Free Radicals in Laser-Irradiated Biological Specimens by Electron-Spin Resonance
V. E. Derr (Martin), E. Klein (Roswell Park Mem. Inst.), and

S. Fine (Northeastern U.)
Appl. Optics, Vol. 3, pp. 786–787, June 1964

23.45 Electro-Optic Amplitude Modulation of Laser-Generated Second Harmonics in KH_2PO_4 (KDP)
J. P. van der Ziel (Harvard U.)
Appl. Phys. Letters, Vol. 5, pp. 27–29, July 15, 1964 (bias voltage changes harmonic amplitude)

23.46 Cooperative Effect in GaAs Lasers
A. B. Fowler (IBM)
J. Appl. Phys., Vol. 35, pp. 2275–2276, July 1964 (short and long units as pair)

23.47 Acoustical Generation by Optical Masers
V. A. Suprynowicz (United Aircraft Corp.)
Proc. IEEE, Vol. 52, p. 849, July 1964 (experiments require care)

23.48 Direct Observation of Optically Induced Generation and Amplification of Sound
A. Korpel, R. Adler, and B. Alpiner (Zenith)
Appl. Phys. Letters, Vol. 5, pp. 86–88, August 15, 1964 (Q-switched ruby, RF modulated acoustic cell)

23.49 Radiation Scattered from the Plasma Produced by a Focused Ruby Laser Beam
S. A. Ramsden and W. E. R. Davies (Natl. Res. Council, Ottawa, Canada)
Phys. Rev. Letters, Vol. 13, pp. 227–229, August 17, 1964 (delayed scattering by plasma electrons)

23.50 Determination of Laser Damage Threshold for Various Glasses
J. H. Cullom and R. W. Waynant (Westinghouse)
Appl. Optics, Vol. 3, pp. 989–990, August 1964 (Q-switched ruby laser focused inside samples)

23.51 Ruby-Laser-Induced Photocurrents and Luminescence in ZnO
D. R. Hotchkiss (MMM)
J. Appl. Phys., Vol. 35, pp. 2455–2457, August 1964 (green luminescence (5250Å) produced)

23.52 Two-Photon Photoelectric Effect in Cs_3Sb
H. Sonnenberg, H. Heffner, and W. Spicer (Stanford U.)
Appl. Phys. Letters, Vol. 5, pp. 95–96, September 1, 1964 (used Nd-doped glass laser)

23.53 Intensity-Dependent Absorption of Light
E. U. Condon (U. of Colorado)
Proc. Natl. Acad. of Sciences, Vol. 52, pp. 635–637, September 15, 1964 (interpretation of Bret and Gires' observations in CdSe glass)

23.54 Mutually Quenched Injection Lasers as Bistable Devices
G. J. Lasher and A. B. Fowler (IBM)
IBM J. Res. and Dev., Vol. 8, pp. 471–475, September 1964 (laser beams are perpendicular)

23.55 Laser-Induced Damage to Transparent Dielectric Materials
C. R. Giuliano (Hughes)
Appl. Phys. Letters, Vol. 5, pp. 137–139, October 1, 1964 (caused by phonons, sparking often on both front and back faces of glass sample, damage occurred on back exit face)

23.56 Low Absorption Measurements by Means of the Thermal Lens Effect Using an He-Ne Laser
R. C. C. Leite, R. S. Moore, and J. R. Whinnery (BTL)
Appl. Phys. Letters, Vol. 5, pp. 141–143, October 1, 1964 (can measure 10^{-3} to 10^{-4} cm^{-1} at 6328 Å)

23.57 Nonlinear Interaction of Light in a Plasma
P. M. Platzman and N. Tzoar (BTL)
Phys. Rev., Vol. 136, pp. A11–A16, October 5, 1964

23.58 Self-Trapping of Optical Beams
R. Y. Chiao, E. Garmire, and C. H. Townes (MIT)
Phys. Rev. Letters, Vol. 13, pp. 479–482, October 12, 1964 (index increases with intensity; self-trapping in water, glass and air at 1, 4, and 80 MW, respectively)

23.59 Erratum: Self-Trapping of Optical Beams
R. Y. Chiao, E. Garmire, and C. H. Townes
Phys. Rev. Letters, Vol. 14, p. 1056, June 21, 1965 (Phys. Rev. Letters, Vol. 13, p. 479; 1964. Refers to prior Soviet work)

23.60 Laser-Induced Emission of Electrons, Ions, and Neutrals from Ti and Ti-D Surfaces
T. Y. Chang and C. K. Birdsall (U. of California)
Appl. Phys. Letters, Vol. 5, pp. 171–172, November 1, 1964 (surface damage due to blow off)

23.61 Double-Quantum Photoelectric Emission from Sodium Metal
M. C. Teich, J. M. Schroeer, and G. J. Wolga (Cornell U.)
Phys. Rev. Letters, Vol. 13, pp. 611–614, November 23, 1964 (pulsed GaAs beam focused on sodium surface)

23.62 Non-Thermal Biological Effects of Laser Beams
V. T. Tomberg (N. Y. School of Medicine)
Nature, Vol. 204, pp. 868–870, November 28, 1964

23.63 Dichroism of the Vacuum
J. J. Klein and B. P. Nigam (State U. of N. Y. at Buffalo)
Phys. Rev., Vol. 136, pp. B1540–B1542, December 7, 1964 (theoretical absorption of light polarized parallel to electric field of 10^{16} V/cm)

23.64 Laser Excitation of Powdered Solids
H. I. S. Ferguson, J. E. Mentall, and R. W. Nichols (U. of Western Ontario, Canada)
Nature, Vol. 204, p. 1295, December 26, 1964 (spectroscopic studies of metals, oxides and other compounds)

23.65 Nanosecond Heating of Aqueous Systems by Giant Laser Pulses
H. Staerk and G. Czerlinski (U. of Pennsylvania)
Nature, Vol. 205, pp. 63–64, January 2, 1965 (one joule Q-sw ruby, 4°C temp rise in liquid, bleaching possibly observed)

23.66 Mechanism of Electron Emission Produced by a Giant-Pulse Laser
J. F. Ready (Honeywell)
Phys. Rev., Vol. 137, pp. A620–A623, January 18, 1965 (thermally induced emission, 30 ns delay)

23.67 Interaction of High Intensity Light Beams with Matter
H. R. Robl (U. S. Army Research Office)
Z. angewandte Mathematik und Physik, Vol. 16, No. 1, p. 33, January 25, 1965

23.68 Free-Radical Production by Laser Photolysis
J. F. Verdieck (U. of Michigan)
Bull. Am. Phys. Soc., Series II, Vol. 10, No. 1, p. 87–GG12, January 1965 (NOH radical produced by ruby laser)

23.69 Second-Order Scattering Studies in Liquids, Using a Laser
P. D. Maker, R. W. Terhune, and C. M. Savage (Ford)
Bull. Am. Phys. Soc., Series II, Vol. 10, No. 1, p. 97–HG5, January 1965 (near second harmonic of ruby laser)

23.70 Perturbation of the Refractive Index of Absorbing Media by a Laser Beam
P. R. Longaker and M. M. Litvak (M.I.T.)
Bull. Am. Phys. Soc., Series II, Vol. 10, No. 1, p. 101–JC4, January 1965 (time-resolved interferometry using argon laser, low loss gas and liquid samples heated by Nd laser pulse)

23.71 Interactions between Closely Coupled GaAs Injection Lasers
C. E. Kelly (IBM)
IEEE Trans. on Electron Devices, Vol. ED-12, pp. 1–4, January 1965 (pulsed, 2° and 77° K, lowers threshold, produces quenching, single block)

23.72 Laser-Induced Temperature Radiation
T. Kushida (Toshiba, Japan)
Japanese J. Appl. Phys., Vol. 4, pp. 73–74, January 1965 (Nd glass laser, focused, generally white light emitted by target)

23.73 Long-Transient Effects in Lasers with Inserted Liquid Samples
J. P. Gordon, R. C. C. Leite, R. S. Moore, S. P. S. Porto, and

J. R. Whinnery (BTL)
J. Appl. Phys., Vol. 36, pp. 3–8, January 1965 (samples within laser cavity, effects possibly due to small absorption)

23.74   Excitation of Hypersonic Vibrations by Means of Photoelastic Coupling of High-Intensity Light Waves to Elastic waves
N. M. Kroll (U. of California)
J. Appl. Phys., Vol. 36, pp. 34–43, January 1965

23.75   Effect of Background Gas on Laser-Induced Electron Emission from Metal Surfaces
N. R. Isenor (U. of Waterloo, Canada)
J. Appl. Phys., Vol. 36, pp. 316–317, January 1965

23.76   Study of Optical Effects due to an Induced Polarization Third Order in the Electric Field Strength
P. D. Maker and R. W. Terhune (Ford)
Phys. Rev., Vol. 137, pp. A801–A818, February 1, 1965 (comprehensive study including ruby laser experiments)

23.77   Plasma Generation by the Laser Illumination of a Tungsten Surface
A. S. Gilmour, Jr. and F. A. Giori (Cornell Aeronautical Lab.)
Bull. Am. Phys. Soc., Series II, Vol. 10, No. 2, p. 189–I7, February 1965 (pulsed ruby laser, efficient process)

23.78   Plasma Production by Laser-Beam Irradiation of Solid Particles
A. F. Haught and R. G. Meyerand, Jr. (United Aircraft)
Bull. Am. Phys. Soc., Series II, Vol. 10, No. 2, p. 227–Q9, February 1965 (particles suspended in vacuum, Q-switched ruby)

23.79   Description of the Plasma Produced by a Laser Pulse Striking an Aluminum Surface
A. W. Ehler (Hughes)
Bull. Am. Phys. Soc., Series II, Vol. 10, No. 2, p. 227–Q10, February 1965 (Q-switched laser)

23.80   The Surface Temperature of Metals Heated with Laser
S. Namba and P. H. Kim (Inst. Physical and Chemical Res., Tokyo, Japan) and S. Nakayama and I. Ida (N.T.T., Tokyo, Japan)
Japanese J. Appl. Phys., Vol. 4, pp. 153–154, February 1965 (to above 5000°K)

23.81   Effects due to Absorption of Laser Radiation
J. F. Ready (Honeywell)
J. Appl. Phys., Vol. 36, pp. 462–468, February 1965 (Q-sw and normal pulse lasers, analyses, experiments with Nd laser, vaporization)

23.82   Laser-Induced Production of Free Radicals in Organic Compounds
Y. Pao and P. M. Rentzepis (BTL)
Appl. Phys. Letters, Vol. 6, pp. 93–95, March 1, 1965 (ruby laser, polymerization, 2 photon absorption)

23.83   Initial Energies of Laser-Induced Electron Emission from W
W. L. Knecht (WPAFB)
Appl. Phys. Letters, Vol. 6, pp. 99–100, March 15, 1965 (14.5 eV first peak, 2 eV delayed peak, focused Q-sw ruby, incident angle dependent)

23.84   Saturation of the Optical Absorption in GaAs
A. E. Michel and M. I. Nathan (IBM)
Appl. Phys. Letters, Vol. 6, pp. 101–102, March 15, 1965 (transmission slope increases with intensity)

23.85   Power Sensitive Optical Switch for the Control of High Brightness Laser Pulses
R. G. Tomlinson (Ohio State U.)
J. Appl. Phys., Vol. 36, pp. 868–870, Part 1, March 1965 (pair of convergent lenses, breakdown at focal point, lower threshold at higher pressures, limit down to 0.2 megawatts with argon)

23.86   Polarization of Anthracene Fluorescence by One and Two Photon Excitation
W. L. Peticolas and K. E. Rieckhoff (IBM)
Physics Letters, Vol. 15, pp. 230–231, April 1, 1965 (experiments with ruby laser)

23.87   Theory of Higher Multipole Contributions to Two-Photon Absorption Processes
R. Guccione and J. van Kranendonk (U. of Toronto, Canada)
Phys. Rev. Letters, Vol. 14, pp. 583–584, April 12, 1965 (higher multipole contributions are negligible)

23.88   Nuclear Superradiance in Solids
J. H. Terhune and G. C. Baldwin (G.E.)
Phys. Rev. Letters, Vol. 14, pp. 589–591, April 12, 1965 (coherent spontaneous gamma-ray emission theory)

23.89   Optical Rectification
M. Bass, P. A. Franken, and J. F. Ward (U. of Michigan)
Phys. Rev., Vol. 138, pp. A534–A542, April 19, 1965 (Q-sw ruby and Nd lasers, coefficients of several crystals, low temperature enhancement)

23.90   Measurements of Nonlinear Light Scattering
R. W. Terhune, P. D. Maker, and C. M. Savage (Ford Motor Co.)
Phys. Rev. Letters, Vol. 14, pp. 681–684, April 26, 1965 (frequency displacements characteristic of material, water, quartz)

23.91   Photocarrier Generation in Anthracene Due to Exciton Interaction of Two-Photon Excited Singlets
K. Hasegawa and S. Yoshimura (Matsushita Res. Inst., Japan)
Phys. Rev. Letters, Vol. 14, pp. 689–690, April 26, 1965 (Q-sw ruby laser)

23.92   Study of Nucleation Induced by Laser Irradiation
C. S. Naiman, M. Y. DeWolf, and J. Schwartz (Mithras)
Bull. Am. Phys. Soc., Series II, Vol. 10, No. 4, p. 493–EH7, April 1965 (supersaturated methanol, nonfocused Q-sw, cloud lasts 2 minutes)

23.93   Laser Induced Photoconductivity in CdS Crystal
K. Yoshino, Y. Watanabe, and Y. Inuishi (Osaka U., Japan)
Japanese J. Appl. Phys., Vol. 4, pp. 312–313, April 1965 (much higher increase in photocurrent with intensity from ruby laser than green light)

23.94   Polarization Dependence of Laser-Induced Fluorescence in Anthracene
M. Iannuzzi (Oxford U., England) and E. Polacco (U. of Pisa, Italy)
Phys. Rev., Vol. 138, pp. A806–A808, May 3, 1965 (maximum and zero fluorescence for linearly and circularly polarized radiation, respectively, from Q-sw ruby laser)

23.95   Optically Induced Magnetization in Ruby
J. P. Van der Ziel (Harvard U.) and N. Bloembergen (U. of California)
Phys. Rev., Vol. 138, pp. A1287–A1292, May 17, 1965 (pumped by Q-sw ruby laser)

23.96   Laser-Induced Thermionic Emission from Tantalum
C. M. Verber and A. H. Adelman (Battelle Mem. Inst.)
J. Appl. Phys., Vol. 36, pp. 1522–1525, May 1965 (focused ruby laser beam)

23.97   On the Production of High Energy Particles by Giant-Pulse Lasers
W. Vali and V. Vali (Boeing)
Proc. IEEE, Vol. 53, pp. 517–518, May 1965 (over 10^{10} eV)

23.98   Effect of a Focused Ruby-Laser Beam on the Ruby
T. P. Belikova and E. A. Sviridenkov (Lebedev Physics Inst., Academy of Sciences, U.S.S.R.)
JETP Letters, Vol. 1, pp. 171–172, June 15, 1965 (internal fractures accompanied by visible flash, blue and orange glow, perhaps due to excitation from ^2E level)

23.99   Electromagnetic Field in the Neighborhood of the Focus of a Coherent Beam
A. Boivin (Laval U., Canada) and E. Wolf (U. of Rochester)
Phys. Rev., Vol. 138, pp. B1561–B1565, June 21, 1965 (contours of energy density)

23.100 Laser Double-Quantum Photodetachment of I⁻
J. L. Hall, E. J. Robinson, and L. M. Branscomb (Joint Inst. for Laboratory Astrophysics, NBS, and U. of Colorado)
Phys. Rev. Letters, Vol. 14, pp. 1013–1016, June 21, 1965 (iodine in high vacuum, Q-sw ruby laser)

23.101 Nonlinear Optical Reflection from a Metallic Boundary
F. Brown, R. E. Parks, and A. M. Sleeper (Williams College)
Phys. Rev. Letters, Vol. 14, pp. 1029–1031, June 21, 1965 (giant pulse ruby laser, second-harmonic reflection by silver surface)

23.102 Quantitative Measurements of Double-Photon Absorption in the Polycyclic Benzene Ring Compounds
D. H. McMahon, R. A. Soref, and A. R. Franklin (Sperry Rand Res. Center)
Phys. Rev. Letters, Vol. 14, pp. 1060–1062, June 28, 1965

23.103 Stimulated Scattering of Light of the Rayleigh-Line Wing
D. I. Mash, V. V. Morozov, V. S. Starunov, and I. L. Fabelinskii (Lebedev Physics Inst., Academy of Sciences, U.S.S.R.)
JETP Letters, Vol. 2, pp. 25–27, July 1, 1965 (passive Q-sw ruby, several liquids, spectra up to 15 cm⁻¹ away)

23.104 Effects of Optical Interaction of Two Diode Lasers
P. G. Eliseev, A. A. Novikov, and V. B. Fedorov (Inst. of Precision Mechanics and Computation Techniques, Academy of Sciences, U.S.S.R.)
JETP Letters, Vol. 2, pp. 36–39, July 15, 1965 (short and long GaAs diodes)

23.105 Observation of Self-Focusing of Light in Liquids
N. F. Pilipetskii and A. R. Rustamov (Moscow State U., U.S.S.R.)
JETP Letters, Vol. 2, pp. 55–56, July 15, 1965 (glowing filaments in organic liquids, focused Q-sw laser beam, threshold of 0.2MW)

23.106 A Crossed-Beam Optical Gate with a Saturable Absorber
F. T. Arecchi, V. Degiorgio, and A. Sona (Laboratori C.I.S.E., Milan, Italy)
Nuovo Cimento, Vol. 38, pp. 1096–1098, July 16, 1965 (Q-sw ruby laser, beam splitter, cell of phthalocyanine in nitrobenzene, rejection factor of 20, variable time delay)

23.107 Electromagnetic Shocks and the Self-Annihilation of Intense Linearly Polarized Radiation in an Ideal Dielectric Material
Gerald Rosen (Southwest Research Inst.)
Phys. Rev., Vol. 139, pp. A539–A543, July 19, 1965 (nonlinear solution of Maxwell's equations, EM shock wave train)

23.108 Reduced Absorption of Light at High Laser Power Densities
Hubertus Staerk and Georg Czerlinski (U. of Pennsylvania)
Nature, Vol. 207, pp. 399–400, July 24, 1965 (photobleaching of methylene blue in solution by narrow beam of 0.63-μ gas laser)

23.109 Cascade Ionization Induced in a Medium by an Intense Light Flash
G. A. Askar'yan and M. S. Rabinovich (Lebedev Physics Institute, Academy of Sciences, U.S.S.R.)
Soviet Physics JETP, Vol. 21, pp. 190–192, July 1965

23.110 Optically-Induced Magnetization Resulting from the Inverse Faraday Effect
J. P. van der Ziel, P. S. Pershan, and L. D. Malmstrom (Harvard U.)
Phys. Rev. Letters, Vol. 15, pp. 190–193, August 2, 1965 (nonabsorbing media, experiments using Q-sw ruby laser)

23.111 Three-Photon Molecular Scattering of Light
S. A. Akhmanov and D. N. Klyshko (Moscow State U., U.S.S.R.)
JETP Letters, Vol. 2, pp. 108–111, August 15, 1965 (estimated cross sections of Rayleigh and Raman scattering)

23.112 Nonlinear Optical Reflection from a Metal Surface
Sudhanshu S. Jha (Tata Inst., India)
Phys. Rev. Letters, Vol. 15, pp. 412–414, August 30, 1965

23.113 Dispersion of the Optical Nonlinearity in Semiconductors
R. K. Chang (Harvard U.), J. Ducuing (M.I.T.), and N. Bloembergen (U. of California)
Phys. Rev. Letters, Vol. 15, pp. 415–418, August 30, 1965 (harmonics reflected from zincblende symmetry semiconductors)

23.114 Role of Flare Images in the Self-Destruction of a Lens in a Strong Laser Beam
G. L. Rogers (College of Advanced Tech., England)
J. Appl. Phys., Vol. 36, p. 2598, August 1965 (suggested optics for elimination of flare images)

23.115 Strong Excitation and Dissociation of Molecules in an Intense Light Field
G. A. Askar'yan (Lebedev Physics Inst., Academy of Sciences, U.S.S.R.)
Soviet Physics JETP, Vol. 21, pp. 439–443, August 1965 (vibrational resonances, energy transfer)

23.116 Self Focusing of Wave Beams in Nonlinear Media
V. I. Talanov (Scientific Research Radio Physics Inst., Gorkii State U., U.S.S.R.)
JETP Letters, Vol. 2, pp. 138–141, September 1, 1965

23.117 Coherent Light Radiation by a Lorentz Field in GaAs
K. K. N. Chang and H. J. Prager (RCA)
Phys. Rev. Letters, Vol. 15, pp. 453–456, September 6, 1965 (3.39-μ gas laser beam through GaAs wafer in transverse magnetic field, alters radiation pattern)

23.118 Measurement of Light-Sound Interaction Efficiencies in Solids
T. M. Smith and A. Korpel (Zenith)
IEEE J. Quantum Electronics, Vol. QE-1, pp. 283–284, September 1965 (42.5 Mc sound wave, He–Ne laser beam at Bragg angle)

23.119 Visual Observation of Infrared Laser Emission
L. S. Vasilenko, V. P. Chebotaev, and Yu. V. Troitskii (Academy of Sciences, U.S.S.R.)
Soviet Physics JETP, Vol. 21, pp. 513–514, September 1965 (0.95 to 1.18 μ radiation observed with unaided eye, 2nd harmonic)

23.120 Influence of Intense Laser Radiation on the Dispersive Properties of "Transparent" Crystals
M. S. Brodin, V. N. Batulev, and S. V. Zakrevskii (Inst. of Physics, Ukrainian Academy of Sciences, U.S.S.R.)
JETP Letters, Vol. 2, pp. 201–203, October 1, 1965 (ruby laser, CdS and ZnS crystals, 10⁻³ increase in index)

23.121 Laser-Induced Damage in Natural White Diamond
P. Whiteman and G. W. Wilson (International Research and Development Co., Ltd., England)
Nature, Vol. 208, pp. 66–67, October 2, 1965 (Q-sw ruby, 0.3 joule, focused, seared front surface, rear surface damage due to acoustic phonons)

23.122 Internal Self-Damage of Ruby and Nd-Glass Lasers
P. V. Avizonis and T. Farrington
Appl. Phys. Letters, Vol. 7, pp. 205–206, October 15, 1965 (Q-switched, oscillator-amplifier system; 10 joules cm⁻³ at 9 ns and 30 joules cm⁻³ at 100 ns for ruby; 3 joules cm⁻³ at 10 ns and 11 joules cm⁻³ at 100 ns for Nd glass)

23.123 Direct Observation of Nonlinear Scattering of Electrons by Laser Beam
H. Schwarz and H. A. Tourtellotte (R.P.I., Hartford Graduate

Center) and W. W. Gaertner (CBS Labs.)
Physics Letters, Vol. 19, pp. 202–203, October 15, 1965 (17.5 keV electron beam affected by 65 joule Nd-glass laser beam)

23.124 Nonlinear Optics of Many-Particle Systems
Gordon Baym (U. of Illinois) and R. W. Hellwarth (Hughes)
IEEE J. Quantum Electronics, Vol. QE-1, pp. 309–320, October 1965

23.125 Laser-Induced Breakdown in Chlorine
John A. Howe (BTL)
J. Appl. Phys., Vol. 36, p. 3363, October 1965 (low breakdown flux, water particles provide free charge)

23.126 Laser Excitation of Modes in X-Cut Quartz
M. Katzman (Electro Optical Systems), F. Molz, Seymour Epstein, W. E. Bicknell, and J. W. Strozyk (U.S. Army Electronics Command, Fort Monmouth)
Proc. IEEE, Vol. 53, pp. 1635–1636, October 1965 (resonances between 40 and 130 kc/s, varied ruby laser spike frequency)

23.127 Electric and Magnetic Free Energy in Nonlinear Media. Quasi-Static Fields
F. Barocchi (Centro Microonde del C.N.R., Firenze, Italy) and M. Mancini and G. Toraldo di Francia (Istituto di Fisica Superiore dell'Università, Firenze, Italy)
Nuovo Cimento, Vol. 40 B, pp. 168–177, November 11, 1965 (lossless, anisotropic)

23.128 Vacuum uv Radiation from Plasmas Produced by a Laser on Metal Surfaces
A. W. Ehler (Hughes) and G. L. Weissler (U. of Southern California)
Bul. Am. Phys. Soc., Series II, Vol. 10, p. 1182–C6, November 1965 (from 4000 to 100 Å; Be, Al, Pt, Ta, and W)

23.129 Time Resolution of Laser Induced Electron and Ion Emission
S. H. Khan, F. A. Richards, and D. Walsh (Oxford U., England)
IEEE J. Quantum Electronics, Vol. QE-1, pp. 359–360, November 1965 (Q-sw ruby laser, tantalum, niobium-tin alloy, targets in vacuum, ion emission delay of 1.5 μs)

23.130 Semiconductor Surface Damage Produced by Ruby Lasers
Milton Birnbaum (Aerospace)
J. Appl. Phys., Vol. 36, pp. 3688–3689, November 1965 (focused beam, regular system of parallel straight lines, also cracks)

23.131 Laser-induced Damage in Diamond
D. J. Bradley, M. Engwell and, H. Komatsu (Royal Holloway College, England)
Nature, Vol. 208, pp. 1081–1082, December 11, 1965 (Q-sw and relaxation oscillation, ruby laser, holes, cleaved diamond, reflection inteferograms of surfaces)

23.132 Effect of Solar Radiation on Atmospheric Laser Returns
W. E. Hoehne and J. L. Karney (Pacific Missile Range, Point Mugu)
Appl. Phys. Letters, Vol. 7, pp. 313–314, December 15, 1965 (Q-sw ruby, amplitude depends on angle between solar and laser radiations, postulate interaction effect)

23.133 Self-Focusing of Optical Beams
P. L. Kelley (Lincoln Lab., M.I.T.)
Phys. Rev. Letters, Vol. 15, pp. 1005–1008, December 27, 1965 (40–100 cm self-focusing distance in CS₂)

23.134 Erratum: Self-Focusing of Optical Beams
P. L. Kelley

Phys. Rev. Letters, Vol. 16, p. 384, February 28, 1966 (Phys. Rev. Letters, Vol. 15, p. 1005, 1965)

23.135 Application of Electron and Optical Microscopy in Studying Laser-Irradiated Metal Surfaces
K. Vogel and P. Backlund (U. of Uppsala, Sweden)
J. Appl. Phys., Vol. 36, pp. 3697–3701, December 1965 (Q-sw ruby laser, several metals, microtopography)

23.136 Erratum: Application of Electron and Optical Microscopy in Studying Laser-Irradiated Metal Surfaces
K. Vogel and P. Backlund (U. of Uppsala, Sweden)
J. Appl. Phys., Vol. 37, p. 2930, June 1966 (J. Appl. Phys., Vol. 36, p. 3697, 1965)

23.137 Influence of the Finite Aperture of a Light Beam on Nonlinear Effects in an Anisotropic Medium
S. A. Akhmanov, A. I. Kovrigin, and N. K. Kulakova (Moscow State U., U.S.S.R.)
Soviet Physics JETP, Vol. 21, pp. 1034–1040, December 1965 (spatial structure of second harmonic radiation, electric field on KDP)

23.138 Heating of Matter by Focused Laser Radiation
R. V. Ambartsumyan, N. G. Basov, V. A. Boĭko, V. S. Zuev, O. N. Krokhin, P. G. Kryukov, Yu. V. Senat-skiĭ, and Yu. Yu Stoĭlov (Lebedev Physics Institute, Academy of Sciences, U.S.S.R.)
Soviet Physics JETP, Vol. 21, pp. 1061–1064, December 1965 (vaporized lithium sample, also high speed photographs of air breakdown)

23.139 Optically Induced Magnetization Resulting from the Inverse Faraday Effect
P. S. Pershan, J. P. van der Ziel, and L. D. Malmstrom (Harvard U.)
Proc. Physics of Quantum Electronics Conf., McGraw-Hill Book Co., New York, pp. 3–12, 1966

23.140 Two-Quantum Absorption and Excitons in Anthracene
John M. Worlock (B.T.L.)
Proc. Physics of Quantum Electronics Conf., McGraw-Hill Book Co., New York, pp. 13–19, 1966

23.141 Nonlinear Effects at Multiples of Laser Frequencies
S. A. Akhmanov, V. G. Dmitriev, R. V. Khokhlov, and A. I. Kovrygin (Moscow U., U.S.S.R.)
Proc. Physics of Quantum Electronics Conf., McGraw-Hill Book Co., New York, pp. 43–48, 1966

23.142 Nonlinear Light Scattering in Methane
P. D. Maker (Ford Motor Company)
Proc. Physics of Quantum Electronics Conf., McGraw-Hill Book Co., New York, pp. 60–66, 1966

23.143 Coupling of Light with Phonons, Magnons, and Plasmons
N. Bloembergen and Y. R. Shen (U. of California)
Proc. Physics of Quantum Electronics Conf., McGraw-Hill Book Co., New York, pp. 119–128, 1966

23.144 Light Waves with Exponential Gain
N. Bloembergen (U. of California), and P. Lallemand (Harvard U.)
Proc. Physics of Quantum Electronics Conf., McGraw-Hill Book Co., New York, pp. 137–154, 1966

23.145 Induced Absorption Spectra at Optical Frequencies
A. K. MacQuillan and B. P. Stoicheff (U. of Toronto, Canada)
Proc. Physics of Quantum Electronics Conf., McGraw-Hill Book Co., New York, pp. 192–199, 1966

23.146 Light-Beating Techniques for the Study of the Rayleigh-Brillouin Spectrum

J. B. Lastovka and G. B. Benedek (M.I.T.)
Proc. Physics of Quantum Electronics Conf., McGraw-Hill Book Co., New York, pp. 231–240, 1966

23.147 Dynamic Effects on the Propagation of Intense Light Pulses in Optical Media
C. L. Tang (Cornell U.) and B. D. Silverman (Raytheon)
Proc. Physics of Quantum Electronics Conf., McGraw-Hill Book Co., New York, pp. 280–293, 1966

23.148 Saturation in Semiconductor Absorbers and Amplifiers of Light
Frank Stern (IBM)
Proc. Physics of Quantum Electronics Conf., McGraw-Hill Book Co., New York, pp. 442–449, 1966

23.149 Field Emission from Atoms in Intense Optical Fields
Eugene R. Peressini (Aerospace Corp.)
Proc. Physics of Quantum Electronics Conf., McGraw-Hill Book Co., New York, pp. 499–508, 1966

23.150 Absorbed Ion Emission from Laser-Irradiated Tungsten
E. Bernal, J. F. Ready and L. P. Levine (Honeywell)
Physics Letters, Vol. 19, pp. 645–647, January 1, 1966

23.151 Frequency Shift of Optical Transition in the Field of a Light Wave
E. B. Aleksandrov, A. M. Bonch-Bruevich, N. N. Kostin, and V. A. Khodovoi (S. I. Vavilov State Optical Institute, U.S.S.R.)
JETP Letters, Vol. 3, pp. 53–55, January 15, 1966 (ruby laser pulse irradiates potassium vapor, optical transition shifts by about 10^9 c/s)

23.152 Complex Intensity-Dependent Index of Refraction, Frequency Broadening of Stimulated Raman Lines, and Stimulated Rayleigh Scattering
N. Bloembergen and P. Lallemand (Harvard U.)
Phys. Rev. Letters, Vol. 16, pp. 81–84, January 17, 1966 (several liquids)

23.153 Optical Design Considerations in Photobleaching with cw Lasers
Hubertus Staerk (U. of Pennsylvania)
Appl. Optics, Vol. 5, pp. 155–158, January 1966 (gas laser, optical systems to alter beam intensity profile)

23.154 Optical Nonlinearities of a Plasma
N. Bloembergen and Y. R. Shen (U. of California)
Phys. Rev., Vol. 141, pp. 298–305, January 1966 (predicts weak stimulated Raman and second harmonic generation effects)

23.155 Erratum: Optical Nonlinearities in a Plasma
N. Bloembergen and Y. R. Shen
Phys. Rev., Vol. 145, p. 390, May 6, 1966 (Phys. Rev., Vol. 141, p. 298, 1966)

23.156 Decay of Laser-Induced Excitations of F Centers
Dietmar Fröhlich and Herbert Mahr (Cornell U.)
Phys. Rev., Vol. 141, pp. 692–695, January 1966 (Q-sw ruby laser, several alkali halides, below 80°K, 0.1–7 μs)

23.157 Bleaching Waves in Two-Level Systems
V. M. Ovchinnikov and V. E. Khartsiev (A. F. Ioffe Physico-technical Institute, U.S.S.R.)
Soviet Physics JETP, Vol. 22, pp. 221–222, January 1966 (predict 42 ns bleaching time in cryptocyanine by ruby)

23.158 Self-Focusing of Light. Role of Kerr Effect and Striction
Ya. B. Zel'dovich and Yu. P. Raizer (Inst. of Mechanics Problems, Academy of Sciences, U.S.S.R.)
JETP Letters, Vol. 3, pp. 86–89, February 1, 1966 (requires 10^{-9} s)

23.159 Quantum Theory of Photon Interaction in a Plasma
H. Cheng (Harvard U. and B.T.L.) and Y. C. Lee (B.T.L.)
Phys. Rev., Vol. 142, pp. 104–114, February 4, 1966 (light-light scattering, optical mixing)

23.160 Vacuum Ultraviolet Radiation from Plasmas Formed by a Laser on Metal Surfaces
A. W. Ehler (Hughes) and G. L. Weissler (U. of Southern California)
Appl. Phys. Letters, Vol. 8, pp. 89–91, February 15, 1966

23.161 Radiative Recombination Lifetimes in Laser-Excited Silicon
N. N. Winogradoff and H. K. Kessler (N.B.S.)
Appl. Phys. Letters, Vol. 8, pp. 99–101, February 15, 1966

23.162 Optocaloric Effect (Amplification of the Atomic Interaction and Cooling of the Medium) in a Laser Beam
G. A. Askar'yan (Lebedev Physics Inst., USSR)
JETP Letters, Vol. 3, pp. 105–110, February 15, 1966

23.163 Length-Dependent Threshold for Stimulated Raman Effect and Self-Focusing of Laser Beams in Liquids
Charles C. Wang (Ford Scientific Lab.)
Phys. Rev. Letters, Vol. 16, pp. 344–346, February 28, 1966 (lower threshold with longer path, benzene, toluene and nitrobenzene, varied beam diameter)

23.164 Dynamics and Characteristics of the Self-Trapping of Intense Light Beams
E. Garmire, R. Y. Chiao, and C. H. Townes (M.I.T.)
Phys. Rev. Letters, Vol. 16, pp. 347–349, February 28, 1966 (50-cm long CS_2 cell, beam splitters every 2 cm, 100 μ diameter spot)

23.165 Higher Order Trapped Light Beam Solutions
H. A. Haus (M.I.T.)
Appl. Phys. Letters, Vol. 8, pp. 128–129, March 1, 1966

23.166 Electrostriction, Optical Kerr Effect and Self-Focusing of Laser Beams
Y. R. Shen (U. of California)
Physics Letters, Vol. 20, pp. 378–380, March 1, 1966 (coefficients of many materials, temperature effects)

23.167 Laser-Induced Damage Thresholds for Various Glasses
John Martinelli (Eastman Kodak)
J. Appl. Phys., Vol. 37, pp. 1939–1940, March 15, 1966 (internal damage, highest threshold of 15×10^3 J/cm²)

23.168 Stimulated Emission Observed from an Organic Dye, Chloroaluminum Phthalocyanine
P. P. Sorokin and J. R. Lankard
IBM J. Res. & Dev., Vol. 10, pp. 162–163, March 1966 (Q-sw ruby, 0.755 μ emission)

23.169 Optical Interactions with Elastic Waves in Lithium Niobate
E. G. Spencer, P. V. Lenzo, and K. Nassau (B.T.L.)
IEEE J. Quantum Electronics, Vol. QE-2, pp. 69–70, March 1966

23.170 Interaction between Light Waves and Spin Waves
Y. R. Shen and N. Bloembergen (U. of California)
Phys. Rev., Vol. 143, pp. 372–384, March 1966 (spin and vibrational Raman effects)

23.171 Absolute Measurement of an Optical-Rectification Coefficient in Ammonium Dihydrogen Phosphate
J. F. Ward (U. of Michigan)
Phys. Rev., Vol. 143, pp. 569–574, March 1966 (Q-sw ruby)

23.172 Theoretical Discussion of the Inverse Faraday Effect, Raman Scattering, and Related Phenomena
P. S. Pershan, J. P. van der Ziel, and L. D. Malmstrom (Harvard U.)
Phys. Rev., Vol. 143, pp. 574–583, March 1966

23.173 cw Measurement of the Optical Nonlinearity of Ammonium Di-hydrogen Phosphate
G. E. Francois (Stanford U.)
Phys. Rev., Vol. 143, pp. 597–600, March 1966 (0.63-μ He–Ne laser)

23.174 Frequency Locking and Dye Spectral Hole Burning in Q-Spoiled Lasers
B. H. Soffer and B. B. McFarland (Korad)
Appl. Phys. Letters, Vol. 8, pp. 166–169, April 1, 1966 (two laser beams intersecting in common cell containing dye)

23.175 Laser-Induced Thermal Etching of Metal and Semi-Conductor Surfaces
R. J. Murphy and G. J. Ritter (Nat. Physical Res. Lab., Council for Scientific and Industrial Res., Pretoria, Republic of South Africa)
Nature, Vol. 210, pp. 191–192, April 9, 1966

23.176 Parametric Interaction of Infrared Waves in a Medium in which Intense Molecular Oscillations are Excited
B. A. Akanaev and Ya. Petselt (Moscow State U.)
JETP Letters, Vol. 3, pp. 211–212, April 15, 1966 (Q-sw ruby, 130 atm hydrogen, Raman effect, 4.50 and 5.16 μ interaction)

23.177 Double Quenching on a Selective Diffused Junction Laser
Akira Kawaji, Hiroo Yonezu, and Yoshihiro Yasuoka (Nippon Electric, Japan)
Japanese J. Appl. Phys., Vol. 5, pp. 340–341, April 1966 (perpendicular beams)

23.178 Effect of a Temperature Gradient on Thermionic Emission
Harold C. Bowers and George J. Wolga (Cornell U.)
J. Appl. Phys., Vol. 37, pp. 2024–2027, April 1966 (laser beam onto metal surface)

23.179 Surface Temperature of Laser Heated Metal
Walter L. Knecht (Wright-Patterson AFB)
Proc. IEEE, Vol. 54, pp. 692–693, April 1966

23.180 Coherent Optical Mixing in Optically Active Liquids
P. M. Rentzepis, J. A. Giordmaine, and K. W. Wecht (B.T.L.)
Phys. Rev. Letters, Vol. 16, pp. 792–794, May 2, 1966 (quadratic polarizability, arabinose solutions, Q-sw ruby plus 2nd harmonic)

23.181 Second-Order Optical Processes and Harmonic Fields in Solids
Sudhanshu S. Jha (Tata Inst. of Fund. Res., India)
Phys. Rev., Vol. 145, pp. 500–506, May 13, 1966

23.182 Laser-Induced Spontaneous Electron Emission from Rear Side of Metal Foils
Walter L. Knecht (W-P Air Force Base)
Appl. Phys. Letters, Vol. 8, pp. 254–256, May 15, 1966

23.183 Effect of Linear Absorption on Self-Focusing of Laser Beam in CS_2
Charles C. Wang and George W. Racette (Ford Sci. Lab.)
Appl. Phys. Letters, Vol. 8, pp. 256–257, May 15, 1966 (threshold power depends on cell length)

23.184 Photoconductivity of Dielectrics under the Influence of Laser Radiation
V. S. Dneprovskii, D. N. Klyshko, and A. N. Penin (Moscow State U., U.S.S.R.)
JETP Letters, Vol. 3, pp. 251–253, May 15, 1966 (NaCl and Al_2O_3 crystals between charged capacitor plates, ruby laser radiation alters charge)

23.185 Photoconductivity of Ruby when Strongly Irradiated by a Ruby Laser
T. P. Belikova and E. A. Sviridenkov (Lebedev Physics Inst., U.S.S.R.)
JETP Letters, Vol. 3, pp. 257–259, May 15, 1966

23.186 Multiphoton Plasma Production and Stimulated Recombination Radiation in Semiconductors
C. K. N. Patel, P. A. Fleury, R. E. Slusher, and H. L. Frisch (B.T.L.)
Phys. Rev. Letters, Vol. 16, pp. 971–974, May 30, 1966 (Q-sw 10.6-μ CO_2 laser, 6.5 μ radiation from PbTe)

23.187 Observation of the Interaction of Plasmons with Longitudinal Optical Phonons in GaAs
A. Mooradian and G. B. Wright (Lincoln Lab., M.I.T.)
Phys. Rev. Letters, Vol. 16, pp. 999–1001, May 30, 1966 (phonons in the Raman spectrum of GaAs)

23.188 Time Resolution of Laser Induced Electron Emission from Cesium at High Laser Power
Walter L. Knecht (Wright-Patterson AFB)
IEEE J. Quantum Electronics, Vol. QE-2, p. 103, May 1966 (cesium diode, Q-sw ruby, photoemission, plasma emission, thermionic emission, up to 2.25 μs delay)

23.189 Elastic Explosions in Solids Caused by Radiation
R. Bullough and J. J. Gilman (U. of Illinois)
J. Appl. Phys., Vol. 37, pp. 2283–2287, May 1966 (short pulse, internal explosion, finite geometry)

23.190 Interaction of Laser Radiation with an Absorbing Semi-Infinite Solid Bar
S. S. Penner and O. P. Sharma (U. of California)
J. Appl. Phys., Vol. 37, pp. 2304–2308, May 1966 (short pulse, temperature profile, ablation, stress wave)

23.191 Effect of Thermal and Ultraviolet Radiation on Thermionic Emission from Tungsten
R. E. Stickney, P. B. Sun, and M. L. Shaw (M.I.T.)
J. Appl. Phys., Vol. 37, pp. 2391–2395, May 1966 (measure radiation density)

23.192 Colour Centres in Borate, Phosphate and Borophosphate Glasses
P. Beekenkamp
Philips Research Rept. Suppl., No. 4, pp. 1–117, 1966 (radiation-induced, optical spectra)

23.193 New Measurements of Intensity-Dependent Changes in the Refractive Index of Liquids
P. D. McWane and D. A. Sealer (Ohio State U.)
Appl. Phys. Letters, Vol. 8, pp. 278–279, June 1, 1966 (CS_2, $CHBr_3$, N_2, and H_2O; unfocused ruby laser, 20 times larger change than previously reported)

23.194 Laser Damage Threshold in NaCl Crystals
D. Olness (U. of California)
Appl. Phys. Letters, Vol. 8, pp. 283–285, June 1, 1966 (Q-sw ruby laser, 2×10^9 W/cm^2)

23.195 Resonant Birefringence in the Electric Field of a Light Wave
A. M. Bonch-Bruevich, N. N. Kostin, and V. A. Khodovoi
JETP Letters, Vol. 3, pp. 279–281, June 1, 1966 (ruby laser electric field, birefringence in potassium vapor)

23.196 Refraction of Electron Beams by Intense Electromagnetic Waves
T. W. B. Kibble (Imperial College, England)
Phys. Rev. Letters, Vol. 16, pp. 1054–1056, June 6, 1966

23.197 Effect of Simultaneous Optical Bleaching and Gamma Radiation on the Room-Temperature Colorability of KCl
J. W. Mathews and W. C. Mallard (Emory U.) and W. A. Sibley (Oak Ridge National Lab.)
Phys. Rev., Vol. 146, pp. 611–614, June 10, 1966

23.198   Conversion of Light to Sound by Electrostrictive Mixing in Solids
D. E. Caddes, C. F. Quate, and C. D. W. Wilkinson (Stanford U.)
Appl. Phys. Letters, Vol. 8, pp. 309–311, June 15, 1966 (Q-sw ruby laser, 720 MHz frequency acoustic translator, mixed two light beams in crystals)

23.199   Fast Overlap of Microwave Radiation by an Ionization Aureole of a Spark in a Laser Beam
G. A. Askar'yan, M. S. Rabinovich, M. M. Savchenko, and V. K. Stepanov (Lebedev Phys. Inst., Acad. of Sci., USSR)
JETP Letters, Vol. 3, pp. 303–305, June 15, 1966 (Q-sw laser, focused, various gases, 8 mm magnetron, spark in antenna beam)

23.200   Filamentary Structure of Light Beams in Nonlinear Liquids
V. I. Bespalov and V. I. Talanov (Radiophys. Sci. Research Inst., USSR)
JETP Letters, Vol. 3, pp. 307–310, June 15, 1966

23.201   Optically-Induced Refractive Index Inhomogeneities in $LiNbO_3$ and $LiTaO_3$
A. Ashkin, G. D. Boyd, J. M. Dziedzic, R. G. Smith, A. A. Ballman, J. J. Levinstein, and K. Nassau (BTL)
Appl. Phys. Letters, Vol. 9, pp. 72–74, July 1, 1966 (detrimental effect produced by focused gas laser beam)

23.202   Stratification of Light Beams in a Nonlinear Medium and the Real Threshold for Self-Focusing
Yu. P. Raizer (Acad. of Sci., USSR)
JETP Letters, Vol. 4, pp. 1–4, July 1, 1966

23.203   Linear Instability Theory of Laser Propagation in Fluids
K. A. Brueckner and S. Jorna (U. of California)
Phys. Rev. Letters, Vol. 17, pp. 78–81, July 11, 9166 (due to interaction, change in induced dipole moment density)

23.204   Erratum: Linear Instability Theory of Laser Propagation in Fluids
K. A. Brueckner and S. Jorna
Phys. Rev. Letters, Vol. 17, p. 279, August 1, 1966 (Phys. Rev. Letters, Vol. 17, p. 78, 1966)

23.205   Light Scattering by Spin Waves in FeF_2
P. A. Fleury, S. P. S. Porto, L. E. Cheesman, and H. J. Guggenheim (BTL)
Phys. Rev. Letters, Vol. 17, pp. 84–87, July 11, 1966 (0.488 μ argon laser, Stokes shifts)

23.206   Self-Induced Divergence of CW Laser Beams in Liquids—A New Nonlinear Effect in the Propagation of Light
Klaus E. Rieckhoff (Simon Fraser U., Canada)
Appl. Phys. Letters, Vol. 9, pp. 87–88, July 15, 1966 (0.63-μ He–Ne laser, several liquids, thermal effect, expands beam diameter)

23.207   Effective Cross Sections of Two-Photon Absorption in Organic Molecules
M. D. Galanin and Z. A. Chizhikova (Lebedev Phys. Inst., Acad. of Sci., USSR)
JETP Letters, Vol. 4, pp. 27–28, July 15, 1966 (Q-sw ruby laser, luminescence induced in anthracene, acridine, and 3-aminophthalimide)

23.208   Possibility of Observing Induced Infrared Radiation in Raman Scattering of Light
V. S. Gorelik, V. A. Zubov, M. M. Sushchinskii, and V. A Chirkov (Lebedev Phys. Inst., Acad. of Sci., USSR)
JETP Letters, Vol. 4, pp. 35–37, July 15, 1966 (population inversion between vibrational or vibrational-rotational levels of molecules)

23.209   Self-Focusing of Optical Beams in Absorbing Media
W. Kaiser, A. Laubereau, M. Maier, and J. A. Giordmaine

(Tech. Hochschule München, Germany)
Physics Letters, Vol. 22, pp. 60–62, July 15, 1966 (distributed absorption, in benzene and nitrobenzene)

23.210   Energies of Ions Produced by Laser Irradiation
Susumu Namba, Pil Hyon Kim, and Akira Mitsuyama (Inst. of Phys. and Chem. Research, Japan)
J. Appl. Phys., Vol. 37, pp. 3330–3331, July 1966 (focused Q-sw lasers, metal targets, about 100 eV for first peak and about 1 eV for second peak)

23.211   Time-of-Flight Spectrometer for Laser Surface Interaction Studies
E. Bernal G., L. P. Levine, and J. F. Ready (Honeywell)
Rev. Sci. Instr., Vol 37, pp. 938–941, July 1966

23.212   Relation Between the Nonlinear Dielectric Constant and the Green's Functions for Electromagnetic Radiation
V. V. Obukhovskii and V. L. Strizhevskii (Kiev State U., USSR)
Soviet Phys. JETP, Vol. 23, pp. 91–93, July 1966

23.213   Reflection of Atoms from Standing Light Waves
S. Altshuler and L. M. Frantz (TRW/Systems), and R. Braunstein (U. of California)
Phys. Rev. Letters, Vol. 17, pp. 231–232, August 1, 1966 (small deflection angle, Bragg scattering)

23.214   Self-Focusing of a Homogeneous Light Beam in a Transparent Medium, Due to Weak Absorption
Yu. P. Raizer (Inst. of Mech. Probs., Acad. of Sci., USSR)
JETP Letters, Vol. 4, pp. 85–88, August 15, 1966

23.215   Nonlinear Effects in a Hypersonic Wave
A. L. Polyakova (Acoustics Inst., USSR)
JETP Letters, Vol. 4, pp. 90–92, August 15, 1966 (in quartz, temperature dependence)

23.216   Self-Focusing and Focusing of Ultrasound and Hypersound
G. A. Askar'yan (Lebedev Phys. Inst., Acad. of Sci., USSR)
JETP Letters, Vol. 4, pp. 99–101, August 15, 1966

23.217   Radiative Coupling Between Two Different Solid-State Lasers
H. Inaba, Y. Isawa, and N. Suda (Tohoku U., Japan)
Physics Letters, Vol. 22, pp. 293–295, August 15, 1966 (Q-sw ruby laser beam into Nd-glass laser, two-photon excitation by ruby, enhanced Nd-laser output)

23.218   Radiation Interactions between Laser Oscillators with Different Active Elements and Different Frequency
H. Inaba, Y. Isawa, and N. Suda (Tohoku U., Japan)
IEEE J. Quantum Electronics, Vol. QE-2, pp. 222–229, August 1966 (Q-sw ruby interacting with Nd-glass laser, double photon absorption)

23.219   On Self-Trapping of a Laser Beam
A. Piekara (Acad. of Sci., Poland)
IEEE J. Quantum Electronics, Vol. QE-2, pp. 249–250, August 1966

23.220   Laser Induced Emission of Electrons, X Rays, and Ions from Solid Targets
G. F. Tonon and P. Langer (CEA, France)
IEEE J. Quantum Electronics, Vol. QE-2, p. 296, August 1966 (abstract)

23.221   Nonlinear Interactions Between Electron and Laser Beams
Helmut Schwarz (Rensselaer Poly. Inst.)
IEEE J. Quantum Electronics, Vol. QE-2, p. 297, August 1966 (abstract)

23.222   Multiphoton Plasma Production and Stimulated Recombination Radiation in Semiconductors
C. K. N. Patel, P. A. Fleury, R. E. Slusher, and H. L. Frisch

(BTL)
IEEE J. Quantum Electronics, Vol. QE-2, p. 298, August 1966 (abstract, Q-sw CO_2 laser, 6.5 μ output from PbTe)

23.223   A Continuum Mechanical Model for Laser-Induced Fracture in Transparent Media
Gary H. Conners and Robert A. Thompson (U. of Rochester)
J. Appl. Phys., Vol. 37, pp. 3434–3440, August 1966 (absorbed energy, thermoelastic wave, stress field)

23.224   Stimulated Emission from Polymethine Dyes
M. L. Spaeth and D. P. Bortfeld (Hughes Aircraft)
Appl. Phys. Letters, Vol. 9, pp. 179–181, September 1, 1966 (pump by Q-sw ruby laser, 7500 to 7900 Å, "hole-burning" effects)

23.225   Direct Observation of the Excess Light Hole Population in Optically Pumped p-Type Germanium
J. M. Feldman and K. M. Hergenrother (Northeastern U.)
Appl. Phys. Letters, Vol. 9, pp. 186–187, September 1, 1966 (Q-sw CO_2 laser pump, 10^{-12} s fast carrier lifetime)

23.226   Excitation of Signals in a Negatively Charged Post of an Antenna Under the Influence of an Unfocused Laser Beam
G. A. Askar'yan, M. S. Rabinovich, A. D. Smirnova, V. K. Stepanov, and V. B. Studenov (Lebedev Phys. Inst., Acad. of Sci., USSR)
JETP Letters, Vol. 4, pp. 122–123, September 1, 1966 (Q-sw ruby laser, few mV change, oxidized antenna surface)

23.227   Variable Focal Length Lenses using Materials with Intensity Dependent Refractive Indices
P. D. McWane (Ohio State U.)
Nature, Vol. 211, pp. 1081–1082, September 3, 1966 (theory, CS_2 example)

23.228   Geometrical Model and Experimental Verification of Two-Photon Absorption in Organic Dye Solutions
F. P. Schäfer and W. Schmidt (U. of Marburg, Germany)
IEEE J. Quantum Electronics, Vol. QE-2, pp. 357–360, September 1966

23.229   Two-Quantum Photoionization of Cs and I⁻
J. L. Hall (NBS, Boulder, Colo.)
IEEE J. Quantum Electronics, Vol. QE-2, pp. 361–363, September 1966

23.230   A New Class of Trapped Light Filaments
R. Y. Chiao, M. A. Johnson, S. Krinsky, H. A. Smith and C. H. Townes (M.I.T.), and E. Garmire (NASA/ERC)
IEEE J. Quantum Electronics, Vol. QE-2, pp. 467–469, September 1966 (about 5 μ diameter, few ergs of energy, about 1 ns, stimulated Raman emission, CS_2)

23.231   Possibility of Self-Focusing Due to Intensity Dependent Anomalous Dispersion
A. Javan (M.I.T.), and P. L. Kelley (M.I.T. Lincoln Lab.)
IEEE J. Quantum Electronics, Vol. QE-2, pp. 470–473, September 1966

23.232   Ion Emission from Laser Irradiated Tungsten
E. Bernal G., J. F. Ready, and L. P. Levine (Honeywell)
IEEE J. Quantum Electronics, Vol. QE-2, pp. 480–482, September 1966 (ion energies up to 180 eV)

23.233   Evaporation and Heating of a Substance Due to Laser Radiation
Yu. V. Afanasyev, O. N. Krokhin, and G. V. Sklizkov (Lebedev Phys. Inst., Acad. of Sci., USSR)
IEEE J. Quantum Electronics, Vol. QE-2, pp. 483–486, September 1966 (heating of solids, blow-off vapor)

23.234   Density and Temperature of a Laser Induced Plasma
C. David, P. V. Avizonis, H. Weichel, and C. Bruce (Kirt-

land AFB), and K. D. Pyatt (General Atomic)
IEEE J. Quantum Electronics, Vol. QE-2, 493–499, September 1966 (blow-off vapor from solid, Q-sw ruby laser, one amplifier, focused beam, 1100 J/cm², electron density of 10^{19} cm⁻³)

23.235   Laser Induced Emission of Electrons, Ions, and X Rays from Solid Targets
P. Langer, G. Tonon, F. Floux, and A. Ducauze (CEA, France
IEEE J. Quantum Electronics, Vol. QE-2, pp. 499–506, September 1966 (spatial distributions of emitted electrons and ions)

23.236   The Self-Focusing of Light of Different Polarizations
D. H. Close, C. R. Giuliano, R. W. Hellwarth, L. D. Hess, F. J. McClung, and W. G. Wagner (Hughes Research)
IEEE J. Quantum Electronics, Vol. QE-2, pp. 553–557, September 1966 (molecular reorientation, circular and linear polarizations)

23.237   Quenching Effects in Coupled Lasers
P. W. Pheneger and R. H. Pantell (Stanford U.)
IEEE J. Quantum Electronics, Vol. QE-2, pp. 644–648, September 1966 (ruby lasers, end coupled)

23.238   Practical Method for the Calculation of Nonlinear Optical Effects in a Plane Parallel Plate
V. L. Strizhevskii
Optics and Spectroscopy, Vol. 21, pp. 196–201, September 1966 (nonlinear crystals, intense radiation)

23.239   Laser-Induced Damage in Copper Crystals
James Murphy and George J. Ritter (Nat'l Phys. Research Lab., South Africa)
Appl. Phys. Letters, Vol. 9, pp. 272–273, October 1, 1966 (Q-sw ruby laser, black spot defects, increased surface hardness)

23.240   Two-Photon Absorption Spectrum in Thallous Chloride
M. Matsuoka and T. Yajima (U. of Tokyo, Japan)
Physics Letters, Vol. 23, pp. 54–55, October 3, 1966 (Nd-glass laser beam and Xe flash lamp excitation)

23.241   Narrow Optical Waveguides and Instabilities Induced in Liquids
R. G. Brewer and J. R. Lifsitz (IBM San Jose)
Physics Letters, Vol. 23, pp. 79–81, October 3, 1966 (self-trapped light beams, 4 μ diameter, over 4 GW/cm²)

23.242   Comment on Photoconduction Induced by Q-Spoiled Lasers in Anthracene Crystals
M. Schott (Ecole Normale Supérieure, France)
Physics Letters, Vol. 23, pp. 92–94, October 3, 1966

23.243   Equations of Ginzburg-Landau Type in Nonlinear Optics
K. Grob and M. Wagner (Tech. Hochschule Stuttgart, Germany)
Phys. Rev. Letters, Vol. 17, pp. 819–821, October 10, 1966 (electrostriction)

23.244   Ionization and Heating of Solid Material by Means of a Laser Pulse
A. Caruso, B. Bertotti, and P. Giupponi, (EURATOM/CNEN, Italy)
Nuovo Cimento, Vol. 45B, pp. 176–189, October 11, 1966 (solid deuterium)

23.245   On the Dynamics of Laser-Induced Damage in Glasses
J. P. Budin and J. Raffy (CGE, France)
Appl. Phys. Letters, Vol. 9, pp. 291–293, October 15, 1966 (internal damage, fractures propagate along beam direction, discussion)

23.246   Performance of a Vibrational H_2-Stokes Oscillator
P. V. Avizonis and A. H. Guenther (AF Weapons Lab.), and

24. TRANSMISSION, PROPAGATION, SCATTERING, REFLECTION, AND FILTERING

Appl. Optics, Vol. 3, pp. 727–729, June 1964 (known added losses increase threshold)

24.43 Parameters for Attenuation in the Atmospheric Windows for Fifteen Wavelengths
L. Elterman (AFCRL)
Appl. Optics, Vol. 3, pp. 745–749, June 1964 (Clear Standard Atmosphere defined, 0.4 to 4.0 μ, up to 30-km altitude)

24.44 A Method for Evaluating the Abrasion Resistance of Optical Surfaces and Thin Films
M. J. Irland and E. B. Schermer (Ford)
Appl. Optics, Vol. 3, pp. 751–754, June 1964 (scattered light measurement technique described)

24.45 Transmission of Polarized Light through a Constant and a Time-Varying Pair of Birefringent Plates
G. B. Thurston (Oklahoma State U.)
Appl. Optics, Vol. 3, pp. 755-759, June 1964

24.46 Radiation Transfer by a Light Pipe Between Media with High Indices of Refraction
M. A. Gilleo (Amelco Semiconductor)
Appl. Optics, Vol. 3, pp. 765–767, June 1964 (GaAs experiment, As_2S_3 light pipes)

24.47 The Laser Interferometer: Application to Plasma Diagnostics
J. B. Gerardo and J. T. Verdeyen (U. of Illinois)
Proc. IEEE, Vol. 52, pp. 690–697, June 1964
(He-Ne gas laser as both source and detector, helium plasma afterglow study)

24.48 Birefringence in Glass Measured by the Scattered-Light Technique with a Laser Source
S. Bateson, J. W. Hunt, D. A. Dalby, and N. Sinha (Duplate Canada Ltd., Canada)
Appl. Optics, Vol. 3, p. 902, July 1964 (used 0.63-μ He-Ne laser, not optimum conditions)

24.49 Attenuation of Laser Light and Efficiency of Optical Diffraction Gratings
G. W. Stroke (U. of Michigan)
Proc. IEEE, Vol. 52, pp. 862–863, July 1964 (discussion of paper by Gerharz, Proc. IEEE, Vol. 52, p. 438; April 1964)

24.50 Instability of Light Beam Due to Lens-Displacements in a Beam Waveguide
J. Hirano (NT&T, Japan)
Proc. IEEE, Vol. 52, pp. 872–873, July 1964 (analytical, periodic displacement)

24.51 Optical Mixing as a Plasma Density Probe
N. M. Kroll, A. Ron, and N. Rostoker (U. of California)
Phys. Rev. Letters, Vol. 13, pp. 83–86, July 20, 1964 (analytical, two crossed laser beams)

24.52 A Lens or Light Guide Using Convectively Distorted Thermal Gradients in Gases
D. W. Berreman (BTL)
Bell Sys. Tech. J., Vol. 43, pp. 1469–1475, July 1964, Pt. 1 (beam passes coaxially through heated helix in moving gas, 5-m focal length)

24.53 A Gas Lens Using Unlike, Counter-Flowing Gases
D. W. Berreman (BTL)
Bell Sys. Tech. J., Vol. 43, pp. 1476–1479, July 1964, Pt. 1 (focal length variable down to 6 meters)

24.54 Directional Control in Light-Wave Guidance
S. E. Miller (BTL)
Bell Sys. Tech. J., Vol. 43, pp. 1727–1739, July 1964, Pt. 2 (considers sequence of lenses, hollow dielectric waveguide, round metallic circular-electric waveguides)

24.55 Alternating-Gradient Focusing and Related Properties of Conventional Convergent Lens Focusing
S. E. Miller (BTL)
Bell Sys. Tech. J., Vol. 43, pp. 1741–1758, July 1964, Pt. 2 (some potential advantages over all-convergent lens systems)

24.56 Analysis of a Tubular Gas Lens
D. Marcuse and S. E. Miller
Bell Sys. Tech. J., Vol. 43, pp. 1759–1782, July 1964, Pt. 2 (cool gas blown into hot tube, calculate 5-feet focal length for 0.325 watt)

24.57 Hollow Metallic and Dielectric Waveguides for Long Distance Optical Transmission and Lasers
E. A. J. Marcatili and R. A. Schmeltzer (BTL)
Bell Sys. Tech. J., Vol. 43, pp. 1783–1809, July 1964, Pt. 2 (low theoretical losses of 1.8 dB/km for 1.2-mm dia. tube at 3 microns)

24.58 Thermal Gas Lens Measurements
A. C. Beck (BTL)
Bell Sys. Tech. J., Vol. 43, pp. 1818–1820, July 1964, Pt. 2 (electrically heated brass tube, up to 5 diopters with 100°C wall temperature rise)

24.59 Gas Mixture Lens Measurements
A. C. Beck (BTL)
Bell Sys. Tech. J., Vol. 43, pp. 1821–1825, July 1964, Pt. 2 (two different index gases, up to 7 diopters)

24.60 Scattering and Attenuation due to Snow at Optical Wave-Lengths
D. C. Hogg (BTL)
Nature, Vol. 203, p. 396, July 25, 1964 (0.63 μ, 2.6-km path)

24.61 Radiation Scattered from the Plasma Produced by a Focused Ruby Laser Beam
S. A. Ramsden and W. E. R. Davies (Natl. Res. Council, Canada)
Phys. Rev. Letters, Vol. 13, pp. 227–229, August 17, 1964 (delayed scattering by plasma electrons)

24.62 A Light Circulator Using the Faraday Effect of Heavy Flint Glass
S. Saito, K. Yokoyama, and Y. Fujii (U. of Tokyo, Japan)
Proc. IEEE, Vol. 52, p. 979, August 1964 (up to 3000 gauss for 6200 Å)

24.63 Display of Infrared Laser Patterns by a Liquid Crystal Viewer
J. R. Hansen, J. L. Fergason, and A. Okaya (Westinghouse)
Appl. Optics, Vol. 3, pp. 987–988, August 1964 (scattered light depends on crystal temperature, He-Ne laser tests)

24.64 Rigorous Calculation of the Electromagnetic Field of Wave Beams
A. G. van Nie (Philips Res. Lab.)
Philips Res. Repts., Vol. 19, pp. 378–394, August 1964 (beam waveguide, iterative lenses, higher order modes)

24.65 Resonant Optical Faraday Rotator
R. Rosenberg, C. B. Rubinstein, and D. R. Herriott (BTL)
Appl. Optics, Vol. 3, pp. 1079–1083, September 1964 (single-pass rotation enhanced by resonant cavity)

24.66 High-Intensity Propagation through Absorptive or Amplifying Media
W. J. Condell (Lab. for Physical Sciences)
J. Opt. Soc. Am., Vol. 54, pp. 1166–1167, September 1964

24.67 Self-Trapping of Optical Beams
R. Y. Chiao, E. Garmire, and C. H. Townes (MIT)
Phys. Rev. Letters, Vol. 13, pp. 479–482, October 12, 1964 (index increases with intensity; self-trapping in water, glass and air at 1, 4, and 80 MW, respectively)

24.68 Light Scattering in a Solution of Europium Benzoylacetonate During Optical Pumping
E. P. Riedel (Westinghouse)
Appl. Phys. Letters, Vol. 5, pp. 162–165, October 15, 1964 (optical homogeneity strongly temperature dependent)

24.69 Mutual Coherence Function Applied to Imaging Through a Random Medium
G. O. Reynolds and T. J. Skinner (Tech. Operations Res.)
J. Opt. Soc. Am., Vol. 54, pp. 1302–1309, November 1964 (turbulent medium)

24.70 Multiple Light Scattering by Spherical Dielectric Particles
D. H. Woodward (U. of Colorado)
J. Opt. Soc. Am., Vol. 54, pp. 1325–1331, November 1964 (5461
Å Hg-arc source, 2-μ diameter polystyrene latex spheres)

24.71 Precision, Absolute Measurement of the Optical Absorption Spectra of Gases
W. H. Culver (IDA)
J. Appl. Phys., Vol. 35, p. 3421, November 1964 (sonic, limit of 3×10^{-11} cm^{-1})

24.72 Electron Density and Temperature Measurements in a 26 kj θ-Pinch by Light Scattering
H. J. Kunze, A. Eberhagen, and E. Fünfer (Inst. f. Plasmaphysik, Germany)
Physics Letters, Vol. 13, pp. 38–39, November 1, 1964 (Q switched ruby laser, 16×10^{16}e/cm^3, 43 eV)

24.73 Stability of a Light Beam in a Beam Waveguide
J. Hirano and Y. Fukatsu (Nippon Tel. & Tel., Japan)
Proc. IEEE, Vol. 52, pp. 1284–1292, November 1964 (Gaussian beam, guide imperfections considered)

24.74 The Minimum Spot Size for a Focused Laser and the Uncertainty Relation
P. W. Carlin (U. of Colorado)
Proc. IEEE, Vol. 52, p. 1371, November 1964

24.75 Scattering Cross Section of Ideal Gases for Narrow Laser Beams
O. Theimer (Los Alamos Scientific Lab.)
Phys. Rev. Letters, Vol. 13, pp. 622–625, November 23, 1964 (interference phenomena produce deviations from Rayleigh scattering, experiments with ruby laser)

24.76 Daytime Thermal Fluctuations in the Lower Atmosphere
E. K. Webb (C.S.I.R.O., Australia)
Appl. Optics, Vol. 3, pp. 1329–1336, December 1964 (conditions given for various seeing conditions, telescope should be higher than Obukhov's scale L, typically several to tens of meters high)

24.77 An Atlas of the Absorption Spectrum of the Lower Atmosphere from 5400A to 8520A
J. A. Curcio, L. F. Drummeter, and G. L. Knestrick (U. S. NRL)
Appl. Optics, Vol. 3, pp. 1401–1409, December 1964 (0.2Å resolution, 16-km path, tungsten source)

24.78 Correlations for Absorption by the 9.4-μ and 10.4-μ CO$_2$ Bands
D. K. Edwards and W. Sun (UCLA)
Appl. Optics, Vol. 3, pp. 1501–1502, December 1964

24.79 Light Depolarizer
C. J. Peters (Sylvania)
Appl. Optics, Vol. 3, pp. 1502–1503, December 1964 (immersed rough-surface retardation plate)

24.80 Loss Measurements with a Beam Waveguide for Long Distance Transmission at Optical Frequencies
G. Goubau and J. R. Christian (U. S. AEL)
Proc. IEEE, Vol. 52, p. 1739, December 1964 (970 m long, 10 iterations, evacuated, 1.3-dB loss at 0.63μ)

24.81 Angular Dependence of the Rayleigh Scattering from Low-Turbidity Molecular Liquids
R. C. C. Leite, R. S. Moore, S. P. S. Porto, and J. E. Ripper (BTL)
Phys. Rev. Letters, Vol. 14, pp. 7–9, January 4, 1965 (CCl$_4$, benzene, toluene, chopped laser beam, Rayleigh scattering about 2×10^{-26} cm^2)

24.82 Erratum: Scattering of Ruby-Laser Beam by Gases
T. V. George, L. Slama, M. Yokoyama, and L. Goldstein
Phys. Rev. Letters, Vol. 14, p. 54, January 11, 1965 (Phys. Rev. Letters, Vol. 11, p. 403; 1963)

24.83 Molecular Scattering of Ruby-Laser Light
T. V. George, L. Goldstein, L. Slama, and M. Yokoyama (U. of Illinois)
Phys. Rev., Vol. 137, pp. A369–A380, January 18, 1965 (argon and xenon, ruby laser, variable angle)

24.84 Absorption and Refractive Index Measurements at a Wavelength of 0.34 mm
H. A. Gebbie, N. W. B. Stone, F. D. Findlay, and E. C. Pyatt (National Physical Lab., England)
Nature, Vol. 205, pp. 377–378, January 23, 1965 (monohalogen substituted benzenes)

24.85 Perturbation of the Refractive Index of Absorbing Media by a Laser Beam
P. R. Longaker and M. M. Litvak (MIT)
Bull. Am. Phys. Soc., Series II, Vol. 10, No. 1, p. 101-JC4, January 1965 (time-resolved interferometry using argon laser, low loss gas and liquid samples heated by Nd laser pulse)

24.86 Theory of Absorption and Scattering within Integrating Spheres
P. J. Richetta (U.S. Army Bio. Labs.)
J. Opt. Soc. Am., Vol. 55, pp. 21–26, January 1965

24.87 Effects of Signal-Dependent Granularity
P. G. Roetling (Cornell Aeronautical Lab.)
J. Opt. Soc. Am., Vol. 55, pp. 67–71, January 1965 (for coherent optical-processing system)

24.88 Broadband Dielectric Mirrors for Multiple Wavelength Laser Operation in the Visible
D. L. Perry (BTL)
Proc. IEEE, Vol. 53, pp. 76–77, January 1965 (over 30 layers, 99.5% reflectance, 0.43 to 0.74μ)

24.89 Variable Phase Shifter for Laser Light Using Birefringent Crystals
J. Hamasaki and H. Noguchi (U. of Tokyo, Japan)
Proc. IEEE, Vol. 53, pp. 80–81, January 1965 (like Fox's microwave phase shifter)

24.90 Absence of Polarization Effects in Diffraction-Attenuated Laser Light
Reinhold Gerharz (USA ERDL)
Proc. IEEE, Vol. 53, pp. 105–106, January 1965

24.91 Transmission of Laser Beams through Various Transparent Rods for Biomedical Applications
J. A. Goldman and R. Meyer (Children's Hosp. Res. Foundation, Cincinnati)
Nature, Vol. 205, pp. 892–894, February 27, 1965 (flexible and rigid rods, some damages, bends, 63 joules into 5 mm rod)

24.92 An Integrating Sphere System for Measuring Average Reflectance and Transmittance
J. M. Davies and W. Zagieboylo (USA Natick Labs.)
Appl. Optics, Vol. 4, pp. 167–174, February 1965

24.93 On an Autocollimation Method of Optical Glass Heterogeneity Measurement
Z. Bodnar and F. Ratajczyk (Warsaw Polytechnic, Poland)
Appl. Optics, Vol. 4, pp. 181–186, February 1965 (accuracy of 10^{-7} cm^{-1})

24.94 Infrared Reflectance and Emittance of Silver and Gold Evaporated in Ultrahigh Vacuum
J. M. Bennett and E. J. Ashley (NOTS)
Appl. Optics, Vol. 4, pp. 221–224, February 1965

24.95 Faraday Effect at Optical Frequencies in Strong Magnetic Fields
N. George (CIT), R. W. Waniek and S. W. Lee (Advanced Kinetics)
Appl. Optics, Vol. 4, pp. 253–254, February 1965 (used 0.63 μ, pulsed 400 kOe field, various substances)

24.96 Losses Suffered by Coherent Light Redirected and Refocused Many Times in an Enclosed Medium
O. E. DeLange (BTL)
Bell Sys. Tech. J., Vol. 44, pp. 283–302, February 1965 (principal loss due to mirror at 1% per reflection, thermal insulation of pipe necessary)

24.97 Modulation of the Reflectivity of Semiconductors
M. Birnbaum (Aerospace)
J. Appl. Phys., Vol. 36, pp. 657–658, February 1965 (reflectivity increases with ruby laser irradiation)

24.98 Interference Structure in the Light Scattered from a Thin Transparent Plate
N. Wainfan and R. Lehr (PIB)
J. Opt. Soc. Am., Vol. 55, pp. 144–145, February 1965 (gas laser beam inclined 10° to surface)

24.99 Absorption of 3.39-Micron Helium-Neon Laser Emission by Methane in the Atmosphere
B. N. Edwards and D. E. Burch (Philco)
J. Opt. Soc. Am., Vol. 55, pp. 174–177, February 1965 (frequencies closely coincide)

24.100 Internal Reflection Barriers as Reflectors in a Modified Fabry-Perot Interferometer
H. A. Daw and J. R. Izatt (New Mexico State U.)
J. Opt. Soc. Am., Vol. 55, pp. 201–202, February 1965

24.101 Focusing of a Light Beam of Gaussian Field Distribution in Continuous and Periodic Lens-like Media
P. K. Tien, J. P. Gordon, and J. R. Whinnery (BTL)
Proc. IEEE, Vol. 53, pp. 129–136, February 1965

24.102 Effects of Atmospheric Turbulence on the Transmission of Visible and Near Infrared Radiation
B. N. Edwards and R. R. Steen (Philco)
Appl. Optics, Vol. 4, pp. 311–316, March 1965 (near ground level, power spectrum, variable diameter, 1500 cps chopper)

24.103 The Reflection and Transmission of Infrared Materials: III, Spectra from 2μ to 50μ
D. E. McCarthy (Beckman)
Appl. Optics, Vol. 4, pp. 317–320, March 1965 (ruby, Al_2O_3, CuCl, etc., 15 materials)

24.104 Simple Construction for Determining the Phase Change of Light Reflected at Normal Incidence
G. B. Wright (M.I.T.)
Appl. Optics, Vol. 4, p. 366, March 1965

24.105 Antireflection Coatings on Glass
L. Young (SRI)
Appl. Optics, Vol. 4, pp. 366–367, March 1965 (three design examples, practical coating indices, wide bandwidth)

24.106 Modulation of Laser Beams by Atmospheric Turbulence
M. Subramanian and J. A. Collinson (BTL)
Bell Sys. Tech. J., Vol. 44, pp. 543–546, March 1965 (120–360 m paths, weather effects, varied aperture sizes)

24.107 Convective Gas Light Guides or Lens Trains for Optical Beam Transmission
D. W. Berreman (BTL)
J. Opt. Soc. Am., Vol. 55, pp. 239–247, March 1965 (straight or slightly curved paths, warm helical light guide, chimney lenses)

24.108 Autocorrelation Function and Power Spectral Density of Laser-Produced Speckle Patterns
L. I. Goldfischer (General Precision)
J. Opt. Soc. Am., Vol. 55, pp. 247–253, March 1965 (diffuse surface, illumination function, single and double beam)

24.109 Coupling of Optical Fibers and Scattering in Fibers
Alan L. Jones (IBM)
J. Opt. Soc. Am., Vol. 55, pp. 261–271, March 1965 (coupling of parallel fibers, sinusoidal oscillatory transfer of energy)

24.110 Zeeman Filter
K. G. Kessler and W. G. Schweitzer, Jr. (NBS)
J. Opt. Soc. Am., Vol. 55, pp. 284–288, March 1965 (very narrow optical band pass filter, Zeeman splitting of resonant absorption line, down to 0.005 cm^{-1} with Hg^{198} at 2537 Å)

24.111 Interferometry with Rotation-Insensitive "Corner-Cube" Systems and Lasers
G. W. Stroke (U. of Michigan)
J. Opt. Soc. Am., Vol. 55, pp. 330–331, March 1965

24.112 Electric Field Effect on the Refractive Index in GaAs
B. O. Seraphin and N. Bottka (Michelson Lab.)
Appl. Phys. Letters, Vol. 6, pp. 134–136, April 1, 1965 (predict index change up to −1.8% at 2×10^5 v/cm)

24.113 The Reflection and Transmission of Infrared Materials: IV, Bibliography
D. E. McCarthy (East Los Angeles College)
Appl. Optics, Vol. 4, pp. 507–511, April 1965 (several materials, 167 references)

24.114 Fluorescence Spectra of Sharp Cut-off Filters
C. S. French (Carnegie Inst. of Washington)
Appl. Optics, Vol. 4, p. 514, April 1965

24.115 New Type of Waveguide for Light and Infrared Waves
A. E. Karbowiak (U. of New South Wales, Australia)
Electronics Letters, Vol. 1, pp. 47–48, April 1965 (thin dielectric supporting the surface E_0 wave mode)

24.116 On the Energy Balance for the Passage of Light Through a Thin Absorbing Film
K. H. Beckmann and B. Caspar (Philips, The Netherlands)
Philips Res. Rept., Vol. 20, pp. 190–205, April 1965 (introduces mixed Poynting vector to conserve energy)

24.117 Refractive Index of Air at 0.377-mm Wave-length
J. E. Chamberlain, F. D. Findlay, and H. A. Gebbie (Natl. Physical Lab., England)
Nature, Vol. 206, pp. 886–887, May 29, 1965 (CN maser, Michelson interferometer, 3-m long cell, water vapor effect)

24.118 Influence of Absorption on Measurement of Refractive Index of Films
O. S. Heavens and H. M. Liddell (U. of London, England)
Appl. Optics, Vol. 4, pp. 629–630, May 1965 (absorption can introduce error)

24.119 Demonstration of Chromatic Aberration in the Eye Using Coherent Light
D. C. Sinclair (U. S. Army Engineer R & D Labs.)
J. Opt. Soc. Am., Vol. 55, pp. 575–576, May 1965 (granularity effect, motion of head)

24.120 Nonlinear Optical Reflection from a Metallic Boundary
F. Brown, R. E. Parks, and A. M. Sleeper (Williams College)
Phys. Rev. Letters, Vol. 14, pp. 1029–1031. June 21, 1965 (giant pulse ruby laser, second-harmonic reflection by silver surface)

24.121 Rayleigh Scattering of 6943 Å Laser Radiation in a Nitrogen Atmosphere
R. D. Watson and M. K. Clark (Natl. Center for Atmospheric Research)
Phys. Rev. Letters, Vol. 14, pp. 1057–1058, June 28, 1965 (Q-sw ruby laser, unfocused, angular distribution)

24.122 Atmospheric Optical Effects—Polarization Fluctuation
D. L. Fried and G. E. Mevers (North American Aviation)
J. Opt. Soc. Am., Vol. 55, pp. 740–741, June 1965 (5-mile path, circularly polarized 0.63-μ gas laser, polarization fluctuation associated with intensity fluctuation)

24.123 Possibility of a Time Delay at Optical Reflections in Metallic Mirrors
B. Sigfridsson (Res. Inst. of Natl. Def., Sweden) and P. Erman (Res. Inst. for Physics, Sweden)
J. Opt. Soc. Am., Vol. 55, p. 742, June 1965 (theoretically less than 5×10^{-14} sec, experimentally less than 10^{-12} sec)

24.124 Efficient Diffraction of Light from Acoustic Waves in Water
W. Kleinhans and D. L. Fried (North American Aviation)
Appl. Phys. Letters, Vol. 7, pp. 19–21, July 1, 1965 (into first-order diffraction mode, possible laser beam steering, 15 Mc/s experiment)

24.125 Double Scattering of Electromagnetic Radiation by a Fluid
H. L. Frisch and J. McKenna (BTL)
Phys. Rev., Vol. 139, pp. A68–A77, July 5, 1965 (with linearly polarized incident light, singly and doubly scattered radiation have different polarizations)

24.126 Transmission of High-Power Laser Light through Tapered Dielectric Tubes and Rods
K. Vogel (U. of Uppsala, Sweden)
Nature, Vol. 207, pp. 281–282, July 17, 1965 (1-joule Q-sw ruby laser, no damage to tapered tubes or rods, beam emerges from small diameter end)

24.127 Incoherent Scattering of Light from Anisotropic Degenerate Plasmas
P. M. Platzman (BTL)
Phys. Rev., Vol. 139, pp. A379–A387, July 19, 1965 (monochromatic incident radiation)

24.128 Coherence in Long-Range Laser Beams
Harald W. Straub (Harry Diamond Labs.)
Appl. Optics, Vol. 4, pp. 875–876, July 1965 (air turbulence effects, source beam diameter vs. distance, beam wandering, spatial coherence)

24.129 FM Laser Communications through a Highly Turbulent Atmosphere
W. M. Doyle, W. D. Gerber, P. M. Sutton, and M. B. White (Philco)
IEEE J. Quantum Electronics, Vol QE-1, pp. 181–182, July 1965 (dual polarization, 0.63-μ He–Ne laser, $\frac{1}{2}$ mile path, varied receiver aperture, low noise with FM)

24.130 An Optical All-Pass Network
A. D. Jacobson and T. R. O'Meara (Hughes)
IEEE Trans. on Microwave Theory and Techniques, Vol. MTT-13, pp. 475–477, July 1965

24.131 Optical Image Quality in a Turbulent Atmosphere
C. E. Coulman (C.S.I.R.O., Australia)
J. Opt. Soc. Am., Vol. 55, pp. 806–812, July 1965 (modulation transfer function, data on horizontal path)

24.132 Wave Propagation Through Quasi-Optical Irregularities
David A. deWolf (RCA)
J. Opt. Soc. Am., Vol. 55, pp. 812–817, July 1965 (compares several simple approximation methods)

24.133 Electric Field Strengths at Totally Reflecting Interfaces
N. J. Harrick (Philips Labs., New York)
J. Opt. Soc. Am., Vol. 55, pp. 851–857, July 1965 (polarization, angle of incidence, experiments)

24.134 p-Polarized Reflectances for Transparent Thin Films on Transparent Substrates
L. A. Catalán (U. of Chile, Chile)
J. Opt. Soc. Am., Vol. 55, pp. 857–859, July 1965 (zero reflectance, single- and two-layer coatings, vary incidence angle)

24.135 Turbulent Backscatter of Light
Raymond J. Munick (North American Aviation)
J. Opt. Soc. Am., Vol. 55, p. 893, July 1965 (clear air laser radar echoes from molecules and aerosol particles, not turbulent backscatter)

24.136 Q-Switched Laser Beam Propagation over a Ten-Mile Path
J. R. Whitten, G. F. Prehmus, and K. Tomiyasu (GE)
Proc. IEEE, Vol. 53, p. 736, July 1965 (sometimes small intense spots received)

24.137 Laser Pulse-Shaping and Mode-Locking with Acoustic Waves
A. J. DeMaria and D. A. Stetser (United Aircraft)
Appl. Phys. Letters, Vol. 7, pp. 71–73, August 1, 1965 (27 Mc/s acoustic waves in quartz within argon ion laser cavity)

24.138 Fine Structure of Spectral Lines of Light Scattered in Cubic Crystals
T. A. Velichkina, O. A. Shustin, and I. A. Yakovlev (Moscow State U., U.S.S.R.)
JETP Letters, Vol. 2, pp. 119–121, August 15, 1965 (0.63-μ He–Ne laser, propagation velocities of elastic waves)

24.139 Nonlinear Optical Reflection from a Metal Surface
Sudhanshu S. Jha (Tata Inst., India)
Phys. Rev. Letters, Vol. 15, pp. 412–414, August 30, 1965

24.140 Low-Loss Multilayer Dielectric Mirrors
D. L. Perry (BTL)
Appl. Optics, Vol. 4, pp. 987–991, August 1965 (preparation, 99.8 percent from 25 layers, broadband)

24.141 Optical Coatings for Laser Use
Milton Laikin (Electro-Optical Systems)
Appl. Optics, Vol. 4, pp. 1032–1033, August 1965 (antireflection films, high reflectivity coatings, other coating materials)

24.142 An Optical Circulator
William B. Ribbens (U. of Michigan)
Appl. Optics, Vol. 4, pp. 1037–1038, August 1965 (proposed, four-port, 45° Faraday rotators)

24.143 Compton Scattering of an Intense Photon Beam
Lee M. Frantz (TRW Space Tech. Labs.)
Phys. Rev., Vol. 139, pp. B1326–B1336, September 6, 1965

24.144 The Apparent Dependence of Terrestrial Scintillation Intensity upon Atmospheric Humidity
Hugh R. Carlon (U.S. Army Edgewood Arsenal Chemical Research and Development Labs.)
Appl. Optics, Vol. 4, pp. 1089–1097, September 1965 (visible to far infrared, experimental data)

24.145 Dichroic Calcite Polarizers for the Infrared
T. J. Bridges and J. W. Kluver (BTL)
Appl. Optics, Vol. 4, pp. 1121–1125, September 1965 (useful in the range of 2.5 to 16 μ)

24.146 A Kerr Cell with Roof Prism
Milton Laikin (Electro-Optical Systems)
Appl. Optics, Vol. 4, pp. 1177–1178, September 1965 (prism apex parallel or perpendicular to Kerr cell plates, phase shift in prism)

24.147 Note on Nonreflective Coatings
G. Lewin (Princeton U.)
Appl. Optics, Vol. 4, p. 1203, September 1965 (erratum)

24.148 Design Considerations for Bent-Beam Waveguides
D. Marcuse (BTL)
IEEE Trans. on Microwave Theory and Techniques, Vol. MTT-13, pp. 647–651, September 1965 (radius of curvature, spot size, lens spacing, focal length)

24.149 Optical Faraday Rotation in Undoped and Doped Sapphire and Glass
Alexander M. Unwin (Boeing)
J. Appl. Phys., Vol. 36, p. 2967, September 1965 (negative results from ruby)

24.150 Laser-Produced Speckle Patterns
L. Allen and D. G. C. Jones (U. of Sussex, England)
J. Opt. Soc. Am., Vol. 55, p. 1188, September 1965 (references in 1963 and 1956)

24.151 Heterodyne Measurements of Light Propagation through Atmospheric Turbulence
I. Goldstein (Raytheon), P. A. Miles (M.I.T.), and A. Chabot (Raytheon)
Proc. IEEE, Vol. 53, pp. 1172–1180, September 1965 (He–Ne laser, 4 and 23.8 km paths, modulated interferometer, temperature fluctuations predominant)

24.152 Granularity in the Angular Spectrum of Scattered Laser-Light
G. Schiffner (Technische Hochschule, Vienna, Austria)

Proc. IEEE, Vol. 53, pp. 1245–1246, September 1965 (discusses Laue's 1914 experiment)

24.153 Simultaneous Measurements of Optical Transmission and Reflection in Thin Films
P. F. Váradi and J. R. Suffredini (Machlett Labs.)
Rev. Sci. Instr., Vol. 36, pp. 1331–1333, September 1965 (modified monochromator)

24.154 Direct Observation of Nonlinear Scattering of Electrons by Laser Beam
H. Schwarz and H. A. Tourtellotte (R.P.I., Hartford Graduate Center), and W. W. Gaertner (CBS Labs.)
Physics Letters, Vol. 19, pp. 202–203, October 15, 1965 (17.5 keV electron beam affected by 65 J Nd-glass laser beam)

24.155 Observation of the Spectrum of Light Scattered from a Pure Fluid Near its Critical Point
N. C. Ford, Jr., and G. B. Benedek (M.I.T.)
Phys. Rev. Letters, Vol. 15, pp. 649–653, October 18, 1965 (SF_6 pure fluid, 0.63-μ He–Ne gas laser, "self-beating" spectrometer)

24.156 0.63 μ Scatter Measurements from Teflon and Various Metallic Surfaces
R. A. Semplak (BTL)
Bell Sys. Tech. J., Vol. 44, pp. 1659–1674, October 1965 (angular and specular scatter measurements)

24.157 Coherent Light Transmitted through Optical Fiber
Ryuichi Hioki and Takeomi Suzuki (U. of Tokyo, Japan)
Japanese J. Appl. Phys., Vol. 4, p. 817, October 1965 (speckle pattern, interference between two close fibers)

24.158 Reflectance of Nonperfect Surfaces in the Integrating Sphere
Bjarne J. Hisdal (Norwegian Council for Scientific and Industrial Research, Norway)
J. Opt. Soc. Am., Vol. 55, pp. 1255–1260, October 1965 (also nonperfect samples in sphere)

24.159 Vacuum Tight Windows with Wide Band Transmission Characteristics
T. P. Vogl, R. O. McIntosh, and M. Garbuny (Westinghouse)
Rev. Sci. Instr., Vol. 36, pp. 1439–1440, October 1965 (barium fluoride, 0.18 to 14 μ range)

24.160 Transmission through a Tapered Quartz Tube in the Laser Near Field
H. J. Caulfield (Texas Instruments)
Nature, Vol. 208, pp. 773–774, November 20, 1965 (sharp transmission peak with intermediate Fresnel number)

24.161 Transmission Characteristics of Fabry-Perot Interferometers and a Related Electrooptic Modulator
V. N. Del Piano, Jr. and A. F. Quesada (Baird-Atomic)
Appl. Optics, Vol. 4, pp. 1386–1390, November 1965 (effects of maladjustments and surface imperfections)

24.162 Lossless Conversion of a Plane Laser Wave to a Plane Wave of Uniform Irradiance
B. Roy Frieden (U. of Rochester)
Appl. Optics, Vol. 4, pp. 1400–1403, November 1965 (Gaussian irradiance beam into constant irradiance beam, plano-aspheric lenses)

24.163 Dependency of Optical Scintillation Frequency on Wind Speed
Edward Ryznar (U. of Michigan)
Appl. Optics, Vol. 4, pp. 1416–1418, November 1965 (frequency of maximum power increases with cross-wind speed, path heights and lengths considered)

24.164 Comparison of Some Recent Experimental Results of Coherent and Incoherent Light Scattering with Theory

L. W. Carrier and L. J. Nugent (Electro-Optical Systems)
Appl. Optics, Vol. 4, pp. 1457–1462, November 1965 (fog, 0.633-μ wavelength sources, angular volume-scattering function ratio)

24.165 Transmittance of Optical Materials at High Temperatures in the 1-μ to 12-μ Range
D. T. Gillespie, A. L. Olsen, and L. W. Nichols (U.S. Naval Ordnance Test Station, China Lake)
Appl. Optics, Vol. 4, pp. 1488–1493, November 1965 (25 to 400°C, glasses, quartz, sapphire, Irtran, etc.)

24.166 Light Propagation in Generalized Lens-Like Media
S. E. Miller
Bell Sys. Tech. J., Vol. 44, pp. 2017–2064, November 1965 (transversally variable media, beamwidth, ray trajectory)

24.167 Statistical Treatment of Light-Ray Propagation in Beam-Waveguides
D. Marcuse
Bell Sys. Tech. J., Vol. 44, pp. 2065–2081, November 1965 (transverse displacement of lenses, lens spacings, tolerances)

24.168 Properties of Periodic Gas Lenses
D. Marcuse
Bell Sys. Tech. J., Vol. 44, pp. 2083–2116, November 1965 (ray optics, focal length, principal surface)

24.169 Growth of Oscillations of a Ray about the Irregularly Wavy Axis of a Lens Light Guide
D. W. Berreman
Bell Sys. Tech. J., Vol. 44, pp. 2117–2132, November 1965

24.170 Theory of a Thermal Gradient Gas Lens
D. Marcuse (BTL)
IEEE Trans. on Microwave Theory and Techniques, Vol. MTT-12, pp. 734–739, November 1965 (ray optics, focal length, principal surface)

24.171 Measurements on a Thermal Gradient Gas Lens
William H. Steier (BTL)
IEEE Trans. on Microwave Theory and Techniques, Vol. MTT-13, pp. 740–748, November 1965 (blow cool gas through hot tube, 0.63-μ interferometric measurements, focal lengths down to 24 cm)

24.172 Optical Transmission in Multidomained KH_2PO_4: Polarization Scattering
R. M. Hill, G. F. Herrmann, and S. K. Ichiki (Lockheed)
J. Appl. Phys., Vol. 36, pp. 3672–3677, November 1965

24.173 Statistics of a Geometric Representation of Wavefront Distortion
D. L. Fried (North American Aviation)
J. Opt. Soc. Am., Vol. 55, pp. 1427–1435, November 1965 (atmospheric turbulence causes significant random tilting of plane-wave front)

24.174 Use of Apophyllite in Polarization Interferometers—Achromatic Half-Wave Plate
Maurice Françon, Shamlal Mallick, and Jacques Vulmière (Institut d'Optique, Paris, France)
J. Opt. Soc. Am., Vol. 55, p. 1553, November 1965 (over visible band)

24.175 Tests for Detecting Self-Trapping in Optical Beams
Bernard L. Lewis (Radiation Inc.)
Proc. IEEE, Vol. 53, pp. 1731–1732, November 1965 (compare beams taken from two widely separated beam splitters)

24.176 Laser Beam "Security"
Selig Kainer (ITT Federal Labs.)
Proc. IEEE, Vol. 53, pp. 1752–1753, November 1965 (secure in space, atmospheric perturbations destroy "security")

24.177 Photon Tunnels: The Waveguides of the Future?
Bernard L. Lewis (Radiation Inc.)
Proc. IEEE, Vol. 53, pp. 1768–1769, November 1965

24.178 Scattering of Intense Light
Walter C. Henneberger (Southern Illinois U.)
Phys. Rev., Vol. 140, pp. A1864–A1866, December 13, 1965
(scattering cross section increases with intensity)

24.179 Radiation Characteristics of Circular Dielectric Waveguides
N. S. Kapany, J. J. Burke, Jr., and K. Frame (Optics Technology, Inc.)
Appl. Optics, Vol. 4, pp. 1534–1543, December 1965 (radiation pattern, mode identification in fibers)

24.180 On the Propagation of Gaussian Beams of Light through Lenslike Media Including Those with a Loss or Gain Variation
Herwig Kogelnik (BTL)
Appl. Optics, Vol. 4, pp. 1562–1569, December 1965

24.181 Time Resolution of Acoustic Mode Patterns in KDP Crystals
G. E. Peterson and P. M. Bridenbaugh (BTL)
Appl. Optics, Vol. 4, pp. 1655–1659, December 1965 (strobe flash lamp, photographic technique, 530 kc/s patterns)

24.182 Measurements of Electromagnetic Backscattering from Known, Rough Surfaces
Jacques Renau and James A. Collinson
Bell Sys. Tech. J., Vol. 44, pp. 2203–2226, December 1965 (He–Ne laser beam, rough aluminum and MgO slab, varied incidence angle, comparison with microwave data)

24.183 Quantum Descriptions of an Infinite Lossless Transmission Line
C. Y. She (U. of Minnesota)
J. Appl. Phys., Vol. 36, pp. 3784–3787, December 1965 (signal plus noise)

24.184 Faraday Rotation of Trivalent Ytterbium
C. B. Rubinstein and S. B. Berger (BTL)
J. Appl. Phys., Vol. 36, pp. 3951–3952, December 1965 (visible wavelengths, also other lanthanides)

24.185 Semiconductor Laser Communications through Multiple-Scatter Paths
E. J. Chatterton (Lincoln Lab., M.I.T.)
Proc. IEEE, Vol. 53, pp. 2114–2115, December 1965 (pulsed GaAs diode, snow, fog, atmospheric shimmer data)

24.186 Propagation Loss in a Distributed Beam Waveguide
Shojiro Kawakami and Jun-ichi Nishizawa (Tohoku U., Japan)
Proc. IEEE, Vol. 53, pp. 2148–2149, December 1965 (iterative positive lenses to approximate radially varying index optical fiber)

24.187 Nonlinear Light Scattering in Methane
P. D. Maker (Ford Motor Company)
Proc. Physics of Quantum Electronics Conf., McGraw-Hill Book Co., New York, pp. 60–66, 1966

24.188 Rayleigh Scattering from Low-Density Gases
A. D. May, E. G. Rawson, and H. L. Welsh (U. of Toronto, Canada)
Proc. Physics of Quantum Electronics Conf., McGraw-Hill Book Co., New York, pp. 260–264, 1966

24.189 Dynamic Effects on the Propagation of Intense Light Pulses in Optical Media
C. L. Tang (Cornell U.) and B. D. Silverman (Raytheon)
Proc. Physics of Quantum Electronics Conf., McGraw-Hill Book Co., New York, pp. 280–293, 1966

24.190 A Light Beam Waveguide Using Hyperbolic-Type Gas Lens
Yasuharu Suematsu, Kenichi Iga, and Shinichi Ito (Tokyo Institute of Technology, Japan)
Digest of Technical Papers, 1966 G-MTT International Symposium, IEEE Cat. No. 17C32, pp. 184–188

24.191 Reflectivity of Atmospheric Shock Waves
William R. Mallory (G.E.)
Nature, Vol. 209, pp. 175–177, January 8, 1966 (laser range finder, 10⁴ greater reflectivity possibly due to condensed water droplets)

24.192 A Laser Output Coupler Using Frustrated Total Internal Reflection
Earl L. Steele, Walter C. Davis, and Robert L. Treuthart (Autonetics)
Appl. Optics, Vol. 5, pp. 5–8, January 1966 (0.483 μ thick air gap for coupler, fused silica prisms)

24.193 The Frustrated Total Reflection Filter. I. Spectral Analysis
Leonard Bergstein and Carl Shulman (Polytechnic Inst. of Brooklyn)
Appl. Optics, Vol. 5, pp. 9–21, January 1966 (also dielectric layered systems)

24.194 Evaporated Inhomogeneous Thin Films
Roland Jacobsson and John Olof Martensson (Institute of Optical Research, Stockholm, Sweden)
Appl. Optics, Vol. 5, pp. 29–34, January 1966 (refractive index profile, simultaneous evaporation of two or more substances)

24.195 Stresses Developed in Optical Film Coatings
Anthony E. Ennos (Perkin-Elmer)
Appl. Optics, Vol. 5, pp. 51–61, January 1966 (many film materials, single and multilayers, measurements)

24.196 Realisations d'empilements de couches minces diélectriques d'épaisseurs quelconques; utilisation d'un calculateur analogique
R. Badoual et P. Giacomo (Faculté des Sciences de Caen, France)
Appl. Optics, Vol. 5, pp. 63–67, January 1966 (in French; nonequal thickness coatings, control problem)

24.197 Multilayer Mirrors with High Reflectance over an Extended Spectral Region
A. F. Turner (Bausch & Lomb) and P. W. Baumeister (U. of Rochester)
Appl. Optics, Vol. 5, pp. 69–76, January 1966

24.198 On a Dielectric Multilayer Fiber by Baumeister
Leo Young and E. G. Cristal (Stanford Research Institute)
Appl. Optics, Vol. 5, pp. 77–80, January 1966 (equal and unequal thicknesses, concept of Herpin equivalent layer, multilayer filter)

24.199 Infrared Modulation by Means of Frustrated Total Internal Reflection
Robert W. Astheimer, Gerald Falbel, and Sheldon Minkowitz (Barnes Engineering Co.)
Appl. Optics, Vol. 5, pp. 87–91, January 1966 (variable air gap, prisms)

24.200 A Tunable Birefringent Filter
F. K. von Willisen (Brown, Boveri & Co., Baden, Switzerland)
Appl. Optics, Vol. 5, pp. 97–104, January 1966 (uses rotating-field phase shifter)

24.201 Factors Affecting the Performance of Commercial Interference Filters
Irving H. Blifford, Jr. (National Center for Atmospheric Research)
Appl. Optics, Vol. 5, pp. 105–111, January 1966 (narrow-band filters, angle of incidence, bandwidth, transmission, temperature)

24.202   Consideration of Atmospheric Turbulence in Laser Systems Design
J. I. Davis (Hughes)
Appl. Optics, Vol. 5, pp. 139–147, January 1966 (many factors)

24.203   The Absorption of Laser Radiation along Atmospheric Slant Paths
Gilbert N. Plass (Southwest Center for Advanced Studies)
Appl. Optics, Vol. 5, pp. 149–154, January 1966 (considers temperature and fractional concentration)

24.204   Light Transmission in a Multiple Dielectric (Gaseous and Solid) Guide
E. A. J. Marcatili
Bell Sys. Tech. J., Vol. 45, pp. 97–103, January 1966 (small diameter thin-walled dielectric tube, CO_2 inside, air outside, 20 atmospheres pressure)

24.205   Ray Propagation in Beam-Waveguides with Redirectors
E. A. J. Marcatili
Bell Sys. Tech. J., Vol. 45, pp. 105–115, January 1966 (to correct adverse effects of displaced lens)

24.206   Power Loss in Propagation through a Turbulent Medium for an Optical-Heterodyne System with Angle Tracking
David M. Chase (TRG)
J. Opt. Soc. Am., Vol. 56, pp. 33–44, January 1966

24.207   Feasibility Model for a Laboratory Simulator of Optical Turbulence
E. Reisman and P. M. Sutton (Philco)
J. Opt. Soc. Am., Vol. 56, pp. 49–50, January 1966 (6-mm dia. Pyrex spheres immersed in glycerine)

24.208   Far-Infrared Spectrum of Liquid Water
David A. Draegert, N. W. B. Stone, Basil Curnutte, and Dudley Williams (Kansas State U.)
J. Opt. Soc. Am., Vol. 56, pp. 64–69, January 1966 (H_2O and D_2O, between 10 and 330 μ)

24.209   Average Transfer Function from Statistics of Wavefront Distortions
Eugene A. Trabka (Cornell Aeronautical Lab.)
J. Opt. Soc. Am., Vol. 56, pp. 128–129, January 1966 (in terms of intensity and phase measurements)

24.210   A Nonreciprocal Electrooptic Device
John L. Wentz (Westinghouse)
Proc. IEEE, Vol. 54, pp. 97–98, January 1966 (2 Gc/s traveling wave interacting in KDP, no magnetic field)

24.211   Laser Scattering from a Bound System
J. H. Eberly and W. M. Frank (U. S. Naval Ord. Lab.)
Nuovo Cimento, Vol. 41 B, pp. 113–122, February 11, 1966

24.212   Backscattering from the Upper Atmosphere (75–160 km) detected by Optical Radar
P. D. McCormick, S. K. Poultney, U. Van Wijk, C. O. Alley, and R. T. Bettinger (U. of Maryland), and J. A. Perschy (Johns Hopkins U.)
Nature, Vol. 209, pp. 798–799, February 19, 1966

24.213   Concepts and Techniques of Microwave Optics
Réal Tramblay and Albéric Boivin (U. Laval, Quebec, Canada)
Appl. Optics, Vol. 5, pp. 249–278, February 1966 (components, beam waveguide, interferometer, 590 references)

24.214   A Laser End Reflector with Spectral Tuning Capability
Robert M. Zoot (Hughes)
Appl. Optics, Vol. 5, pp. 349–350, February 1966 (Pellin-Broca type prism with reflective interface)

24.215   Geometrical Representation of Gaussian Beam Propagation
T. S. Chu

Bell Sys. Tech. J., Vol. 45, pp. 287–299, February 1966 (beam matching problem, Smith chart)

24.216   The Attenuation of 3.392 μ He–Ne Laser Radiation by Methane in the Atmosphere
T. S. Chu and D. C. Hogg
Bell Sys. Tech. J., Vol. 45, pp. 301–306, February 1966 (also 3.508 μ He–Xe laser, 2.6-km atmospheric path, 5.5 dB/km due to methane)

24.217   Optics of General Guiding Media
J. P. Gordon
Bell Sys. Tech. J., Vol. 45, pp. 321–332, February 1966 (weakly focusing transparent media, scalar wave equation)

24.218   Low-Pass and High-Pass Filters Consisting of Multilayer Dielectric Stacks
L. Young and E. G. Cristal (S.R.I.)
IEEE Trans. on Microwave Theory and Techniques, Vol. MTT-14, pp. 75–80, February 1966 (equal and unequal thicknesses)

24.219   Effect of Internal Reflection on Optical Faraday Rotation
Herbert Piller (U. S. Naval Ordnance Laboratory)
J. Appl. Phys., Vol. 37, pp. 763–767, February 1966

24.220   Staggered Broad-Band Reflecting Multilayers
O. S. Heavens (U. of York, England) and Heather M. Liddell (U. of London, England)
Appl. Optics, Vol. 5, pp. 373–376, March 1966 (symmetric and asymmetric filters, 15 to 35 layers)

24.221   Precision Mapping of Pairs of Uncoated Optical Flats
F. L. Roesler and W. Traub (U. of Wisconsin)
Appl. Optics, Vol. 5, pp. 463–468, March 1966 (accuracy within $\lambda/500$)

24.222   Detection of Low Angle Optical Scattering by Fabry-Perot Resonance
D. G. Peterson (Lockheed) and Amnon Yariv (Calif. Inst. of Tech.)
Appl. Optics, Vol. 5, pp. 469–470, March 1966

24.223   A Hybrid Narrow Band Filter
Harold Zirin (Calif. Inst. of Tech.)
Appl. Optics, Vol. 5, pp. 474–475, March 1966 (combined Lyot type and multilayer filters)

24.224   The Statistical Effects of Random Variations in the Components of a Beam Waveguide
William H. Steier
Bell Sys. Tech. J., Vol. 45, pp. 451–471, March 1966 (Gaussian beam, exponential degradation)

24.225   Atmospheric Attenuation at Submillimetre Wavelengths
K. H. Breeden, W. K. Rivers, and A. P. Sheppard (Georgia Inst. of Tech.)
Electronics Letters, Vol. 2, p. 88, March 1966 (500 to 1500 Gc/s, several water vapor absorption lines)

24.226   Laser Wave Propagation Through the Atmosphere
H. Hodara (National Engineering Science Co.,)
Proc. IEEE, Vol. 54, pp. 368–375, March 1966 (thermal turbulences, beam quivering, breathing, depolarization)

24.227   Comments on "Laser Wave Propagation Through the Atmosphere"
Leonard S. Taylor (Case Inst. of Tech.)
Proc. IEEE, Vol. 54, pp. 1461–1462, October 1966
(Proc. IEEE, Vol. 54, pp. 368–375, March 1966, reply by H. Hodara)

24.228 Transmission Filters for Visible Light
Sukumar Maiti, Ashok Ghosh and Mihir K. Saha (Indian Assoc. for the Cultivation of Science, India)
Nature, Vol. 210, pp. 513–514, April 30, 1966 (dye solutions, tunable by changing pH, high transmission, photo-stable using tungsten lamp)

24.229 The Elliptical Polarization of Light Scattered by a Volume of Atmospheric Air
Reiner Eiden (Johannes Gutenberg U., Mainz, Germany)
Appl. Optics, Vol. 5, pp. 569–575, April 1966

24.230 Multiple Reflection Effects in the Faraday Rotation in Thin-Film Semiconductors
E. D. Palik and J. R. Stevenson (U. S. Naval Res. Lab), and J. Webster (U. of London, England)
J. Appl. Phys., Vol. 37, pp. 1982–1988, April 1966 (PbS, 3–30 μ infrared, oscillatory behavior)

24.231 Anomalous Surface Heating Rates
R. E. Harrington (Carbon Products Div., Union Carbide)
J. Appl. Phys., Vol. 37, pp. 2028–2034, April 1966 (apparently lower thermal conductivity, metals, semiconductors and insulators, MgO reflectivity saturates with xenon flashtube, hysteresis effect)

24.232 Coefficient for Self-Trapping of Optical Beams
Bernard L. Lewis (Radiation Inc.)
Proc. IEEE, Vol. 54, p. 688, April 1966

24.233 Deterioration of the Coherence Properties of a Laser Beam by Atmospheric Turbulence and Molecular Scattering
A. Consortini, L. Ronchi, A. M. Scheggi, and G. Toraldo di Francia (Centro Microonde del Consiglio Nazionale delle Ricerche, Firenze, Italy)
Radio Science, Vol. 1, pp. 523–530, April 1966 (optimum beam diameter, line width broadening)

24.234 The Slumping of Optical Surfaces during Coating
A. S. De Vany (Northrop Corp.)
Appl. Optics, Vol. 5, pp. 735–736, May 1966 (at 205°C for MgF$_2$)

24.235 Selection and Processing of Infrared Materials
P. W. Collyer (Barnes Engineering)
Appl. Optics, Vol. 5, pp. 765–770, May 1966 (nine materials)

24.236 Note on Reflectance Measurements on Metals
W. E. Müller (IBM, Switzerland)
Appl. Optics, Vol. 5, pp. 876–877, May 1966 (absolute measurement)

24.237 Far Infrared Transmittance of Irtrans 1 to 5 in the 250–10 cm^{-1} Spectral Region
G. M. Ressler and K. D. Möller (Fairleigh Dickinson U.)
Appl. Optics, Vol. 5, pp. 877–879, May 1966

24.238 Humidity Effects in the 8–13 μ Infrared Window
Hugh R. Carlon (U. S. Army, Edgewood Arsenal)
Appl. Optics, Vol. 5, p. 879, May 1966

24.239 Some Characteristics of Alternating Gradient Optical Transmission Lines
William H. Steier (B.T.L.)
IEEE Trans. on Microwave Theory and Techniques, Vol. MTT-14, pp. 228–233, May 1966 (alternate negative and positive lenses)

24.240 Infrared Reflectance Spectra of Igneous Rocks, Tuffs, and Red Sandstone from 0.5 to 22 μ
W. A. Hovis, Jr., and William R. Callahan (Goddard Space Flight Center)
J. Opt. Soc. Am., Vol. 56, pp. 639–643, May 1966

24.241 Excitation of Waveguide Modes in Retinal Receptors
Allan W. Snyder (Sylvania)
J. Opt. Soc. Am., Vol. 56, pp. 705–706, May 1966

24.242 On Strong Fluctuations of Light Wave Parameters in a Turbulent Medium
V. I. Tatarskiĭ (Institute of Physics of the Atmosphere, Academy of Sciences, U.S.S.R.)
Soviet Physics JETP, Vol. 22, pp. 1083–1088, May 1966

24.243 Optical Properties and Applications of Photochromic Glass
G. K. Megla (Corning Glass Works)
Appl. Optics, Vol. 5, pp. 945–960, June 1966 (usable as Q switch)

24.244 Interferometry through the Turbulent Atmosphere at an Optical Path Difference of 354 m
R. Bruce Herrick (Corning Glass Works) and Jurgen R. Meyer-Arendt (Indiana State U.)
Appl. Optics, Vol. 5, pp. 981–983, June 1966 (modified Michelson interferometer, He–Ne laser)

24.245 Perfect Match in Antireflection Systems
K. C. Park (Honeywell)
Appl. Optics, Vol. 5, pp. 1082–1083, June 1966 (considers absorbing films)

24.246 Far Infrared Transmission through Metal Light Pipes with Low Thermal Conductance
R. E. Harris, R. L. Cappelletti, and D. M. Ginsberg (U. of Illinois)
Appl. Optics, Vol. 5, pp. 1083–1084, June 1966

24.247 Optical Waveguide Modes in a Bisected Dielectric Slab
D. W. Wilmot and E. R. Schineller (Wheeler Labs.)
J. Opt. Soc. Am., Vol. 56, pp. 839–840, June 1966

24.248 Optical Waveguides
A. E. Karbowiak (U. of New South Wales, Australia)
Advances in Microwaves, Vol. 1, Edited by L. Young, Academic Press, New York, 1966, pp. 75–113. (principles, properties of materials)

24.249 The Effect of Internal Reflection in a Corner Cube Upon the Polarization of a Reflected Beam
Peter J. Walsh, and Irvin Krause (Fairleigh Dickinson U.)
Proc. 8th Annual Electron and Laser Beam Symposium, April 6–8, 1966. Sponsored by U. of Michigan and IEEE, pp. 139–155

24.250 Light Scattering by Spin Waves in FeF$_2$
P. A. Fleury, S. P. S. Porto, L. E. Cheesman, and H. J. Guggenheim (BTL)
Phys. Rev. Letters, Vol. 17, pp. 84–87, July 11, 1966 (0.488 μ argon laser, Stokes shifts)

24.251 Spectrum of Light Scattered from Thermal Fluctuations in Gases
T. J. Greytak and G. B. Benedek (M.I.T.)
Phys. Rev. Letters, Vol. 17, pp. 179–182, July 25, 1966 (0.63 μ He–Ne laser, five gases, one atmosphere, up to 1.5 GHz shift)

24.252 Faraday Rotators for High Power Laser Cavities
Nicholas George (California Inst. of Tech.), and R. W. Waniek (Advanced Kinetics)
Appl. Optics, Vol. 5, pp. 1183–1185, July 1966 (quartz, up to 400 kOe magnetic field, cascaded with cryptocyanine cell, ruby laser)

24.253 Wavefronts and Construction Tolerances for a Cat's-Eye Retroreflector
Reinhard Beer and Darwin Marjaniemi (California Inst. of Tech.)
Appl. Optics, Vol. 5, pp. 1191–1197, July 1966

24.254 The Ray Packet Equivalent of a Gaussian Light Beam
William H. Steier (BTL)
Appl. Optics, Vol.. 5, pp. 1229–1233, July 1966

24.255 Faraday Effect Measurements with Pulsed Magnetic Fields
K. Dismukes, S. H. Lott, Jr., and J. P. Barach (Vanderbilt U.)
Appl. Optics, Vol. 5, pp. 1246–1247, July 1966 (small hysteresis effect, Verdet constants of ten materials at 0.63 μ)

24.256 Acoustic Scattering of Light in a Fabry-Perot Resonator
M. G. Cohen and E. I. Gordon
Bell Sys. Tech. J., Vol. 45, pp. 945–966, July-August 1966 (0.63 μ beam, 200 to 500 MHz acoustic waves, enhancement by 50 in fused quartz resonator)

24.257 Loss Measurement of Organic Materials at 6328 Å
Domenico Solimini (U. of California)
J. Appl. Phys., Vol. 37, pp. 3314–3315, July 1966 (27 liquid samples within laser resonator)

24.258 Phase Control by Polarization in Coherent Spatial Filtering
Thomas M. Holladay and John D. Gallatin (Cornell Aeronautical Lab.)
J. Opt. Soc. Am., Vol. 56, pp. 869–872, July 1966 (0 or 180° phase generation)

24.259 Polarization, Directional Distribution, and Off-Specular Peak Phenomena in Light Reflected from Roughened Surfaces
K. E. Torrance and E. M. Sparrow (U. of Minnesota), and R. C. Birkebak (Georgia Inst. of Tech.)
J. Opt. Soc. Am., Vol. 56, pp. 916–925, July 1966 (polarized 0.5-μ light source, some departure from Lambert's cosine law)

24.260 Optical Network Synthesis Using Birefringent Crystals. III. Some General Properties of Lossless Birefringent Networks
E. O. Ammann (Sylvania)
J. Opt. Soc. Am., Vol. 56, pp. 943–951, July 1966

24.261 Optical Network Synthesis Using Birefringent Crystals. IV. Synthesis of Lossless Double-Pass Networks
E. O. Ammann (Sylvania)
J. Opt. Soc. Am., Vol. 56, pp. 952–955, July 1966

24.262 A Close Look at Optical Waveguides
A. E. Karbowiak (U. of New South Wales, Australia)
Microwaves, Vol. 5, No. 7, pp. 37–46, July 1966 (review)

24.263 Metal Mirrors for Infrared Cells in Corrosive Atmospheres
Dale E. Armstrong and Eugene S.. Robinson (Los Alamos)
Rev. Sci. Instr., Vol. 37, pp. 965–966, July 1966

24.264 The Pressure Dependence of Absorption Coefficient in Organic Gases at 3.3913 μm and 3.5070 μm
Katsumi Sakurai and Koichi Shimoda (U. of Tokyo, Japan)
Japanese J. Appl. Phys., Vol. 5, p. 744, August 1966 (several gases, 0.1 to 100 Torr)

24.265 Validity of the Rytov Approximation in Optical Propagation Calculations
W. P. Brown, Jr. (Hughes)
J. Opt. Soc. Am., Vol. 56, pp. 1045–1052, August 1966

24.266 Depolarization of the Components of Rayleigh Scattering in Liquids
D. H. Rank, Amos Hollinger, and D. P. Eastman (Penn. State U.)
J. Opt. Soc. Am., Vol. 56, pp. 1057–1058, August 1966 (eight liquids, He–Ne laser source, Brillouin components strongly polarized)

24.267 Synthesis of Electro-Optic Shutters having a Prescribed Transmission vs Voltage Characteristic
E. O. Ammann (Sylvania)
J. Opt. Soc. Am., Vol. 56, pp. 1081–1088, August 1966

24.268 Effect of Precipitation on Transmission through the Atmosphere at 10 Microns
R. W. Wilson and A. A. Penzias (BTL)
Nature, Vol. 211, p. 1081, September 3, 1966 (fog, snow, rain, 50 dB fade)

24.269 Effects of Atmospheric Turbulence on the Transmission of a Laser Beam at 6328 Å. I-Distribution of Intensity
D. H. Höhn (U. of Tübingen, Germany)
Appl. Optics, Vol. 5, pp. 1427–1431, September 1966

24.270 Effects of Atmospheric Turbulence on the Transmission of a Laser Beam at 6328 Å. II-Frequency Spectra
D. H. Höhn (U. of Tübingen, Germany)
Appl. Optics, Vol. 5, pp. 1433–1436, September 1966

24.271 Underwater Transmission Characteristics for Laser Radiation
Howard J. Okoomian (RCA)
Appl. Optics, Vol. 5, pp. 1441–1446, September 1966 (5300 Å radiation, irradiance versus distance)

24.272 A Field-Sensitive Spectrometer for Quasi-Coherent Scattering Studies
L. R. Wilcox (Columbia U.)
IEEE J. Quantum Electronics, Vol. QE-2, pp. 557–562, September 1966 (spectral power density, scattered light from rotating screen, CO_2 scattering)

24.273 Investigation of the Scattering of Laser Light by a Plasma
W. H. McMahan and J. R. Bowen (Martin)
IEEE J. Quantum Electronics, Vol. QE-2, pp. 567–579, September 1966 (argon ion laser, argon plasma by theta pinch)

24.274 Laser Operation with Liquid Semiconductor Mirrors
M. Birnbaum and T. L. Stocker (Aerospace)
IEEE J. Quantum Electronics, Vol. QE-2, pp. 632–635, September 1966 (ruby laser, one liquid selenium mirror at 500°C, reflectivity measurements)

24.275 Coherence of Light from Random Medium
Takeomi Suzuki and Ryuichi Hioki (U. of Tokyo, Japan)
Japanese J. Appl. Phys., Vol. 5, pp. 807–813, September 1966 (speckled diffraction patterns of diffused coherent light)

24.276 Phase Compensation of Total Internal Reflection
Paul Mauer (Eastman Kodak)
J. Opt. Soc. Am., Vol. 56, pp. 1219–1221, September 1966 (symmetrical 3-layer thin-film combinations, zero phase difference between S and P components beyond critical angle)

24.277 Radiation Characteristics of Light Beams Transmitted Through Straight Dielectric Tubes
K. Vogel (U. of Uppsala, Sweden)
J. Opt. Soc. Am., Vol. 56, pp. 1222–1226, September 1966

24.278 Dependence of Image Quality on Horizontal Range in a Turbulent Atmosphere
C. E. Coulman (NSL, Australia)
J. Opt. Soc. Am., Vol. 56, pp. 1232–1238, September 1966

24.279 Spectrum of Light Scattered by Density and Anisotropy Fluctuations in Liquid Nitrobenzene
V. S. Starunov, E. V. Tiganov, and I. L. Fabelinskii (Lebedev Phys. Inst., Acad. of Sci., USSR)
JETP Letters, Vol. 4, pp. 176–179, October 1, 1966 (narrow diffuse wing and fine-structure lines, Mandel'shtam-Brillouin components)

24.280 Transmission of Sub-millimetre Waves in Fog
W. J. Burroughs, E. C. Pyatt, and H. A. Gebbie (Nat'l Phys. Lab., England)
Nature, Vol. 212, pp. 387–388, October 22, 1966 (0.337 mm

wavelength CN maser, about 50 dB/km loss at 0°C and saturated vapor pressure)

24.281 Optical Transmission Research
S. E. Miller and L. C. Tillotson (BTL)
Appl. Optics, Vol. 5, pp. 1538–1549, October 1966 (summary, modulation, atmospheric effects, guided propagation, detection)

24.282 Optical Activity and Electrooptic Effect in Bismuth Germanium Oxide ($Bi_{12}GeO_{20}$)
P. V. Lenzo, E. G. Spencer, and A. A. Ballman (BTL)
Appl. Optics, Vol. 5, pp. 1688–1689, October 1966 (modulation up to 500 MHz)

24.283 Comparison Between a Gas Lens and Its Equivalent Thin Lens
D. Marcuse
Bell Sys. Tech. J., Vol. 45, pp. 1339–1344, October 1966 (good agreement, use warranted)

24.284 Deformation of Fields Propagating Through Gas Lenses
D. Marcuse
Bell Sys. Tech. J., Vol. 45, pp. 1345–1368, October 1966 (essential to keep beam on axis)

24.285 Propagation Characteristics of a Partially Filled Cylindrical Waveguide for Light Beam Modulation
D. Chen and T. C. Lee (Honeywell)
IEEE Trans. Microwave Theory and Techniques, Vol. MTT-14, pp. 482–486, October 1966 (partial dielectric, electromagnetic wave, and optical beam propagation)

24.286 Optical Transmittance of Fused Silica at Elevated Temperatures
Oliver J. Edwards (Lewis Research Ctr.)
J. Opt. Soc. Am., Vol. 56, pp. 1314–1319, October 1966 (up to 982°C, wavelength range of 0.17 to 3.5 μ)

24.287 Optical Resolution Through a Randomly Inhomogeneous Medium for Very Long and Very Short Exposures
D. L. Fried (Autonetics)
J. Opt. Soc. Am., Vol. 56, pp. 1372–1379, October 1966

24.288 Limiting Resolution Looking Down Through the Atmosphere
D. L. Fried (Autonetics)
J. Opt. Soc. Am., Vol. 56, pp. 1380–1384, October 1966

24.289 Far-Infrared Spectrum of Liquid Water
J. A. Lane (Sci. Research Council, England)
J. Opt. Soc. Am., Vol. 56, pp. 1398–1399, October 1966 (10 to 2000 μ)

24.290 Optical Transmission Research
S. E. Miller and L. C. Tillotson (BTL)
Proc. IEEE, Vol. 54, pp. 1300–1311, October 1966 (summary, modulation, atmospheric effects, guided propagation, detection)

24.291 Induced Scattering of Light by Light
A. A. Varfolomeev
Soviet Phys. JETP, Vol. 23, pp. 681–688, October 1966

24.292 Angular Dependence of Transmission Characteristics of Interference Filters and Application to a Tunable Fluorometer
S. A. Pollack (TRW/Systems)
Appl. Optics, Vol. 5, pp. 1749–1756, November 1966

24.293 The Stability of Interference Filters
J. Meaburn (U. of Manchester, England)
Appl. Optics, Vol. 5, pp. 1757–1759, November 1966 (thermal treatment, passband position changes due to crystal growth, up to one year)

24.294 Phase and Amplitude Measurements of Coherent Optical Wavefronts

Joseph T. Ruscio
Bell Sys. Tech. J., Vol. 45, pp. 1583–1597, November 1966 (phase-locked laser loop, resolution of 1° in phase and 1 dB in amplitude)

24.295 Statistical Distribution of the Envelope of an Amplitude-Modulated Laser Signal After Passage Through a Turbulent Atmosphere
M. Chomát and F. Hoff (Acad. of Sci., Czechoslovakia)
Electronics Letters, Vol. 2, pp. 409–410, November 1966 (1 kHz modulation, 0.63-μ He–Ne laser, 0.8 to 12 km path length, noise increases with path length)

24.296 Interferometric Uses of Optical Fiber
Takeomi Suzuki (U. of Tokyo, Japan)
Japanese J. Appl. Phys., Vol. 5, pp. 1065–1074, November 1966 (up to 4 fibers, coherence effects)

24.297 Spatial Coherence in Periodic Systems
William Streifer (U. of Rochester)
J. Opt. Soc. Am., Vol. 56, pp. 1481–1489, November 1966 (van Cittert-Zernike theorem, apertures and lenses, improvement in coherence)

24.298 Diffraction of a Plane Wave at a Sinusoidally Stratified Dielectric Grating
C. B. Burckhardt (BTL)
J. Opt. Soc. Am., Vol. 56, pp. 1502–1509, November 1966 (maximum amplitude at Bragg angle)

24.299 Equivalent Layers in Multilayer Filters
Alfred Thelen (Optical Coating Lab.)
J. Opt. Soc. Am., Vol. 56, pp. 1533–1538, November 1966

24.300 Effect of Molecular Redistribution on the Nonlinear Refractive Index of Liquids
R. W. Hellwarth (Hughes Research)
Phys. Rev., Vol. 152, pp. 156–165, December 2, 1966 (as important as molecular reorientation)

24.301 Rotierende Mattscheiben zur Änderung der Kohärenz von Laserlicht
K. Goetz and D. Unangst (U. of Jena)
Physics Letters, Vol. 23, pp. 667–668, December 12, 1966 (laser beam on rotating ground glass, used for optical transforms)

24.302 Thermal Expansion and Other Physical Properties of the Newer Infrared-Transmitting Optical Materials
Stanley S. Ballard and James Steve Browder (U. of Florida)
Appl. Optics, Vol. 5, pp. 1873–1876, December 1966

24.303 Light Rays in Lens-Like Media
Yoshinao Aoki (Hokkaido U., Japan)
J. Opt. Soc. Am., Vol. 56, pp. 1648–1651, December 1966

24.304 Multiple Scattering of Light in a Turbulent Atmosphere
P. M. Livingston (Inst. for Defense Analysis)
J. Opt. Soc. Am., Vol. 56, pp. 1660–1667, December 1966

24.305 Propagation of an Infinite Plane Wave in a Randomly Inhomogeneous Medium
D. L. Fried and J. D. Cloud (Autonetics)
J. Opt. Soc. Am., Vol. 56, pp. 1667–1676, December 1966 (log-amplitude covariance, phase-structure functions)

24.306 Effect of a Turbulent Medium on the Power Pattern of a Wavefront-Tracking Circular Aperture
G. R. Heidbreder and R. L. Mitchell (Aerospace)
J. Opt. Soc. Am., Vol. 56, pp. 1677–1684, December 1966

24.307 Optical Network Synthesis Using Birefringent Crystals. V. Synthesis of Lossless Networks Containing Equal-Length Crystals and Compensators

E. O. Ammann and J. M. Yarborough (Sylvania)
J. Opt. Soc. Am., Vol. 56, pp. 1746–1754, December 1966

24.308 Reflectance and Phase Envelopes of an Iterated Multilayer
Joseph Arndt and Philip Baumeister (U. of Rochester)
J. Opt. Soc. Am., Vol. 56, pp. 1760–1762, December 1966 (minimum and maximum reflectance curves)

24.309 Electromagnetic Wave Propagation in Birefringent Multilayers

D. A. Holmes (AF Weapons Lab.), and D. L. Feucht (Carnegie Inst. of Tech.)
J. Opt. Soc. Am., Vol. 56, pp. 1763–1769, December 1966

24.310 Simple Relation Between Reflectances of Polarized Components of a Beam When the Angle of Incidence is 45 Degrees
D. W. Berreman (BTL)
J. Opt. Soc. Am., Vol. 56, p. 1784, December 1966

25. HOLOGRAPHY AND WAVEFRONT RECONSTRUCTION

25.1 Wavefront Reconstruction with Continuous-Tone Objects
Emmett N. Leith and Juris Upatnieks (U. of Michigan)
J. Opt. Soc. Am., Vol. 53, pp. 1377–1381, December 1963

25.2 Wavefront Reconstruction with Diffused Illumination and Three-Dimensional Objects
E. N. Leith and J. Upatnieks (U. of Michigan)
J. Opt. Soc. Am., Vol. 54, pp. 1295–1301, November 1964 (holograms, 3-D scenes, also multicolor possible)

25.3 An Application of Wavefront Reconstruction to Interferometry
M. H. Horman (Boeing)
Appl. Optics, Vol. 4, pp. 333–336, March 1965 (comparison with holograms, limitations)

25.4 Attainment of High Resolutions in Holography by Multi-Directional Illumination and Moving Scatterers
G. W. Stroke and D. G. Falconer (U. of Michigan)
Physics Letters, Vol. 15, pp. 238–240, April 1, 1965 (0.63-μ gas laser)

25.5 Three-Dimensional Wavefront Reconstruction Using a Phase Hologram
W. T. Cathey, Jr. (Autonetics)
J. Opt. Soc. Am., Vol. 55, p. 457, April 1965 (instead of spatial amplitude modulation)

25.6 Lensless Fourier-Transform Method for Optical Holography
George W. Stroke (U. of Michigan)
Appl. Phys. Letters, Vol. 6, pp. 201–203, May 15, 1965

25.7 Fresnel Holograms: Their Imaging Properties and Aberrations
J. A. Armstrong (IBM)
IBM J. Res. & Dev., Vol. 9, pp. 171–178, May 1965

25.8 Microscopy by Wavefront Reconstruction
E. N. Leith and J. Upatnieks (U. of Michigan)
J. Opt. Soc. Am., Vol. 55, pp. 569–570, May 1965

25.9 Multicolor Wavefront Reconstruction
K. S. Pennington and L. H. Lin (BTL)
Appl. Phys. Letters, Vol. 7, pp. 56–57, August 1, 1965 (three-dimensional holograms)

25.10 Holographic Photography of High-Speed Phenomena with Conventional and Q-switched Ruby Lasers
R. E. Brooks, L. O. Heflinger, R. F. Wuerker, and R. A. Briones (TRW Systems)
Appl. Phys. Letters, Vol. 7, pp. 92–94, August 15, 1965 (Kerr cell, bullet, water droplet)

25.11 Holograms on Thick Emulsions
Albert A. Friesem (U. of Michigan)
Appl. Phys. Letters, Vol. 7, pp. 102–103, August 15, 1965 (Kodak type-649 emulsion)

25.12 Optical Image Synthesis (Complex Amplitude Addition and Subtraction) by Holographic Fourier Transformation
D. Gabor (Imperial College of Science and Technology, England) and G. W. Stroke, R. Restrick, A. Funkhouser, and D. Brumm (U. of Michigan)
Physics Letters, Vol. 18, pp. 116–118, August 15, 1965

25.13 Microscopy by Wavefront Reconstruction
Emmett N. Leith, Juris Upatnieks, and Kenneth A. Haines (U. of Michigan)
J. Opt. Soc. Am., Vol. 55, pp. 981–986, August 1965 (two-beam Gabor microscope, aberrations discussed)

25.14 Magnification and Third-Order Aberrations in Holography
Reinhard W. Meier (Xerox)
J. Opt. Soc. Am., Vol. 55, pp. 987–992, August 1965

25.15 Resolution-Retrieving Compensation of Source Effects by Correlative Reconstruction in High-Resolution Holography
G. W. Stroke, R. Restrick, A. Funkhouser, and D. Brumm (U. of Michigan)
Physics Letters, Vol. 18, pp. 274–275, September 1, 1965

25.16 Resolution-Retrieving Source-Effect Compensation in Holography with Extended Sources
G. W. Stroke, R. Restrick, A. Funkhouser, and D. Brumm (U. of Michigan)
Appl. Phys. Letters, Vol. 7, pp. 178–179, September 15, 1965

25.17 Phase-Contrast Holograms
G. L. Rogers (College of Advanced Technology, Birmingham, England)
J. Opt. Soc. Am., Vol. 55, p. 1181, September 1965 (earlier reference in 1952)

25.18 Application of Moiré Techniques to Holography
R. J. Collier, E. T. Doherty, and K. S. Pennington (BTL)
Appl. Phys. Letters, Vol. 7, pp. 223–225, October 15, 1965

25.19 Reconstruction of Wavefronts in All Directions
Ryuichi Hioki and Takeomi Suzuki (U. of Tokyo, Japan)
Japanese J. Appl. Phys., Vol. 4, p. 816, October 1965 (object illuminated 360° from laser beam, reconstructed images)

25.20 Three-Dimensional Holography with "Lensless" Fourier-Transform Holograms and Coarse P/N Polaroid Film
G. W. Stroke, D. Brumm, and A. Funkhouser (U. of Michigan)
J. Opt. Soc. Am., Vol. 55, pp. 1327–1328, October 1965

25.21 Holography with Spatially Noncoherent Light
George W. Stroke and Robert C. Restrick III (U. of Michigan)
Appl. Phys. Letters, Vol. 7, pp. 229–231, November 1, 1965

25.22 Interferometry with a Holographically Reconstructed Comparison Beam
R. E. Brooks, L. O. Heflinger, and R. F. Wuerker (TRW Systems)
Appl. Phys. Letters, Vol. 7, pp. 248–249, November 1, 1965

25.23 Holograms Produced with Pulsed Laser Illumination
A. D. Jacobson and F. J. McClung (Hughes)
Appl. Optics, Vol. 4, pp. 1509–1510, November 1965 (30 ns exposure, 60 mJ output from mode controlled ruby laser)

25.24 Wavefront Reconstruction for Incoherent Objects
A. W. Lohmann (IBM)
J. Opt. Soc. Am., Vol. 55, pp. 1555–1556, November 1965

25.25   Inverted Reference-Beam Hologram
        A. S. Hoffman, J. G. Doidge, and D. G. Mooney (Autonetics)
        J. Opt. Soc. Am., Vol. 55, p. 1559, November 1965

25.26   Some Curious Properties of Holograms
        W. E. Kock and J. Rendeiro (NASA Electronics Research
        Center)
        Proc. IEEE, Vol. 53, p. 1787, November 1965 (rotated hologram)

25.27   Comments on "Some Curious Properties of Holograms"
        T. S. Huang (M.I.T.), R. A. Becker (Altadena, Calif.), W. E.
        Kock (NASA Electronics Research Center), and J. Rendeiro
        (Merrimac College)
        Proc. IEEE, Vol. 54, p. 716, April 1966 (W. E. Kock and J.
        Rendeiro, "Some Curious Properties of Holograms," Proc. IEEE,
        Vol. 53, p. 1787, November 1965)

25.28   Reconstruction of Vectorial Wavefronts
        A. W. Lohmann (IBM)
        Appl. Optics, Vol. 4, pp. 1667–1668, December 1965 (three-color
        holography)

25.29   Holographic Image Projection through Inhomogeneous Media
        H. Kogelnik
        Bell Sys. Tech. J., Vol. 44, pp. 2451–2455, December 1965

25.30   Interferometric Vibration Analysis by Wavefront Reconstruction
        Robert L. Powell and Karl A. Stetson (U. of Michigan)
        J. Opt. Soc. Am., Vol. 55, pp. 1593–1598, December 1965 (vi-
        brating objects, resonant modes detected)

25.31   Wavefront Reconstruction by Reflection
        A. K. Rigler (Westinghouse)
        J. Opt. Soc. Am., Vol. 55, p. 1693, December 1965 (aluminized
        the emulsion side of hologram)

25.32   Depth of Focus and Depth of Field in Holography
        Reinhard W. Meier (Xerox)
        J. Opt. Soc. Am., Vol. 55, pp. 1693–1694, December 1965

25.33   Interferometric Hologram Evaluation and Real-Time Vibration
        Analysis of Diffuse Objects
        Karl A. Stetson and Robert L. Powell (U. of Michigan)
        J. Opt. Soc. Am., Vol. 55, pp. 1694–1695, December 1965

25.34   Color Imagery by Wavefront Reconstruction
        L. Mandel (U. of Rochester)
        J. Opt. Soc. Am., Vol. 55, pp. 1697–1698, December 1965

25.35   Hologram-Generated Ghost Image Experiments
        K. S. Pennington and R. J. Collier (B.T.L.)
        Phys. Rev. Letters, Vol. 8, pp. 14–16, January 1, 1966

25.36   Two–Beam Interferometry by Successive Recording of Intensities
        in a Single Hologram
        George W. Stroke and Antoine E. Labeyrie (U. of Michigan)
        Appl. Phys. Letters, Vol. 8, pp. 42–44, January 15, 1966

25.37   Ghost Imaging by Holograms Formed in the Near Field
        R. J. Collier and K. S. Pennington (B.T.L.)
        Appl. Phys. Letters, Vol. 8, pp. 44–46, January 15, 1966 (solid
        objects)

25.38   Interferometric Measurements Using the Wavefront Reconstruc-
        tion Technique
        B. P. Hildebrand and K. A. Haines (U. of Michigan)
        Appl. Optics, Vol. 5, pp. 172–173, January 1966 (real time de-
        formation of 3-dimensional objects)

25.39   Influence of Photographic Film on Wavefront Reconstruction. I.
        Plane Wavefronts
        Raoul F. van Ligten (American Optical)
        J. Opt. Soc. Am., Vol. 56, pp. 1–9, January 1966

25.40   Holography and Its Crystallographic Equivalent
        Patrick Tollin, Peter Main, and Michael G. Rossman (Purdue
        U.), and George W. Stroke and Robert C. Restrick (U. of
        Michigan)
        Nature, Vol. 209, pp. 603–604, February 5, 1966

25.41   On the Use of Moving Scatterers in Conventional Holography
        D. J. DeBitetto (Philips, New York)
        Appl. Phys. Letters, Vol. 8, pp. 78–80, February 15, 1966

25.42   Hologram Transmission via Television
        L. H. Enloe, J. A. Murphy and C. B. Rubinstein
        Bell Sys. Tech. J., Vol. 45, pp. 335–339, February 1966

25.43   Polarization Selection for Reconstructed Wavefronts and Applica-
        tion to Polarizing Microholography
        W. H. Carter, P. D. Engeling, and A. A. Dougal (U. of Texas)
        IEEE J. Quantum Electronics, Vol. QE-2, pp. 44–46, February
        1966

25.44   Holographic Interferometry
        L. O. Heflinger, R. F. Wuerker, and R. E. Brooks (TRW Sys-
        tems)
        J. Appl. Phys., Vol. 37, pp. 642–649, February 1966

25.45   Cardinal Points and the Novel Imaging Properties of a Holo-
        graphic System
        Reinhard W. Meier (Xerox)
        J. Opt. Soc. Am., Vol. 56, pp. 219–223, February 1966

25.46   Hologram Television
        Winston E. Kock (NASA Electronics Research Center)
        Proc. IEEE, Vol. 54, p. 331, February 1966

25.47   White-Light Reconstruction of Holographic Images Using the
        Lippmann-Bragg Diffraction Effect
        G. W. Stroke and A. E. Labeyrie (U. of Michigan)
        Physics Letters, Vol. 20, pp. 368–370, March 1, 1966

25.48   Dual- and Multiple-beam Interferometry by Wavefront Recon-
        struction
        J. M. Burch, A. E. Ennos, and R. J. Wilton (National Physical
        Lab., England)
        Nature, Vol. 209, pp. 1015–1016, March 5, 1966

25.49   Holograms with Nonpseudoscopic Real Images
        F. B. Rotz and A. A. Friesem (U. of Michigan)
        Appl. Phys. Letters, Vol. 8, pp. 146–148, March 15, 1966

25.50   Incoherent Holograms with Mercury Light Source
        P. J. Peters (Conductron)
        Appl. Phys. Letters, Vol. 8, pp. 209–210, April 15, 1966

25.51   Correction of Lens Aberrations by Means of Holograms
        J. Upatnieks, A. Vander Lugt, and E. Leith (U. of Michigan)
        Appl. Optics, Vol. 5, pp. 589–593, April 1966

25.52   Surface-Deformation Measurement Using the Wavefront Recon-
        struction Technique
        K. A. Haines and B. P. Hildebrand (U. of Michigan)
        Appl. Optics, Vol. 5, pp. 595–602, April 1966

25.53   Copying Holograms
        Franklin S. Harris, Jr., George C. Sherman, and Bruce H. Bill-
        ings (Aerospace)
        Appl. Optics, Vol. 5, pp. 665–666, April 1966

25.54   Thermoplastic Xerographic Holography
        John C. Urbach and Reinhard W. Meier (Xerox)
        Appl. Optics, Vol. 5, pp. 666–667, April 1966

25.55   Generation of a Hologram from a Moving Target
        V. J. Corcoran, R. W. Herron, Jr., and J. G. Jaramillo (Martin
        Co.)
        Appl. Optics, Vol. 5, pp. 668–669, April 1966

25.56 Multicolor Holographic Image Reconstruction with White-Light Illumination
L. H. Lin and K. S. Pennington (B.T.L.) and G. W. Stroke and A. E. Labeyrie (U. of Michigan)
Bell. Sys. Tech. J., Vol. 45, pp. 659–661, April 1966

25.57 On the Absence of Phase-Recording or "Twin-Image" Separation Problems in "Gabor" (in-line) Holography
G. W. Stroke, D. Brumm, A. Funkhouser, A. Labeyrie and R. C. Restrick (U. of Michigan)
British J. Appl. Phys., Vol. 17, pp. 497–500, April 1966

25.58 Schlieren Photographs from Holograms
J. B. Story, G. S. Ballard, and R. H. Gibbons (U. of Arkansas)
J. Appl. Phys., Vol. 37, pp. 2183–2184, April 1966

25.59 Image Reconstruction with Fraunhofer Holograms
John B. DeVelis, George B. Parrent, Jr., and Brian J. Thompson (Technical Operations Res.)
J. Opt. Soc. Am., Vol. 56, pp. 423–427, April 1966

25.60 Photographic Recording of Spatially Modulated Coherent Light
Adam Kozma (U. of Michigan)
J. Opt. Soc. Am., Vol. 56, pp. 428–432, April 1966

25.61 Image Luminance and Ray Tracing in Holography
Carl W. Helstrom (Westinghouse)
J. Opt. Soc. Am., Vol. 56, pp. 433–441, April 1966

25.62 Holographic Imagery Through Diffusing Media
Emmett N. Leith and Juris Upatnieks (U. of Michigan)
J. Opt. Soc. Am., Vol. 56, p. 523, April 1966

25.63 Wide-Angle Holography
E. P. Supertzi and A. K. Rigler (Westinghouse)
J. Opt. Soc. Am., Vol. 56, pp. 524–525, April 1966

25.64 Hologram Visual Displays
E. N. Leith, J. Upatnieks, A. Kozma and N. Massey (U. of Michigan)
J. Soc. Motion Picture and Television Engineers, Vol. 75, pp. 323–326, April 1966 (storing sequence of frames, readout and projection problems)

25.65 Bibliography on Holograms
R. P. Chambers and J. S. Courtney-Pratt (B.T.L.)
J. Soc. Motion Picture and Television Engineers, Vol. 75, pp. 373–435, April 1966 (over 180 papers, 1948 to 1966, many abstracts)

25.66 Hologram Illumination with a Flashlight
D. W. Wilmot, E. R. Schineller, and R. W. Heuman (Wheeler Labs.)
Proc. IEEE, Vol. 54, pp. 690–691, April 1966

25.67 Reactive Processing of Phase Objects
R. V. Pole, H. Wieder, and R. A. Myers (IBM)
Appl. Phys. Letters, Vol. 8, pp. 229–231, May 1, 1966 (phase object within laser cavity to process image)

25.68 Holographische Interferometrie Diffus Reflektierender Objekte
H. Nassenstein (Farbenfabriken Bayer AG, Germany)
Physics Letters, Vol. 21, pp. 290–291, May 15, 1966

25.69 Obtaining Increased Focal Depth in Bubble Chamber Photography by an Application of the Hologram Principle
W. T. Welford (Imperial College, England)
Appl. Optics, Vol. 5, pp. 872–873, May 1966

25.70 Color Holograms for White Light Reconstruction
J. Upatnieks, J. Marks, and R. Fedorowicz (U. of Michigan)
Appl. Phys. Letters, Vol. 8, pp. 286–287, June 1, 1966

25.71 The Generation of Three-Dimensional Contour Maps by Wavefront Reconstruction
B. P. Hildebrand and K. A. Haines (U. of Michigan)
Physics Letters, Vol. 21, pp. 422–423, June 1, 1966

25.72 Wavefront-Reconstruction Imaging through Random Media
J. W. Goodman, W. H. Huntley, Jr., D. W. Jackson, and M. Lehmann (Stanford U.)
Appl. Phys. Letters, Vol. 8, pp. 311–313, June 15, 1966 (improves resolution capability)

25.73 Complex Spatial Filtering with Binary Masks
B. R. Brown and A. W. Lohmann (IBM)
Appl. Optics, Vol. 5, pp. 967–969, June 1966 (Fraunhofer holography, character recognition)

25.74 Recent Advances in Multicolor Wavefront Reconstruction
A. A. Friesem and R. J. Fedorowicz (U. of Michigan)
Appl. Optics, Vol. 5, pp. 1085–1086, June 1966

25.75 On Some Properties of Photographically Produced Diffraction Gratings
A. K. Rigler and T. P. Vogl (Westinghouse)
Appl. Optics, Vol. 5, pp. 1086–1087, June 1966 (diffractive properties of holograms)

25.76 Polarization Effects in Holography
G. L. Rogers (College of Advanced Technology, Birmingham, England)
J. Opt. Soc. Am., Vol. 56, p. 831, June 1966 (shadow effect, internal strains)

25.77 Wavefront Reconstruction or "Holography"
Dennis Gabor (U. of London, England)
Proc. 8th Annual Electron and Laser Beam Symposium, April 6–8, 1966. Sponsored by U. of Michigan and IEEE, pp. 1–19 (fundamentals, applications)

25.78 Recent Results in Holography
E. N. Leith (U. of Michigan)
Proc. 8th Annual Electron and Laser Beam Symposium, April 6–8, 1966. Sponsored by U. of Michigan and IEEE, pp. 21–37

25.79 On Some Possible Properties of the Hologram
G. I. Kopylov (Joint Inst. of Nuclear Research, USSR)
Physics Letters, Vol. 21, pp. 645–646, July 1, 1966

25.80 Holographic Microscopy
Raoul F. vanLigten and Harold Osterberg (American Optical)
Nature, Vol. 211, pp. 282–283, July 16, 1966 (12 μ lower level of detectability, He–Ne laser)

25.81 Simplification of Holographic Procedures
Joseph T. Carcel, Alfred H. Rodemann, Edward Florman, and S. Domeshek (USN Training Device Ctr.)
Appl. Optics, Vol. 5, pp. 1199–1201, July 1966

25.82 Demonstration of the Application of Wavefront Reconstruction to Interferometry
Henry H. M. Chau and Melvin H. Horman (Boeing)
Appl. Optics, Vol. 5, pp. 1237–1239, July 1966

25.83 Some Current Views on Holography
Robert J. Collier (BTL)
IEEE Spectrum, Vol. 3, No. 7, pp. 67–74, July 1966 (fundamentals, experiments)

25.84 Interference Microscope with Total Wavefront Reconstruction
D. Gabor and W. P. Goss (Imperial Coll. of Sci. and Tech., England)
J. Opt. Soc. Am., Vol. 56, pp. 849–858, July 1966

25.85 Geometrical Relationships Between the Original Object and the Two Images of a Hologram Reconstruction
Don B. Neumann (Wright-Patterson AFB)
J. Opt. Soc. Am., Vol. 56, pp. 858–861, July 1966

25.86 Magnification and Observation of a Holographic Interference Pattern
Sten Walles (Inst. of Optical Research, Sweden)
Optica Acta, Vol. 13, pp. 241–246, July 1966

25.87 Reconstruction of an Image from a Hologram in Nonmonochromatic Light
G. I. Kosourov, I. N. Kalinkina, and M. P. Golovei (Acad. of Sci., USSR)
JETP Letters, Vol. 4, pp. 57–58, August 1, 1966

25.88 Holographic Investigation of a Laser Spark
G. V. Ostrovskaya and Yu. I. Ostrovskii (Ioffe Phys. Tech. Inst., USSR)
JETP Letters, Vol. 4, pp. 83–85, August 15, 1966 (three holograms during one spark, structure of spark)

25.89 Holographic Data Storage in Three-Dimensional Media
E. N. Leith, A. Kozma, J. Upatnieks, J. Marks, and N. Massey (U. of Michigan)
Appl. Optics, Vol. 5, pp. 1303–1311, August 1966

25.90 Pulsed Laser Holograms
R. E. Brooks, L. O. Heflinger, and R. F. Wuerker (TRW/Systems)
IEEE J. Quantum Electronics, Vol. QE-2, pp. 275–279, August 1966

25.91 Influence of Photographic Film on Wavefront Reconstruction. II: "Cylindrical" Wavefronts
Raoul F. vanLigten (American Optical)
J. Opt. Soc. Am., Vol. 56, pp. 1009–1014, August 1966

25.92 Coffee-Table Holography
John Landry (U. of California, Santa Barbara)
J. Opt. Soc. Am., Vol. 56, p. 1133, August 1966 (solid table, 10 s exposure)

25.93 Underwater Holography
R. M. Grant, R. L. Lillie, and N. E. Barnett (U. of Michigan)
J. Opt. Soc. Am., Vol. 56, p. 1142, August 1966 (He–Ne gas laser, three minute exposure, good results)

25.94 Holographic Diffraction Gratings
Nicholas George and J. W. Matthews (California Inst. of Tech.)
Appl. Phys. Letters, Vol. 9, pp. 212–215, September 1, 1966

25.95 White-Light Reconstruction of Color Images from Black-and-White Volume Holograms Recorded on Sheet Film
George W. Stroke and Richard G. Zech (U. of Michigan)
Appl. Phys. Letters, Vol. 9, pp. 215–217, September 1, 1966

25.96 Noise Limitations on the Reconstruction of Three-Dimensional Pictures
A. L. Mikaelyan and V. I. Bobrinev
JETP Letters, Vol. 4, pp. 118–120, September 1, 1966

25.97 Speckled Diffraction Pattern and Source Effect on Resolution Limit in Holography
Takeomi Suzuki and Ryuichi Hioki (U. of Tokyo, Japan)
Japanese J. Appl. Phys., Vol. 5, pp. 814–817, September 1966

25.98 Hologram Interferometry
Karl A. Stetson and Robert L. Powell (U. of Michigan)
J. Opt. Soc. Am., Vol. 56, pp. 1161–1166, September 1966

25.99 Spatial Phase Modulation of Wavefronts in Spatial Filtering and Holography
Wade T. Cathey, Jr. (Autonetics)
J. Opt. Soc. Am., Vol. 56, pp. 1167–1171, September 1966

25.100 360° Holography
Tung H. Jeong, Paul Rudolf, and Arleigh Luckett (Lake Forest College)
J. Opt. Soc. Am., Vol. 56, pp. 1263–1264, September 1966

25.101 Achromatization of Holograms
Henri Paques (Wright-Patterson AFB)
Proc. IEEE, Vol. 54, pp. 1195–1196, September 1966

25.102 Interferometry Holographic Investigation of a Laser Spark
A. Kakos, G. V. Ostrovskaya, Yu. I. Ostrovskii, and A. N. Zaidel (Ioffe Phys. Tech. Inst., USSR)
Physics Letters, Vol. 23, pp. 81–83, October 3, 1966 (three holograms during spark development, two beam holograms, electron density of 2×10^{19} cm^{-3})

25.103 Copying Holograms
M. J. Landry (Sandia)
Appl. Phys. Letters, Vol. 9, pp. 303–304, October 15, 1966 (He–Ne laser light source for copying process)

25.104 White-Light Reconstruction of Holographic Images Using Transmission Holograms Recorded with Conventionally-Focused Images and 'In-Line' Background
G. W. Stroke (U. of Michigan)
Physics Letters, Vol. 23, pp. 325–327, October 31, 1966

25.105 Hologram on Photochromic Glass
J. P. Kirk (IBM Endicott)
Appl. Optics, Vol. 5, pp. 1684–1685, October 1966 (used 4880 Å argon laser, glass sensitive to wavelengths shorter than 5500 Å)

25.106 Fresnel-Transform Representation of Holograms and Hologram Classification
John T. Winthrop and C. R. Worthington (U. of Michigan)
J. Opt. Soc. Am., Vol. 56, pp. 1362–1368, October 1966

25.107 Effects of Partial Coherence on Holography with Diffuse Illumination
M. Lurie (Newark Coll. of Engrg.)
J. Opt. Soc. Am., Vol. 56, pp. 1369–1372, October 1966

25.108 Production of Holograms with Incoherent Illumination
H. R. Worthington, Jr. (Sci. & Engrg. Inst.)
J. Opt. Soc. Am., Vol. 56, pp. 1397–1398, October 1966

25.109 Space-Bandwidth Theorem for Holograms
George B. Parrent, Jr., and George O. Reynolds (Technical Operations)
J. Opt. Soc. Am., Vol. 56, pp. 1400–1401, October 1966

25.110 Les Transformations de l'Information en Optique
D. Gabor (Imperial Coll. of Sci. and Tech., England)
Optica Acta, Vol. 13, pp. 299–310, October 1966 (review)

25.111 Sound Holograms and Optical Reconstruction
R. K. Mueller and N. K. Sheridon (Bendix)
Appl. Phys. Letters, Vol. 9, pp. 328–329, November 1, 1966 (underwater, 7 MHz sound waves, water surface photographed)

25.112 Focused-Image Holography with Extended Sources
Lowell Rosen (NASA/ERC)
Appl. Phys. Letters, Vol. 9, pp. 337–339, November 1, 1966

25.113 Ray Tracing Through a Holographic System
Abe Offner (Perkin-Elmer)
J. Opt. Soc. Am., Vol. 56, pp. 1509–1512, November 1966 (diffraction grating, aberrations)

25.114 New Method of Making Fresnel Transforms with Incoherent Light
Gary Cochran (Conductron)
J. Opt. Soc. Am., Vol. 56, pp. 1513–1517, November 1966 (two- and three-dimensional scenes, triangular interferometer, afocal optical system, experiments)

25.115 Wavefront Reconstruction with Light of Finite Coherence Length

L. Mandel (U. of Rochester)
J. Opt. Soc. Am., Vol. 56, pp. 1636–1637, November 1966

25.116   Holograms and Zone Plates
W. E. Kock, L. Rosen, and J. Rendeiro (NASA/ERC)
Proc. IEEE, Vol. 54, pp. 1599–1601, November 1966

25.117   Three-Color Hologram Zone Plates
Winston E. Kock (NASA/ERC)
Proc. IEEE, Vol. 54, pp. 1610–1612, November 1966

25.118   Holographic Image Synthesis Utilizing Theoretical Methods
James P. Waters (United Aircraft)
Appl. Phys. Letters, Vol. 9, pp. 405–407, December 1, 1966

25.119   Rotierende Mattscheiben zur Änderung der Kohärenz von Laserlicht
K. Goetz and D. Unangst (U. of Jena)
Physics Letters, Vol. 23, pp. 667–668, December 12, 1966 (laser beam on rotating ground glass, used for optical transforms)

25.120   Holography with a Scatter-plate as Beam Splitter and a Pulsed Ruby Laser as Light Source
J. M. Burch, J. W. Gates, R. G. N. Hall, and L. H. Tanner (Nat'l Phys. Lab., England)
Nature, Vol. 212, pp. 1347–1348, December 17, 1966

25.121   Copying Holograms
D. B. Brumm (U. of Michigan)
Appl. Optics, Vol. 5, pp. 1946–1947, December 1966 (noncontacting)

25.122   Display of Holograms in White Light
C. B. Burckhardt
Bell Sys. Tech. J., Vol. 45, pp. 1841–1844, December 1966 (reasonably good reconstruction)

25.123   Holographic Interferometry of the Distortion of Thermoelectric Cooling Modules
R. Wolfe and E. T. Doherty (BTL)
J. Appl. Phys., Vol. 37, pp. 5008–5009, December 1966

25.124   Theory and Practice of Image Formation
A. Maréchal (Inst. of Optics, France)
J. Opt. Soc. Am., Vol. 56, pp. 1645–1648, December 1966

25.125   Field Range and Resolution in Holography
William H. Carter and Arwin A. Dougal (U. of Texas)
J. Opt. Soc. Am., Vol. 56, pp. 1754–1759, December 1966

25.126   Visual Display of Incoherent Wave Fields by Planar Arrays
W. Duane Montgomery (Inst. for Defense Analyses)
J. Opt. Soc. Am., Vol. 56, pp. 1769–1774, December 1966

25.127   Realism of Lens Action in Holograms
W. E. Kock, L. Rosen, and J. Rendeiro (NASA/ERC)
Proc. IEEE, Vol. 54, p. 1985, December 1966

25.128   The Hologram—Properties and Applications
E. G. Ramberg
RCA Rev., Vol. 27, pp. 467–499, December 1966 (review)

26. APPLICATIONS

26.1   Rotation Rate Sensing with Traveling-Wave Ring Lasers
W. M. Macek and D. T. M. Davis, Jr. (Sperry Gyroscope)
Appl. Phys. Letters, Vol. 2, pp. 67–68, February 1, 1963

26.2   Analysis of Receiving Techniques and Devices for Microwave Bandwidth Coherent Light Communication Systems
Monte Ross (Hallicrafters)
1963 IEEE Int. Conv. Record, Vol. 11, Part 8, pp. 145–152, March 1963

26.3   Quantum Effects and Noise in Optical Communications
Monte Ross (Hallicrafters)
Proc. IEEE, Vol. 51, pp. 602–603, April 1963

26.4   Optical Communications Employing Infrared Emitting Diodes and FM Techniques
E. J. Chatterton (MIT)
Proc. IEEE, Vol. 51, p. 612, April 1963

26.5   Laser-Alloyed Tunnel Diodes for Microwave Applications
L. Wandinger and K. Klohn (U.S. Army)
Proc. IEEE, Vol. 51, pp. 938–939, June 1963

26.6   Application of a Ruby Laser to High-Speed Photography
A. T. Ellis and M. E. Fourney (Calif. Inst. of Tech.)
Proc. IEEE, Vol. 51, pp. 942–943, June 1963

26.7   The Use of a Laser Amplifier in a Laser Communication System
H. Steinberg (Tech. Research Group)
Proc. IEEE, Vol. 51, p. 943, June 1963

26.8   Will Lasers Weld Circuit Components?
K. W. Dunlap and D. L. Williams (G.E.)
Electronics, Vol. 36, No. 28, pp. 54–57, July 12, 1963

26.9   On the Minimum Detectable Rotation Rate for a Laser Rotation Sensor
E. K. Proctor (Stanford Res. Inst.)
Proc. IEEE, Vol. 51, p. 1035, July 1963

26.10   Gatling-Gun Laser
Michael F. Wolff
Electronics, Vol. 36, No. 38, pp. 25–29, September 20, 1963

26.11   The Laser as a Machine Tool
David L. Williams (G.E.)
Proc. Nat'l Electronics Conf., Vol. 19, pp. 574–587, 1963

26.12   Laser-Formed Apertures for Electron Beam Instruments
J. F. Norton and J. G. McMullen (G.E.)
J. Appl. Phys., Vol. 34, pp. 3640–3641, December 1963

26.13   A Pulse Modulation Application of the Gallium Arsenide Light Emitting Diode
C. R. Seashore (Minneapolis-Honeywell)
Proc. IEEE, Vol. 51, p. 1781, December 1963

26.14   Quantum Noise in Communication Channels
J. P. Gordon (U. of California)
Proc. Third Int. Conf. Quantum Electronics, Columbia U. Press, N. Y., Vol. 1, pp. 55–64, 1964

26.15   Laser-Pumped Maser
D. P. Devor, I. J. D'Haenens, and C. K. Asawa (Hughes)
Proc. Third Int. Conf. Quantum Electronics, Columbia U. Press, N. Y., Vol. 1, pp. 931–936, 1964

26.16   Generation of Millimeter Waves in Optically Pumped Ruby
G. M. Zverev, A. M. Prokhorov, and A. K. Shevchenko (Moscow State U., U.S.S.R.)
Proc. Third Int. Conf. Quantum Electronics, Columbia U. Press, N. Y., Vol. 1, pp. 963–966, 1964

26.17 An Experiment for the Observation of the "Coriolis-Zeeman"
Effect for Photons
C. V. Heer (Ohio State U.)
Proc. Third Int. Conf. Quantum Electronics, Columbia U. Press,
N. Y., Vol. 2, pp. 1305-1311, 1964

26.18 Ring Laser Rotation Rate Sensor
W. M. Macek, D. T. M. Davis, Jr., R. W. Olthuis, J. R.
Schneider, and G. R. White (Sperry Gyroscope)
Proc. Third Int. Conf. Quantum Electronics, Columbia U. Press,
N. Y. Vol. 2, pp. 1313-1317, 1964

26.19 The S-66 Laser Satellite Tracking Experiment
H. H. Plotkin (NASA, Goddard Space Flight Center, Greenbelt
Md.)
Proc. Third Int. Conf. Quantum Electronics, Columbia U. Press,
N. Y., Vol. 2, pp. 1319-1332, 1964

26.20 Usinage Photonique Avec Generateur Laser
M. S. Bruma (Lab. de Chimie Physique, Faculté des Sciences
de Paris, France)
Proc. Third Int. Conf. Quantum Electronics, Columbia U. Press,
N. Y., Vol. 2, pp. 1333-1337, 1964

26.21 Distance Measurement by Means of a Light Beam Polarization-
Modulated at a Microwave Frequency
R. H. Bradsell (Natl. Physical Lab., England)
Proc. Third Int. Conf. Quantum Electronics, Columbia U. Press,
N. Y., Vol. 2, pp. 1541-1547, 1964

26.22 Optically-Pumped Parametric Oscillators at Microwave and In-
frared Frequencies
J. H. Dennis, P. R. Longaker, and R. H. Kingston (Lincoln
Lab., MIT)
Proc. Third Int. Conf. Quantum Electronics, Columbia U. Press,
N. Y., Vol. 2, pp. 1627-1633, 1964

26.23 Interferometric Phase Shift Technique for Measuring Short
Fluorescent Lifetimes
R. J. Carbone and P. R. Longaker (MIT)
Appl. Phys. Letters, Vol. 4, pp. 32-34, January 15, 1964

26.24 Analysis and Optimization of Laser Ranging Techniques
Graham W. Flint (Martin)
IEEE Trans. on Military Electronics, Vol. MIL-8, pp. 22-
28, January 1964

26.25 On the Narrow Beam Communication System Acquisition
Problem
Joel S. Greenberg (RCA)
IEEE Trans. on Military Electronics, Vol. MIL-8, pp. 28-39,
January 1964

26.26 Meteor Dust Detected by Laser Radar
(Technical notes)
J. Opt. Soc. Am., Vol. 54, p. 135, January 1964

26.27 A High Repetition Rate Laser System
W. T. Haswell, III, J. S. Hitt, and J. M. Feldman (Carnegie
Inst. of Tech.)
Proc. IEEE, Vol. 52, p. 93, January 1964

26.28 Solid-State Light Sources for Distance-Measuring Equipment
A. E. Harrison (U. of Washington)
Proc. IEEE, Vol. 52, p. 101, January 1964

26.29 A Light-Modulated Data Link
B. A. Boerschig (G.E.)
IEEE Trans. on Broadcasting, Vol. BC-10, pp. 4-7, February
1964 (GaAs incoherent diode, 12 Mc video)

26.30 Detection of the Transverse Doppler Effect with Laser Light
Reinhold Gerharz
Proc. IEEE, Vol. 52, p. 218, February 1964

26.31 Communication Channel Model of a Photoelectric Detector
L. P. Bolgiano, Jr., and L. F. Jelsma (U. of Delaware)
Proc. IEEE, Vol. 52, pp. 218-219, February 1964

26.32 Interconnection Techniques for Microcircuits
F. Z. Keister, R. D. Engquist, and J. H. Holley (Hughes)
IEEE Trans. on Component Parts, Vol. CP-11, pp. 33-41,
March 1964 (laser welding also)

26.33 Noise-Modulated Optical Radar
Hans E. Band (Concord Radiance Lab.)
Proc. IEEE, Vol. 52, pp. 306-307, March 1964

26.34 Transmitter for Coherent Light Communication System
M. Ito (Nippon Electric)
IEEE Int. Conv. Record, Vol. 12, Pt. 2, pp. 59-66, 1964 (dis-
cussion of components, cooled KDP suggested)

26.35 A High Resolution, Microwave Modulated Optical Doppler Radar
R. B. Hankin (Hallicrafters) and A. C. Todd (Ill. Inst. Tech.)
IEEE Int. Conv. Record, Vol. 12, Pt. 7, pp. 246-251, 1964 (anal-
yses, DCF photomultiplier)

26.36 Amplitude Modulation with a Noise Carrier
Harrison E. Rowe (BTL)
Proc. IEEE, Vol. 52, pp. 389-395, April 1964

26.37 Frequency or Phase Modulation with a Noise Carrier
Harrison E. Rowe (BTL)
Proc. IEEE, Vol. 52, pp. 396-408, April 1964

26.38 Measurement of Continental Drift and Earth Movement with
Lasers
William Honig (Honig Labs.)
Proc. IEEE, Vol. 52, p. 430, April 1964

26.39 Localized Fluid Flow Measurements with an He-Ne Laser
Spectrometer
Y. Yeh and H. Z. Cummins (Columbia U.)
Appl. Phys. Letters, Vol. 4, pp. 176-178, May 15, 1964
(suspended polystyrene spheres, heterodyne, profile velocity
range 5×10^{-2} to 0.8×10^{-2} cm/sec)

26.40 Index of Refraction Measured by Double-Slit Diffraction of Co-
herent Light from a Gas Laser
R. L. Aagard, D. Chen, and G. N. Otto (Honeywell Res. Ctr.)
Appl. Optics, Vol. 3, pp. 643-644, May 1964

26.41 Gases from Flash and Laser Irradiation of Coal
A. G. Sharkey, Jr., J. L. Shultz, and R. A. Friedel (U. S. Bureau
of Mines)
Nature, Vol. 202, pp. 988-989, June 6, 1964 (different gases compared to high-temperature carbonization)

26.42 Beat Frequency Between Two Traveling Waves in Fabry-Perot
Square Cavity
P. K. Cheo and C. V. Heer (Ohio State U.)
Appl. Opt., Vol. 3, pp. 788-789, June 1964 (He-Ne ring laser,
3.39μ, sharp edge near beam produces nonreciprocal frequency
bias)

26.43 Determination of the Velocity of Light Using the Laser as a
Source
D. Sinclair and M. P. Givens (U. of Rochester)
J. Opt. Soc. Am., Vol. 54, pp. 795-797, June 1964 (axial modes
of 0.63μ He-Ne laser produce 300-Mc beats, 1 per cent accuracy
can be improved)

26.44 An Optically Pumped Rb Laser Oscillator
P. Davidovits (Columbia U.)
Appl. Phys. Letters, Vol. 5, pp. 15-16, July 1, 1964

26.45 Measurement of Earth Tides and Continental Drift with Laser
Interferometer
V. Vali, R. S. Krogstad (Boeing), and W. Vali (Stanford U.)

Proc. IEEE, Vol. 52, pp. 857-858, July 1964 (Michelson interferometric strain measuring device proposed)

26.46 Birefringence in Glass Measured by the Scattered-Light Technique with a Laser Source
S. Bateson, J. W. Hunt, D. A. Dalby, and N. Sinha (Duplate Canada Ltd., Canada)
Appl. Optics, Vol. 3, p. 902, July 1964 (used 0.63-μ He-Ne laser, not optimum conditions)

26.47 Laser Radar Echoes from the Clear Atmosphere
R. T. H. Collis and M. G. H. Ligda (SRI)
Nature, Vol. 203, p. 508, August 1, 1964

26.48 Measurement of Fresnel Drag with the Ring Laser
W. M. Macek, J. R. Schneider, and R. M. Salamon (Sperry)
J. Appl. Phys., Vol. 35, pp. 2556-2557, August 1964 (moving solid, liquid and gas; can detect 10 cm/s)

26.49 Detection of the Transverse Doppler Effect with Laser Light
D. Censor (Technion-Israel Inst. of Tech., Israel)
Proc. IEEE, Vol. 52, p. 987, August 1964

26.50 Laser Beam Detection of Electron-Photon Interaction
I. R. Gatland, L. Gold, and J. W. Moffatt (Res. Inst. for Adv. Studies)
Physics Letters, Vol. 12, pp. 105-106, September 15, 1964

26.51 Laser Radar Echoes from a Stratified Clear Atmosphere
R. T. H. Collis, F. G. Fernald, and M. G. H. Ligda (SRI)
Nature, Vol. 203, pp. 1274-1275, September 19, 1964

26.52 Electron and Laser Beam Processing
S. Namba and P. H. Kim (Inst. Phys. and Chem. Res., Japan)
Japanese J. Appl. Phys., Vol. 3, pp. 536-545, September 1964 (critical comparison, laser beam more efficient probably due to its spiking)

26.53 Spectrum Properties of Pulse Modulated Lasers
S. Karp (Douglas Aircraft)
Proc. IEEE, Vol. 52, pp. 1264-1265, October 1964

26.54 Laser-Excited Photoluminescence of Overcompensated P⁺ GaAs and the Band-Filling Model
R. C. C. Leite, J. E. Ripper, and P. A. Guglielmi (BTL)
Appl. Phys. Letters, Vol. 5, pp. 188-190, November 1, 1964 (gas laser, many experimental advantages, frequency shift with excitation intensity)

26.55 Measurement of Absolute Optical Collision Diameters in Methane Using Tuned-Laser Spectroscopy
H. J. Gerritsen and S. A. Ahmed (RCA)
Physics Letters, Vol. 13, pp. 41-42, November 1, 1964 (Zeeman tuned HeNe at 3.39 μ)

26.56 Modes in a Triangular Ring Optical Resonator
S. A. Collins, Jr. and D. T. M. Davis, Jr. (Sperry Gyroscope)
Appl. Optics, Vol. 3, pp. 1314-1315, November 1964

26.57 Dual Polarization FM Laser Communications
W. M. Doyle and M. B. White (Philco)
Proc. IEEE, Vol. 52, p. 1353, November 1964

26.58 Room-Temperature GaAs Laser Voice-Communication System
D. Karlsons, C. W. Reno, and W. J. Hannan (RCA)
Proc. IEEE, Vol. 52, pp. 1354-1355, November 1964 (30 A., 30-ns pulses, 1.5 watts out, 20 kc average rep rate)

26.59 Laser Excitation of Powdered Solids
H. I. S. Ferguson, J. E. Mentall, and R. W. Nichols (U. of Western Ontario, Canada)
Nature, Vol. 204, p. 1295, December 26, 1964 (spectroscopic studies of metals, oxides and other compounds)

26.60 A Simple Optical Filter for Chirp Radar
J. S. Gerig and H. Montague (Scope)
Proc. IEEE, Vol. 52, p. 1753, December 1964 (laser, ultrasonic cell, first-order light pattern)

26.61 Long-Distance Interferometry with an He-Ne Laser
F. T. Arecchi and A. Sona (Lab. CISE, Italy)
Proc. Symp. on Quasi-Optics, Polytech. Inst. of Bklyn., New York, MRI Symp. Series, Vol. 14, pp. 623-633, 1964

26.62 Optics in Laser Research
H. de Lang (Philips Res. Labs., The Netherlands)
Z. angewandte Mathematik und Physik, Vol. 16, No. 1, pp. 7-14, January 25, 1965

26.63 Laser Applications in the Field of Physics and Optics
J. M. Burch (National Physical Lab., Great Britain)
Z. angewandte Mathematik und Physik, Vol. 16, No. 1, pp. 111-119, January 25, 1965

26.64 Application of a C. W. Laser as a Light Source in an Optical Alignment Method
H. J. Raterink (Technisch Physische Dienst T.N.O.—T. H. Delft, The Netherlands)
Z. angewandte Mathematik und Physik, Vol. 16, No. 1, pp. 126-128, January 25, 1965 (used zone plate)

26.65 Laser Application in the Field of Testing Materials
W. D. Hagenah (Institut fur Spektrochemie und angewandte Spektroskopie, Dortmund, Germany)
Z. angewandte Mathematik und Physik, Vol. 16, No. 1, pp. 130-138, January 25, 1965 (Raman effect, plasma diagnostics, spectrochemical analysis)

26.66 Laser Applications in the Field of Treating Materials
S. Panzer (Carl Zeiss, Oberkochen, Germany)
Z. angewandte Mathematik und Physik, Vol. 16, No. 1, pp. 138-155, January 25, 1965 (welding, drilling)

26.67 Laser Applications to Communication
D. Sette (U. of Rome, Rome, Italy)
Z. angewandte Mathematik und Physik, Vol. 16, No. 1, pp. 156-169, January 25, 1965 (general discussion, comparison of modulation and detection methods)

26.68 Laser Application in the Field of Medicine
M. M. Zaret (Zaret Foundation)
Z. angewandte Mathematik und Physik, Vol. 16, No. 1, pp. 178-181, January 25, 1965

26.69 Vacuum Deposited Thin Films Using a Ruby Laser
Howard M. Smith (U. of Rochester) and A. F. Turner (Bausch & Lomb)
Appl. Optics, Vol. 4, pp. 147-148, January 1965

26.70 Laser Benioff Strain Seismometers
W. Honig (Loral)
Proc. IEEE, Vol. 53, pp. 101-102, January 1965

26.71 Transmission of Laser Beams through Various Transparent Rods for Biomedical Applications
J. A. Goldman and R. Meyer (Children's Hosp. Res. Foundation, Cincinnati)
Nature, Vol. 205, pp. 892-894, February 27, 1965 (flexible and rigid rods, some damages, bends, 63 joules into 5 mm rod)

26.72 The Surface Temperature of Metals Heated with Laser
S. Namba and P. H. Kim (Inst. Physical and Chemical Res., Tokyo, Japan) and S. Nakayama and I. Ida (N.T.T., Tokyo, Japan)
Japanese J. Appl. Phys., Vol. 4, pp. 153-154, February 1965 (to above 5000°K)

26.73 Laser Welding—Where It Stands Today
J. E. Anderson and J. E. Jackson (Union Carbide)
Materials in Design Engineering, Vol. 61, No. 2, pp. 92-96, February 1965 (tiny, precise joints in many dissimilar metals)

26.74 Some Parameters of a Laser-Type Beyond-the-Horizon Communication Link
M. King and S. Kainer (ITT)
Proc. IEEE, Vol. 53, pp. 137-141, February 1965 (cloud and haze scattering)

26.75 Electro-Optic TV Communication System
W. J. Hannan, J. Bordogna, and T. E. Penn (RCA)
Proc. IEEE, Vol. 53, pp. 171–172, February 1965 (GaAs electro-optic birefringent modulator, 400 volts for 13% modulation, Nd:YAG laser oscillator)

26.76 Internal Modulation of a He–Ne Laser with a Television Signal
G. Schiffner and O. Hintringer (Technische Hochschule Wien, Austria)
Proc. IEEE, Vol. 53, pp. 172–173, February 1965 (KDP inside HeNe laser cavity, 120 volts for modulator)

26.77 Motion Sensing by Optical Heterodyne Doppler Detection from Diffuse Surfaces
R. D. Kroeger (Sperry)
Proc. IEEE, Vol. 53, pp. 211–212, February 1965 (0.63μ gas laser, from surfaces up to 180 feet away)

26.78 Laser-Induced Production of Free Radicals in Organic Compounds
Yoh-Han Pao and P. M. Rentzepis (BTL)
Appl. Phys. Letters, Vol. 6, pp. 93–95, March 1, 1965 (ruby laser, polymerization, 2 photon absorption)

26.79 Gas-Phase Laser as a Source of Light for an Optical Diffractometer
G. Harburn, K. Walkley, and C. A. Taylor (U. of Manchester, England)
Nature, Vol. 205, pp. 1095–1096, March 13, 1965 (considerable improvement in resolution)

26.80 Progress in Optical Computer Research
O. A. Reimann (RADC) and W. F. Kosonocky (RCA)
IEEE Spectrum, Vol. 2, No. 3, pp. 181–195, March 1965 (laser digital devices, ruby laser experiments)

26.81 Laser Beam Machining
N. Forbes (Ferranti Ltd., England)
Microelectronics and Reliability, Vol. 4, pp. 105–108, March 1965 (solid state and gas lasers, 80 watt pulsed gas laser at 2.5 kc/s, general discussion and results)

26.82 Measurement of Differential Earth Tides
N. D. Newby, Jr. (Autonetics)
Proc. IEEE, Vol. 53, p. 296, March 1965 (long baseline, measure relative velocity)

26.83 Satellite Laser Ranging Experiment
G. L. Snyder, S. R. Hurst, A. B. Grafinger, and H. W. Halsey (G.E.)
Proc. IEEE, Vol. 53, pp. 298–299, March 1965 (Beacon Explorer-B satellite, out to 1379 km)

26.84 Reflection of Ruby Laser Radiation From Explorer XXII
H. H. Plotkin, T. S. Johnson, P. Spadin, and J. Moye (NASA)
Proc. IEEE, Vol. 53, pp. 301–302, March 1965 (Beacon Explorer-B satellite, 6.25 ms round trip, over 1000 km)

26.85 An Evaluation of Pulsed Laser Welding
J. E. Anderson and J. E. Jackson (Union Carbide Corp., Linde Div.)
Proc. Electron and Laser Beam Symposium, March 31-April 2, 1965, edited by A. B. El-Kareh, sponsored by Pennsylvania State U. and Alloyd General Corp.; pp. 17–50.

26.86 Laser Beam Micro-Processing
P. H. Kim and S. Namba (Inst. of Physical and Chemical Research, Japan) and I. Ida and S. Nakayama (N.T.T., Japan)
Proc. Electron and Laser Beam Symposium, March 31–April 2, 1965; edited by A. B. El-Kareh, sponsored by Pennsylvania State U. and Alloyd General Corp., pp. 335–342 (metal surface temperature, machining efficiency)

26.87 Laser Welding of Microcircuits
T. A. Osial, K. B. Steinbruegge, and P. Scharf (Westinghouse)
Proc. Electron and Laser Beam Symposium, March 31–April 2, 1965; edited by A. B. El-Kareh, sponsored by Pennsylvania State U. and Alloyd General Corp., pp. 343–368

26.88 Acoustic Beam Probing Using Optical Techniques
M. G. Cohen and E. I. Gordon (BTL)
Bell Sys. Tech. J., Vol. 44, pp. 693–721, April 1965 (sonic waves in media deflect or scatter optical beams)

26.89 Fundamentals of Laser Beam Machining and Drilling
C. M. Adams, Jr. (M.I.T.) and G. A. Hardway (Applied Lasers, Inc.)
IEEE Trans. on Industry and General Applications, Vol. IGA-1, pp. 90–96, March/April 1965 (thermal features, laser operational characteristics)

26.90 Solid-State Modulation and Demodulation of Light with Information from Five Television Channels Simultaneously
K. M. Johnson and D. D. Eden (Texas Instruments)
Proc. IEEE, Vol. 53, pp. 402–403, April 1965 (KDP modulator, 0.63μ HeNe laser, laser mode beat noise).

26.91 Laser Retinal Photocoagulator
N. S. Kapany, N. Silbertrust, and N. A. Peppers (Optics Technology)
Appl. Optics, Vol. 4, pp. 517–522, May 1965 (clinical data)

26.92 A Laser Microscope
N. A. Peppers (Optics Technology)
Appl. Optics, Vol. 4, pp. 555–558, May 1965 (pulsed ruby laser, bio-optical research)

26.93 Optical System for Coupling a Point Radiation Source to an Aircraft Beacon and a Laser
M. Dalton (Electro-Optical Systems)
Appl. Optics, Vol. 4, pp. 628–629, May 1965

26.94 Vignetting Test for Catadioptric Systems Using a cw Laser
R. M. Zoot (Hughes)
Appl. Optics, Vol. 4, p. 755, June 1965

26.95 Linac Alignment Techniques
W. B. Herrmannsfeldt (Stanford Linear Accelerator Center)
IEEE Trans. on Nuclear Science, Vol. NS-12, No. 3 pp. 9–18, June 1965 (requirement of ± 0.5 mm over length of 3050 m, 0.63-μ gas laser, rectangular Fresnel-zone plates)

26.96 A Technique for the Transmission of Digital Information over Short Distances Using Infra-red Radiation
A. T. Davies (National Physical Lab., England)
The Radio and Electronic Engineer, Vol. 29, pp. 369–373, June 1965 (GaAs diode source, system considerations, 110 words per second, 800-foot path)

26.97 Production Laser Welding for Specialized Applications
K. J. Miller and J. D. Nunnikhoven (AiResearch)
Welding Journal, Vol. 44, pp. 480–485, June 1965 (maximum 10 pulses per sec., 10 J per pulse)

26.98 Laser Beam Welding Electronic-Component Leads
A. R. Pfluger and P. M. Maas (Lockheed)
Welding Journal, Vol. 44, pp. 264(s)–269(s), June 1965 (welds between wires and ribbons)

26.99 Oncolysis with Laser Energy Combined with Chemotherapy
John Peter Minton, George H. Weiss, and Marvin Zelen (National Cancer Inst.)
Nature, Vol. 207, pp. 140–141, July 10, 1965 (destructive effect on tumor systems)

26.100 Schlieren Interferometry. An Optical Method for Determining Temperature and Velocity Distributions in Liquids
J. Bruce Brackenridge and W. Paul Gilbert (Lawrence U.)
Appl. Optics, Vol. 4, pp. 819–821, July 1965 (gas laser source)

26.101 FM Laser Communications through a Highly Turbulent Atmosphere
W. M. Doyle, W. D. Gerber, P. M. Sutton, and M. B. White (Philco)

IEEE J. Quantum Electronics, Vol. QE-1, pp. 181–182, July 1965 (dual polarization, 0.63-μ He–Ne laser, $\frac{1}{2}$ mile path, varied receiver aperture, low noise with FM)

26.102 Measurement of Localized Flow Velocities in Gases with a Laser Doppler Flowmeter
J. W. Foreman, Jr., E. W. George, and R. D. Lewis (Brown Engineering Co.)
Appl. Phys. Letters, Vol. 7, pp. 77–78, August 15, 1965 (smoke added, optical heterodyne, 14.4 m/s produced 5 Mc/s Doppler shift)

26.103 Effects of Laser Radiation
John F. Ready (Honeywell)
Industrial Research, Vol. 7, No. 8, pp. 44–50, August 1965 (review, long pulses for welding, advantages and disadvantages, plumes, electron emission)

26.104 Laser Welding
K. J. Miller and J. D. Ninnikhoven (AiResearch)
Machine Design, Vol. 37, pp. 120–125, August 1965 (several joint designs, materials, costs)

26.105 A Proposed First-Order Relativity Test Using Lasers
J. Shamir and R. Fox (Technion, Israel)
Proc. IEEE, Vol. 53, pp. 1155–1156, August 1965 (rebuttal by Chalon W. Carnahan)

26.106 New Ether Drift Experiment Using Lasers
A. Szöke (Weizmann Inst. of Science, Israel)
Physics Letters, Vol. 18, pp. 267–268, September 1, 1965 (vacuum interferometer, modulated laser)

26.107 Comment on "Lasers and Ether Drift"
J. Shamir and R. Fox (Technion, Israel, Israel Inst. of Tech.)
Physics Letters, Vol. 18, pp. 277–278, September 1, 1965 (negative comment on Cristescu and Guirgea, Physics Letters, Vol. 5, p. 128, 1963)

26.108 Laser Interferometer for Earth-Strain Measurements
R. W. Moss, V. Vali, and R. S. Krogstad (Boeing Scientific Research Labs.)
Bull. Am. Phys. Soc., Series II, Vol. 10, p. 704–CF12, September 2, 1965 (interferometer constructed, capable of detecting strains of less than 10^{-10})

26.109 Computing at the Speed of Light
Kendall Preston, Jr. (Perkin-Elmer)
Electronics, Vol. 38, No. 18, pp. 72–83, September 6, 1965 (optical analog computers, correlators, limitations)

26.110 Laser Interferometer for Earth Strain Measurements
V. Vali, R. S. Krogstad, and R. W. Moss (Boeing)
Rev. Sci. Instr., Vol. 36, pp. 1352–1355, September 1965 (10-m long interferometer arm, 0.63-μ gas laser, earthquake recordings)

26.111 Interferometer for Aerodynamic and Heat Transfer Measurements
R. J. Goldstein (U. of Minnesota)
Rev. Sci. Instr., Vol. 36, pp. 1408–1410, October 1965 (modification of Jamin interferometer, diverging lenses and paraboloidal mirrors, gas laser)

26.112 Switching with Light
T. E. Bray (GE)
Electronics, Vol. 38, No. 22, pp. 58–65, November 1, 1965 (optoelectronics, coherent optical logic devices)

26.113 Spatially and Temporally Resolved Electron and Atom Concentrations in an Afterglow Gas Discharge
J. B. Gerardo, J. T. Verdeyen, and M. A. Gusinow (U. of Illinois)
J. Appl. Phys., Vol. 36, pp. 3526–3534, November 1965 (laser interferometer, detect 2×10^{13} e/cm³, also 4 and 8 mm microwaves)

26.114 Incorporation of a Laser into the Arm of an Interferometer for Measurement of Transient Phase Changes
Edward Thornton (IIT Research Inst.)
J. Appl. Phys., Vol. 36, pp. 3539–3541, November 1965 (0.63-μ laser)

26.115 Laser Beam "Security"
Selig Kainer (ITT Federal Labs.)
Proc. IEEE, Vol. 53, pp. 1752–1753, November 1965 (secure in space, atmospheric perturbations destroy "security")

26.116 Optimum Detection Thresholds in Optical Communications
T. F. Curran and M. Ross (Hallicrafters)
Proc. IEEE, Vol. 53, pp. 1770–1771, November 1965

26.117 Search via Laser Receivers for Interstellar Communications
Monte Ross (Hallicrafters)
Proc. IEEE, Vol. 53, p. 1780, November 1965

26.118 Life Performance of Prism Q Switched Laser
W. L. Knecht (AF Avionics Lab., Wright Patterson AFB)
Proc. IEEE, Vol. 53, pp. 1785–1786, November 1965 (ruby laser, 12 000 pulses, component breakdown, performance data)

26.119 Investigation of a Laser Triggered Spark Gap
Winston K. Pendleton (AF Inst. of Tech.) and Arthur H. Guenther (AF Weapons Lab.)
Rev. Sci. Instr., Vol. 36, pp. 1546–1550, November 1965 (less than 10 ns delay time in SF₆ atmosphere, various gases, pressures, etc.)

26.120 Laser Photography of Hypervelocity Projectiles
Warren V. Trammell (GM Defense Research Labs.)
Rev. Sci. Instr., Vol. 36, pp. 1551–1553, November 1965 (Q-sw laser, 7.1 km/s velocity)

26.121 Interferometer Technique for Measuring the Dynamic Mechanical Properties of Materials
L. M. Barker and R. E. Hollenbach (Sandia Lab.)
Rev. Sci. Instr., Vol. 36, pp. 1617–1620, November 1965 (gas laser, surface motion due to impact)

26.122 Optical Crystallographic Orientation Determination Using a He–Ne Gas Laser
Robert E. Green, Jr. (Johns Hopkins U.)
Rev. Sci. Instr., Vol. 36, pp. 1668–1669, November 1965

26.123 Effect of Solar Radiation on Atmospheric Laser Returns
W. E. Hoehne and J. L. Karney (Pacific Missile Range, Point Mugu)
Appl. Phys. Letters, Vol. 7, pp. 313–314, December 15, 1965 (Q-sw ruby, amplitude depends on angle between solar and laser radiations, postulate interaction effect)

26.124 Electrodeless Excitation of He–Ne Gas Lasers
D. S. Smith, K. M. Baird, and W. E. E. Berger (National Research Council, Canada)
Appl. Optics, Vol. 4, pp. 1673–1674, December 1965 (application to voice transmission)

26.125 Moon Distance Measurement by Laser
A. Orszag (École Polytechnique, Paris, France)
NBS J. Research, Section D, Radio Science Vol. 69D, pp. 1681–1689, December 1965 (discussion of proposed system parameters)

26.126 Semiconductor Laser Communications through Multiple-Scatter Paths
E. J. Chatterton (Lincoln Lab., M.I.T.)
Proc. IEEE, Vol. 53, pp. 2114–2115, December 1965 (pulsed GaAs diode, snow, fog, atmospheric shimmer data)

26.127 Pulse-Code Modulation Multiplex Transmission over an Injection Laser Transmission System

E. J. Schiel, E. C. Bullwinkel and R. B. Weimer (U.S. Army Electronics Command, Fort Monmouth)
Proc. IEEE, Vol. 53, pp. 2140–2141, December 1965 (overcomes air turbulence effects, GaAs, 1.1×10^6 pulses/sec, 13-km path)

26.128   Observation of Time-Dependent Density Fluctuations in Carbon Dioxide Near the Critical Point Using an He–Ne Laser
S. S. Alpert, D. Balzarini, R. Novick, L. Seigel, and Y. Yeh (Columbia U.)
Proc. Physics of Quantum Electronics Conf., McGraw-Hill Book Co., New York, pp. 253–259, 1966

26.129   Spectroscopy with Gas Lasers
M. S. Feld, J. H. Parks, H. R. Schlossberg, and A. Javan (M.I.T.)
Proc. Physics of Quantum Electronics Conf., McGraw-Hill Book Co., New York, pp. 567–580, 1966

26.130   Tuned-Laser Spectroscopy of Organic Vapors
Hendrik J. Gerritsen (RCA)
Proc. Physics of Quantum Electronics Conf., McGraw-Hill Book Co., New York, pp. 581–590, 1966

26.131   Laser Probing the Lower Atmosphere
B. R. Clemesha, G. S. Kent, and R. W. H. Wright (U. of the West Indies, Jamaica)
Nature, Vol. 209, pp. 184–185, January 8, 1966 (Q-sw ruby laser, extra scattering at 18 and 25 km heights)

26.132   Optical Design Considerations in Photobleaching with cw Lasers
Hubertus Staerk (U. of Pennsylvania)
Appl. Optics, Vol. 5, pp. 155–158, January 1966 (gas laser, optical systems to alter beam intensity profile)

26.133   Micromachining with a Pulsed Gas Laser
H. A. H. Boot, D. M. Clunie, and R. S. A. Thorn (Services Electronics Res. Lab., England)
Electronics Letters, Vol. 2, p. 1, January 1966 (200 W 0.25 μs pulses, 500 pps)

26.134   Optical Simulation of Microwave Antennas
Arthur L. Ingalls
IEEE Trans. on Antennas and Propagation, Vol. AP-14, pp. 2–6, January 1966 (gas laser, 96-element array, near and far fields)

26.135   Laser-Stimulated Nucleation in a Bubble Chamber
R. C. Stamberg and D. E. Gillespie (U. of Michigan)
J. Appl. Phys., Vol. 37, pp. 459–460, January 1966 (contains $CBrF_3$, ruby laser)

26.136   Images of Coherently Illuminated Edged Objects Formed by Scanning Optical Systems
Robert E. Kinzly (Cornell Aeronautical Lab.)
J. Opt. Soc. Am., Vol. 56, pp. 9–11, January 1966 (microdensitometer system)

26.137   Communication by Laser
Stewart E. Miller (B.T.L.)
Scientific American, Vol. 214, No. 1, pp. 19–27, January 1966 (general principles, comparison with microwave)

26.138   Observation of Earth Tides using a Laser Interferometer
V. Vali, R. S. Krogstad, and R. W. Moss (Boeing)
J. Appl. Phys., Vol. 37, pp. 580–582, February 1966

26.139   Measurement of the Distance to the Moon by Optical Radar
Yu. L. Kokurin, V. V. Kurbasov, V. F. Lobanov, V. M. Mozhzherin, A. N. Sukhanovskii, and N. S. Chernykh (Lebedev Physics Institute, U.S.S.R.)
JETP Letters, Vol. 3, pp. 139–141, March 1, 1966 (ruby laser, 50 ns 5 J output pulse)

26.140   Recent Applications of Lasers
O. S. Heavens (U. of York, England)
British J. Appl. Phys., Vol. 17, pp. 287–309, March 1966 (general review)

26.141   A Laser Radar Ranging System Using Pseudo-Random-Code Modulation
Jesse B. Sherman (Cooper Union)
IEEE Trans. on Education, Vol. E-9, pp. 2–9, March 1966 (eliminates range ambiguity)

26.142   Laser Doppler Velocimeter for Measurement of Localized Flow Velocities in Liquids
J. W. Foreman, Jr., R. D. Lewis, J. R. Thornton, and H. J. Watson (Brown Engineering Co.)
Proc. IEEE, Vol. 54, pp. 424–425, March 1966 (gas laser)

26.143   The Two-Way Transmission of a Ruby-Laser Beam Between Earth and a Retroreflecting Satellite
P. H. Anderson, C. G. Lehr, and L. A. Maestre (Smithsonian Inst. Astrophysical Observatory) and H. W. Halsey and G. L. Snyder (G.E.)
Proc. IEEE, Vol. 54, pp. 426–427, March 1966 (36 J laser output, up to 2.1 Mm range)

26.144   Solar-Pumped Modulated Laser
C. W. Reno (RCA)
RCA Review, Vol. 27, pp. 149–157, March 1966 (Dy and Nd lasers, transmitted television picture)

26.145   Operating Features of the Ring Laser
I. L. Bershteĭn and Yu. I. Zaĭtsev (Radiophysics Institute, Gorkiĭ State U.)
Soviet Phys. JETP, Vol. 22, pp. 663–667, March 1966 (3-mirrors, 0.63 μ)

26.146   Laser Welding of Aerospace Structural Alloys
L. P. Earvolino and J. R. Kennedy (Grumman)
Welding J., Vol. 45, pp. 127-s–134-s, March 1966 (ruby laser, up to 9 ms and 50 J pulses, experimental procedure, results)

26.147   The Use of Lasers in Signal Processing for Radar and Communications
L. J. Cutrona (U. of Michigan and Conductron)
Proc. 8th Annual Electron and Laser Beam Symposium, April 6–8, 1966. Sponsored by U. of Michigan and IEEE, pp. 39–85

26.148   Airborne Investigations of Clear Air Turbulence with Laser Radars
P. A. Franken, J. A. Jenney, and D. M. Rank (U. of Michigan)
Proc. 8th Annual Electron and Laser Beam Symposium, April 6–8, 1966. Sponsored by U. of Michigan and IEEE, pp. 87–103

26.149   The Application of Lasers in Thermophysical Properties Measurements
M. M. Nakata (Atomics International)
Proc. 8th Annual Electron and Laser Beam Symposium, April 6–8, 1966. Sponsored by U. of Michigan and IEEE, pp. 107–115 (thermal diffusivity, heat capacity)

26.150   Laser Machining Study
Warren V. Trammell (GM Defense Research Labs.)
Proc. 8th Annual Electron and Laser Beam Symposium, April 6–8, 1966. Sponsored by U. of Michigan and IEEE, pp. 157–186 (several metals, various holes)

26.151   The Laser Now a Production Tool
J. P. Epperson, R. W. Dyer, and J. C. Grzywa (Western Electric)
Proc. 8th Annual Electron and Laser Beam Symposium, April 6–8, 1966. Sponsored by U. of Michigan and IEEE, pp. 187–205 (drilling diamond, wire dies, multiple pulses per hole)

26.152   Design of a Production-Worthy Laser Microwelder
Jon H. Myer (Hughes Aircraft)
Proc. 8th Annual Electron and Laser Beam Symposium, April 6–8, 1966. Sponsored by U. of Michigan and IEEE, pp. 207–215

26.153 High-Resolution Absorption Measurement in CO_2 with a Tuned Laser
B. F. Jacoby and R. K. Long (Ohio State U.)
Appl. Phys. Letters, Vol. 8, pp. 202–204, April 15, 1966 (He–Xe laser, near 2.026 μ)

26.154 A New Technique for Measuring Magnitudes of Photoelastic Tensors and its Application to Lithium Niobate
R. W. Dixon and M. G. Cohen (B.T.L.)
Appl. Phys. Letters, Vol. 8, pp. 205–207, April 15, 1966 (pulsed ultrasound on material, scatters 0.63 μ laser beam)

26.155 Further Evidence of Enhanced Tumour Destruction with Combined Laser Energy and Chemotherapy
Robert C. Hoye and George H. Weiss (National Cancer Institute)
Nature, Vol. 210, pp. 432–433, April 23, 1966 (ruby and Nd glass lasers produce similar effects, up to 1,100 J in 2 msec, focused)

26.156 Digital Laser Ranging and Tracking Using a Compound Axis Servomechanism
Thomas W. Barnard and Carroll R. Fencil (Perkin-Elmer)
Appl. Optics, Vol. 5, pp. 497–505, April 1966

26.157 Precision Laser Automatic Tracking System
R. F. Lucy, C. J. Peters, E. J. McGann, and K. T. Lang (Sylvania)
Appl. Optics, Vol. 5, pp. 517–524, April 1966 (accuracy of 25μ rads)

26.158 The FM Laser and Optical Communication Systems
J. Richard Kerr (Sylvania)
Appl. Optics, Vol. 5, pp. 671–672, April 1966

26.159 Increasing the Memory Capacity of the Digital Light Deflector by "Color Coding"
J. G. Skinner
Bell. Sys. Tech. J., Vol. 45, pp. 597–608, April 1966 (potential increase by 20, KTN optical switches)

26.160 Microperforation with Laser Beam in the Preparation of Microelectrodes
M. J. Mela (U. of Pennsylvania)
IEEE Trans. on Bio-Medical Engineering, Vol. BME-13, pp. 70–76, April 1966 (in vinyl lacquer insulation of tungsten microelectrodes)

26.161 High-Speed Stroboscopic Photography Using a Kerr-Cell Modulated Laser Source
George J. Hecht, Gerald B. Steel, and Antoni K. Oppenheim (U. of California)
ISA Trans., Vol. 5, pp. 133–138, April 1966 (10 ns pulses with rate up to 1 Mc/s)

26.162 Yield-Point Phenomenon in Impact-Loaded 1060 Aluminum
L. M. Barker, B. M. Butcher, and C. H. Karnes (Sandia)
J. Appl. Phys., Vol. 37, pp. 1989–1991, April 1966 (elastic and plastic waves, laser beam, Michelson interferometer)

26.163 A Proposal for a Magneto-Optical Variable Memory
F. Forlani and N. Minnaja (Olivetti-General Electric, Italy)
Proc. IEEE, Vol. 54, pp. 711–712, April 1966

26.164 Laser Radar (Lidar) for Meteorological Observations
C. A. Northend, R. C. Honey, and W. E. Evans (Stanford Res. Inst.)
Rev. Sci. Instr., Vol. 37, pp. 393–400, April 1966 (noise considerations, ruby laser, system analysis)

26.165 Alignment of Fastie-Ebert Spectrometers Using He–Ne Laser
R. C. Ohlmann and A. Mego (Westinghouse)
Rev. Sci. Instr., Vol. 37, pp. 530–531, April 1966

26.166 Laser Lesions: Changes in Retinal Excitability
A. N. Nicholson and M. J. Allwood (Royal Air Force Institute of Aviation Medicine, England)
Nature, Vol. 210, pp. 637–638, May 7, 1966

26.167 Observations of Power Station Plumes using a Pulsed Ruby Laser Rangefinder
P. M. Hamilton, K. W. James and D. J. Moore (Central Electricity Res. Labs., England)
Nature, Vol. 210, pp. 723–724, May 14, 1966

26.168 Backscattering from the Upper Atmosphere (75–160 km) Detected by Optical Radar
W. C. Bain and M. C. W. Sandford (Radio and Space Res. Station, England)
Nature, Vol. 210, p. 826, May 21, 1966 (data different from McCormick et al., Nature, Vol. 209, p. 798, 1966)

26.169 p-n Junction Lasers for Communication Systems
Lothar Wandinger and Kenneth L. Klohn (U. S. Army Electronics Command Labs.)
IEEE Trans. on Aerospace and Electronic Systems, Vol. AES-2, pp. 271–277, May 1966 (pulsed, one audio channel, eight mile range)

26.170 Technique for Streak Camera Writing Rate Calibration Using Pulsed Laser
A. B. Christensen and W. M. Isbell (Stanford Res. Inst.)
Rev. Sci. Instr., Vol. 37, pp. 559–561, May 1966

26.171 Laser Interferometry of a Dropping Mercury Electrode
J. Leja (U. of British Columbia, Canada), and R. N. O'Brien (U. of Alberta, Canada)
Nature, Vol. 210, pp. 1217–1219, June 18, 1966

26.172 The Measurement of Homogeneity of Optical Materials in the Visible and Near Infrared
F. W. Rosberry (N.B.S.)
Appl. Optics, Vol. 5, pp. 961–966, June 1966 (gas laser source)

26.173 The Laser as a Light Source for Ultramicroscopy and Light Scattering by Imperfections in Crystals. Investigation of Imperfections in LiF, MgO, and Ruby
V. Vand, K. Vedam, and R. Stein (Penn. State U.)
J. Appl. Phys., Vol. 37, pp. 2551–2557, June 1966

26.174 Technique for the Measurement of the Transit Time of an Optical Signal
L. E. Wood and M. C. Thompson, Jr. (Environmental Science Services Admin.)
Nature, Vol. 211, pp. 173–174, July 9, 1966 (each of two oppositely traveling collimated light beams modulated before and after transit, about 10 GHz modulation, proposal)

26.175 Determination of Small Wedge-Angles Using a Gas Laser
Viktor Met (Electro-Optics)
Appl. Optics, Vol. 5, pp. 1242–1244, July 1966

26.176 Optical Communications in the Earth's Atmosphere
Bernard Cooper (ITT)
IEEE Spectrum, Vol. 3, No. 7, pp. 83–88, July 1966 (noise, detection efficiency, atmospheric turbulence)

26.177 Velocity-Profile Measurement in Plasma Flows using Tracers Produced by a Laser Beam
Chen Jen Chen (Jet Propulsion Lab.)
J. Appl. Phys., Vol. 37, pp. 3092–3095, July 1966 (focused Q-sw ruby laser produces small highly luminous plasma)

26.178 Apparent Illuminance as a Function of Range in Gated, Laser Night-Viewing Systems
Lester F. Gillespie (Fort Belvoir)
J. Opt. Soc. Am., Vol. 56, pp. 883–889, July 1966 (pulsed laser illuminator, various pulse shapes, shuttered viewer, analytical)

26.179 Cyclotron Resonance in Semiconductors with Far Infrared Laser
K. J. Button (M.I.T. Nat'l Magnet Lab.), H. A. Gebbie (Nat'l Phys. Lab., England), and B. Lax (Nat'l Magnet Lab.)
IEEE J. Quantum Electronics, Vol. QE-2, pp. 202–207, August 1966 (337 μ cyanide laser, up to 180 000 gauss)

26.180 Laser Experiments for Determining Satellite Orbits
P. H. Anderson, C. G. Lehr, and L. A. Maestre (Smithsonian Astrophys. Obs.), and G. L. Snyder (GE)
IEEE J. Quantum Electronics, Vol. QE-2, pp. 215–219, August 1966 (60 ns 8MW pulsed ruby laser)

26.181 The Laser Current Transformer for EHV Power Transmission Lines
S. Saito, Y. Fujii, K. Yokoyama, and J. Hamasaki (U. of Tokyo, Japan) and Y. Ohno (Tokyo Electric Power Co., Japan)
IEEE J. Quantum Electronics, Vol. QE-2, pp. 255–259, August 1966 (Faraday rotation in flint glass, He–Ne gas laser)

26.182 Fluid Flow Measurements with a Laser Doppler Velocimeter
J. W. Foreman, Jr., E. W. George, J. L. Jetton, R. D. Lewis, J. R. Thornton, and H. J. Watson (Brown Engrg.)
IEEE J. Quantum Electronics, Vol. QE-2, pp. 260–266, August 1966 (up to 6000 f/s)

26.183 Alternative Interpretation of Rotation Rate Sensing by Ring Laser
E. O. Schulz-DuBois (BTL)
IEEE J. Quantum Electronics, Vol. QE-2, pp. 299–305, August 1966

26.184 Hypersonic Velocity Measurements in Aqueous Alkali Halide Solutions
Julian Stone (Columbia U.)
J. Opt. Soc. Am., Vol. 56, pp. 1136–1137, August 1966 (0.63-μ He–Ne laser, Brillouin scattering, 1500–1600 m/s)

26.185 On the Application of Coherent Optical Processing Techniques to Synthetic-Aperture Radar
L. J. Cutrona, E. N. Leith, L. J. Porcello, and W. E. Vivian (U. of Michigan)
Proc. IEEE, Vol. 54, pp. 1026–1032, August 1966 (sidelooking radar)

26.186 Laser Used for Mass Analysis
N. C. Fenner and N. R. Daly (UKAEA, England)
Rev. Sci. Instr., Vol. 37, pp. 1068–1070, August 1966 (vaporization and ionization of solid materials)

26.187 Zeeman Laser Interferometer
J. A. Dahlquist, D. G. Peterson, and W. Culshaw (Lockheed)
Appl. Phys. Letters, Vol. 9, pp. 181–183, September 1, 1966 (He–Ne laser, left- and right-hand circularly polarized modes, detect target motions)

26.188 Characteristics of a Traveling-Wave Ruby Single-Mode Laser as a Laser Radar Transmitter
I. Goldstein and A. Chabot (Raytheon)
IEEE J. Quantum Electronics, Vol. QE-2, pp. 519–523, September 1966 (20 kW, 1 μs, 1 pps master oscillator, 24 dB gain power amplifier)

26.189 High Resolution Spectroscopy of Formaldehyde by a Tunable Infrared Maser
Katsumi Sakurai, Koichi Shimoda, and Michio Takami (U. of Tokyo, Japan)
J. Phys. Soc. Japan, Vol. 21, p. 1838, September 1966

26.190 Optical Collision Diameters of H_2CO and CH_4 Measured by the Infrared Maser Spectrograph
Katsumi Sakurai and Koichi Shimoda (U. of Tokyo, Japan)
J. Phys. Soc. Japan, Vol. 21, pp. 1842–1843, September 1966

26.191 Continuously-Variable Ultrasonic-Optical Delay Line
M. J. Brienza and A. J. DeMaria (United Aircraft)
Appl. Phys. Letters, Vol. 9, pp. 312–314, October 15, 1966 (argon ion laser, CdS transducer, acoustic wave in quartz diffracts beam)

26.192 Computer Applications of Lasers
William V. Smith (IBM Watson)
Appl. Optics, Vol. 5, pp. 1533–1538, October 1966 (memory, interconnection, input-output)

26.193 A Television Display Using Acoustic Deflection and Modulation of Coherent Light
A. Korpel, R. Adler, P. Desmares, and W. Watson (Zenith)
Appl. Optics, Vol. 5, pp. 1667–1675, October 1966 (Bragg reflection)

26.194 Single-Beam-Laser Rotation-Rate Sensor
Paul Fenster and Walter K. Kahn (Poly. Inst. Brooklyn)
Electronics Letters, Vol. 2, pp. 380–381, October 1966 (lower locking frequency than for dual beam system)

26.195 Computer Applications of Lasers
William V. Smith (IBM Watson)
Proc. IEEE, Vol. 54, pp. 1295–1300, October 1966 (memory, interconnection, input-output)

26.196 A Television Display Using Acoustic Deflection and Modulation of Coherent Light
A. Korpel, R. Adler, P. Desmares, and W. Watson (Zenith)
Proc. IEEE, Vol. 54, pp. 1429–1437, October 1966 (Bragg reflection)

26.197 Measurement of Optical Surface Roughness Using Coherent Radiation
S. Karp, R. B. Hankin, S. Spinak, E. J. Pisa, and P. P. Barron (Douglas)
Proc. IEEE, Vol. 54, pp. 1484–1485, October 1966 (0.63-μ He–Ne laser, 1–5 μ rms surface roughness of aluminum)

26.198 Velocity Aberration and Atmospheric Refraction in Satellite Laser Communication Experiments
L. J. Nugent and R. J. Condon (Electro-Optical)
Appl. Optics, Vol. 5, pp. 1832–1837, November 1966

26.199 Ring-Laser Accuracy
A. F. H. Thomson and P. G. R. King (Services Elec. Research Lab., England)
Electronics Letters, Vol. 2, p. 417, November 1966 (triangular resonator, minimized inaccuracies)

26.200 Mode Interaction in a Gas Laser
B. L. Zhelnov, A. P. Kazantsev, and V. S. Smirnov (Siberian Semiconductor Phys. Inst., USSR)
Soviet Phys. JETP, Vol. 23, pp. 858–860, November 1966 (traveling waves in ring resonator, mode coupling by mirror reflection)

26.201 Frequency Spectrum of Giant Acoustic Wave Packets Generated in CdS by High Electric Fields
J. Zucker and S. Zemon (GT&E Labs.)
Appl. Phys. Letters, Vol. 9, pp. 398–400, December 1, 1966 (laser beam probe)

26.202 Amplitude and Frequency Characteristics of a Ring Laser
T. J. Hutchings, J. Winocur, R. H. Durrett, E. D. Jacobs, and W. L. Zingery (Autonetics)
Phys. Rev., Vol. 152, pp. 467–473, December 2, 1966 (isotopes)

26.203 Hyperfine Structure and Paramagnetic Properties of Excited States of Xenon Studied with a Gas Laser
H. R. Schlossberg and A. Javan (M.I.T.)
Phys. Rev. Letters, Vol. 17, pp. 1242–1244, December 19, 1966

27. MISCELLANEOUS

27.29 Cross Modulation of Optical, RF, and Microwave Signals with Optical Pumping
B. W. Harned, L. B. Leder, and M. E. Lasser (Philco) and D. L. Carter (U. of Pennsylvania)
Proc. IEEE, Vol. 52, p. 632, May 1964 (cesium vapor)

27.30 Stimulated Phonon Emission by Supersonic Electrons and Collective Phonon Propagations
E. W. Prohofsky (Sperry Rand Res. Center)
Phys. Rev., Vol. 134, pp. A1302-A1312, June 1, 1964 (phonon frequencies of 10^{10} to 10^{11} c/s)

27.31 Astronomical Image-Integration System using a Television Camera Tube
E. Luedicke, A. D. Cope, and L. E. Flory (RCA)
Appl. Optics, Vol. 3, pp. 677-689, June 1964 (image orthicon camera 3-4 stellar magnitudes more sensitive than 103a0 film in same exposure interval)

27.32 Quartz Optical Phonon-Masers
C. H. Becker (Precision Instrument)
IEEE Trans. on Sonics and Ultrasonics, Vol. SU-11, pp. 34-40, June 1964 (Stokes and Anti-Stokes lines, experiments, forward scattering)

27.33 Theory of Boil-Off Calorimetry
R. B. Jacobs (NBS)
Rev. Sci. Instr., Vol. 35, pp. 828-832, July 1964

27.34 Flash and CW Methods for Laser Potentiality Measurements
H. H. Theissing, P. J. Caplan, T. Ewanizky, and G. deLhery (U. S. Army ERDL)
Appl. Optics, Vol. 3, pp. 951-955, August 1964 (test measurements on $CaF_2:Sm^{2+}$)

27.35 Improved Laser Angular Brightness through Diffraction Coupling
J. T. LaTourette, S. F. Jacobs, and P. Rabinowitz (TRG)
Appl. Optics, Vol. 3, pp. 981-982, August 1964 (single-mode output)

27.36 Laser Interferometry of Small Windows
R. M. Zoot (Litton Systems)
Appl. Optics, Vol. 3, pp. 985-986; August 1964 (surface flatness, homogeneity and parallelism)

27.37 Display of Infrared Laser Patterns by a Liquid Crystal Viewer
J. R. Hansen, J. L. Fergason, and A. Okaya (Westinghouse)
Appl. Optics, Vol. 3, pp. 987-988, August 1964 (scattered light depends on crystal temperature, He-Ne laser tests)

27.38 Measurement of Microwave Shot-Noise Reduction Factor by Laser Light Induced Photoemission
S. Saito and Y. Fujii (U. of Tokyo, Japan)
Proc. IEEE, Vol. 52, p. 980, August 1964

27.39 On the Feasibility of Flamelasers
R. Bleekrode and W. C. Neiuwpoort (Philips Res. Lab., Netherlands)
Physics Letters, Vol. 12, pp. 204-205, October 1, 1964

27.40 Flux Growth, Czochralski Growth, and Hydrothermal Synthesis of Lithium Metagallate Single Crystals
J. P. Remeika and A. A. Ballman (BTL)
Appl. Phys. Letters, Vol. 5, pp. 180-181, November 1, 1964 (hydrothermal at 10000 psi and 350°C)

27.41 High-Power Light Flux Resulting from Bunching of Light Wave Packets
A. Bramley (Bramley Consultants)
Appl. Phys. Letters, Vol. 5, pp. 210-212, November 15, 1964 (proposed method)

27.42 Large Energy Transfer from Uranyl to Europium Ions in Glass
L. G. DeShazer and A. Y. Cabezas (Hughes)
Proc. IEEE, Vol. 52, p. 1355, November 1964 (uranyl fluoresces at 5250 Å)

27.43 The Minimum Spot Size for a Focused Laser and the Uncertainty Relation
P. W. Carlin (U. of Colorado)
Proc. IEEE, Vol. 52, p. 1371, November 1964

27.44 A Scanning Spherical Mirror Interferometer for Spectral Analysis of Laser Radiation
R. L. Fork, D. R. Herriott, and H. Kogelnik (BTL)
Appl. Optics, Vol. 3, pp. 1471-1484, December 1964 (improved resolution, some spatial dependence)

27.45 Comparison of Monochromatic Semiconductor Radiation Sources with Tungsten Lamps
O. A. Weinreich and S. Mayburg (GT&E)
Appl. Optics, Vol. 3, pp. 1489-1493, December 1964 (efficiency, source life, radiant emittance)

27.46 Active Image Formation in Lasers
W. A. Hardy (IBM)
IBM J. Res. & Dev., Vol. 9, pp. 31-46, January 1965 (modes controlled by imaging masks, can be projected)

27.47 Making of Spacers for Fabry-Perot Etalons
F. M. Phelps, III (U. of Michigan)
J. Opt. Soc. Am., Vol. 55, pp. 293-295, March 1965 (inexpensive, Invar spacer, fabrication and testing techniques)

27.48 Effect of Laser on Carcinoma in Man
Hubert L. Rosomoff, Richard Hellstrom, Jerry Brown, and Fred Carroll (U. of Pittsburgh School of Medicine and Veterans Admin. Hospital, Pittsburgh)
J. Am. Med. Assoc., Vol. 192, pp. 167-168, April 12, 1965 (preliminary, 1-3 joule, ruby laser, will destroy cancer, limited volume, dosage tolerated)

27.49 The Dark Side of the Laser
J. J. Schlickmann and R. H. Kingston (M.I.T.)
Electronics, Vol. 38, No. 8, pp. 93-98, April 19, 1965 (laser dosimeter, critical levels)

27.50 On the Visibility of 8400 Å Light from GaAs Laser Diodes
Y. Nannichi (Nippon Electric, Japan)
Japanese J. Appl. Phys., Vol. 4, p. 308, April 1965 (eye excited by intense infrared)

27.51 Note on Laser Monitors
S. E. Schwarz (U. of California)
Proc. IEEE, Vol. 53, pp. 414-415, April 1965 (photodiode more reliable than photomultiplier)

27.52 Eye Protection against Lasers
C. H. Swope and C. J. Koester (American Optical)
Appl. Optics, Vol. 4, pp. 523-526, May 1965 (damaged filters, suggestions to personnel)

27.53 New Oxy-Hydrogen Burner for Flame Fusion
J. A. Adamski (Amer. Sci. Engineering)
J. Appl. Phys., Vol. 36, pp. 1784-1786, May 1965 (larger diameter crystals by using three-tube postmix burner)

27.54 Demonstration of Chromatic Aberration in the Eye Using Coherent Light
D. C. Sinclair (U.S. Army Engineer R & D Labs.)
J. Opt. Soc. Am., Vol. 55, pp. 575-576, May 1965 (granularity effect, motion of head)

27.55 Nanosecond Pulses of Very Low Impedance
Heinz Fischer (AFCRL)
Proc. IEEE, Vol. 53, pp. 545-546, May 1965 (13 ns pulse, 1300 amps, 5 kV)

27.56 Calcite Prisms as High-Power Laser Beam Combiners
C. C. Wang and G. W. Racette (Philco)
Appl. Optics, Vol. 4, pp. 759-761, June 1965 (collinearized 6328 Å and 3470 Å beams, latter is second harmonic of ruby laser)

27.57 Circulators for Optical Radar Systems
Paul C. Fletcher and David L. Weisman (Electro-Optical Systems)

Appl. Optics, Vol. 4, pp. 867–873, July 1965 (Faraday rotation, Lummer-Gehrcke plate, 1.1 dB insertion loss, over 40 dB isolation)

27.58 Simple Device for Leveling Collimated Light Beams
Melvin D. Daybell (New Mexico State U.)
Appl. Optics, Vol. 4, pp. 877–878, July 1965 (45° prism, horizontal surface of liquid)

27.59 Silicon-Controlled-Rectifier Long-Pulse Driver for Injection Lasers
M. C. Teich, D. A. Berkley, and G. J. Wolga (Cornell U.)
Rev. Sci. Instr., Vol. 36, pp. 973–974, July 1965 (longer than 30 μs, up to 20 amps)

27.60 Cavity Resonances in Accelerated Systems
E. J. Post (AFCRL) and Asim Yildiz (Raytheon)
Phys. Rev. Letters, Vol. 15, pp. 177–178, August 2, 1965 (electromagnetic waves in rotating systems)

27.61 Fundamental Theorem in Quantum Optics
I. R. Senitzky (U.S. Army Electronics Command)
Phys. Rev. Letters, Vol. 15, pp. 233–235, August 9, 1965 (interaction of field and detector)

27.62 Disproof of Senitzky's "Fundamental Theorem in Quantum Optics"
A. E. Glassgold (New York U.) and Dennis Holliday (RAND)
Phys. Rev. Letters, Vol. 15, pp. 741–742, November 8, 1965 (I. R. Senitzky, Phys. Rev. Letters, Vol. 15, pp. 233–235, August 9, 1965)

27.63 Thermal Effects in Optically Pumped Laser Rods
R. L. Townsend, C. M. Stickley, and A. D. Maio (AFCRL)
Appl. Phys. Letters, Vol. 7, pp. 94–96, August 15, 1965 (distortion during and after pumping, ruby and Nd glass, beam divergence)

27.64 Folded Optical Delay Lines
Donald R. Herriott and Harry J. Schulte (BTL)
Appl. Optics, Vol. 4, pp. 883–889, August 1965 (up to 10 μs delay)

27.65 Measurement of Radiant Power Output with a Solar Cell
S. Deb and M. K. Mukherjee (Jadavpur U., India)
IEEE J. Quantum Electronics, Vol. QE-1, pp. 219–220, August 1965 (expect limit of 10^{-9}W)

27.66 Gas Effects Produced by a Pulsed Laser Beam in a Ballistic Torsional Pendulum
M. Stimler and Z. I. Slawsky (U. of Maryland and N.O.L.)
J. Appl. Phys., Vol. 36, pp. 2598–2599, August 1965

27.67 Alignment of Cr^{3+} in Ruby
G. F. Hull, Jr., J. T. Smith, and A. F. Quesada (Baird-Atomic)
Appl. Optics, Vol. 4, pp. 1117–1120, September 1965 (pump with circularly polarized 6943 Å, 10^{-7} seconds spin-lattice relaxation time)

27.68 Proof of the Manley-Rowe Relations from Quantum Considerations
B. D. Anderson (Stanford Electronics Labs.) and J. Brown (University College, London, England)
Electronics Letters, Vol. 1, p. 199, September 1965

27.69 Proposed Gas Maser Pumping Scheme for the Far Infrared
Willard H. Wells (Jet Propulsion Lab., California Inst. Tech.)
J. Appl. Phys., Vol. 36, pp. 2838–2843, September 1965 (rotational state interaction between HCl and HF, HF emission at 81.25 μ)

27.70 Visual Observation of Infrared Laser Emission
L. S. Vasilenko, V. P. Chebotaev, and Yu. V. Troitskiĭ (Academy of Sciences, U.S.S.R.)
Soviet Physics JETP, Vol. 21, pp. 513–514, September 1965 (0.95 to 1.18 μ radiation observed with unaided eye, 2nd harmonic)

27.71 On the Possibility of Stimulated Emission in the Far Ultraviolet
P. A. Bazhulin, I. N. Knyazev, and G. G. Petrash (Academy of Sciences, U.S.S.R.)
Soviet Physics JETP, Vol. 21, pp. 649–650, September 1965 (considers H_2 molecule)

27.72 On the Equivalence between Semiclassical and Quantum Descriptions of Light
J. Perina (Palacký U., Olomouc, Czechoslovakia)
Physics Letters, Vol. 19, pp. 195–197, October 15, 1965

27.73 Measurements of the Longitudinal Component of the Electromagnetic Field at the Focus of a Coherent Beam
A. I. Carswell (RCA, Canada)
Phys. Rev. Letters, Vol. 15, pp. 647–649, October 18, 1965 (microwave results can be applied to laser beams)

27.74 Infrared Laser Interferometry Utilizing Quantum-Electronic Cross Modulation
R. J. Freiberg, L. A. Weaver, and J. T. Verdeyen (U. of Illinois)
J. Appl. Phys., Vol. 36, p. 3352, October 1965 (detection of cross-modulation sidelight at visible wavelength)

27.75 Nanosecond Background Source for Flash Photolysis
Heinz Fischer (AFCRL), and Emma Duchane and Alfred Büchler (Arthur D. Little)
J. Opt. Soc. Am., Vol. 55, pp. 1275–1277, October 1965 (20–30 ns halfwidth, UV from argon-air mixture, 2000–4000 amp discharge)

27.76 Interaction of Laser Modes
L. A. Ostrovskiĭ (Gorkiĭ State U., U.S.S.R.)
Soviet Physics JETP, Vol. 21, pp. 727–732, October 1965 (steady-state and transient processes in 2-mode laser)

27.77 The Optical Whispering Mode of Polished Cylinders and its Implications in Laser Technology
Franklin G. Reick (ITT)
Appl. Optics, Vol. 4, pp. 1395–1399, November 1965 (clockwise and counterclockwise trapped light in cylinder)

27.78 Mach-Zehnder Interferometer with Adjustable Compensation
A. Stein and T. Shultz (Control Data Corp.)
Appl. Optics, Vol. 4, pp. 1510–1511, November 1965 (dual prism adjustable compensator)

27.79 Information Content of a Beam of Photons
Ellen Hisdal (U. of Oslo, Norway)
J. Opt. Soc. Am., Vol. 55, pp. 1446–1454, November 1965 (entropy and information, photon fluctuations)

27.80 Rayleigh's Integral in the Near Fresnel Region
Harold Osterberg (American Optical)
J. Opt. Soc. Am., Vol. 55, pp. 1467–1471, November 1965

27.81 Laser Beam "Security"
Selig Kainer (ITT Federal Labs.)
Proc. IEEE, Vol. 53, pp. 1752–1753, November 1965 (secure in space, atmospheric perturbations destroy "security")

27.82 Doppler Shift of Ruby Laser Light by Means of a Kerr Cell Traveling Wave Line
F. P. Küpper and E. Fünfer (Institut für Plasmaphysik, Germany)

Physics Letters, Vol. 19, pp. 486–487, December 1, 1965 (60-cm-long cell, theoretical shift of 2.3×10^{10} c/s)

27.83 Quantum Descriptions of an Infinite Lossless Transmission Line
C. Y. She (U. of Minnesota)
J. Appl. Phys., Vol. 36, pp. 3784–3787, December 1965 (signal plus noise)

27.84 Thin Spacers for Fabry-Perot Étalons
R. Mewe and R. F. de Vries (FOM-Instituut voor Plasma-Fysica, Netherlands)
J. Opt. Soc. Am., Vol. 55, p. 1697, December 1965 (thin tungsten wires, precision bearing balls)

27.85 A Proposed First-Order Relativity Test using Lasers
R. Fox and J. Shamir (Technion, Israel)
Proc. IEEE, Vol. 53, pp. 2141–2142, December 1965 (further comments, also by C. W. Carnahan)

27.86 Multiple Watt Submicrosecond High Repetition Rate Light Source and Its Application
R. C. Mackey (U. of California) and S. A. Pollack and R. S. Witte (TRW)
Rev. Sci. Instr., Vol. 36, pp. 1715–1718, December 1965 (20 ns at 5 kc/s rate, $0.3 - 0.5 \mu$ wavelength)

27.87 Generation and Detection of Subnanosecond Light Pulses: Application to Luminescence Studies
Juan Yguerabide (Sandia Lab.)
Rev. Sci. Instr., Vol. 36, pp. 1734–1742, December 1965

27.88 Hydrofluorination Unit for Purification of Fluoride Laser Materials
Stanley I. Warshaw and Robert E. Jackson (Raytheon)
Rev. Sci. Instr., Vol. 36, pp. 1774–1776, December 1965

27.89 Photon Echoes in Ruby
N. A. Kurnit, I. D. Abella, and S. R. Hartmann (Columbia U.)
Proc. Physics of Quantum Electronics Conf., McGraw-Hill Book Co., New York, pp. 267–279, 1966

27.90 Nonlinear Quantum Effect in Solid-State Masers
Humio Inaba and Hideo Morita (Tohoku U., Japan)
Proc. Physics of Quantum Electronics Conf., McGraw-Hill Book Co., New York, pp. 294–304, 1966 (X-band microwave)

27.91 Far-Infrared Solid-State Masers: A Speculative Account with Some Related Experiments
F. Varsanyi (B.T.L.)
Proc. Physics of Quantum Electronics Conf., McGraw-Hill Book Co., New York, pp. 370–375, 1966

27.92 Stimulated Phonon Emission from an Inverted Spin System to a Bottlenecked Lattice
Peter E. Wagner and William J. Brya (John Hopkins U.)
Proc. Physics of Quantum Electronics Conf., McGraw-Hill Book Co., New York, pp. 376–384, 1966

27.93 Stimulated Phonon-Photon Double-Quantum Emission
N. S. Shiren (IBM)
Proc. Physics of Quantum Electronics Conf., McGraw-Hill Book Co., New York, pp. 385–394, 1966 (9 Gc/s)

27.94 Light Shift, Light Modulation, and Phase Pulling in the Optically Pumped Rubidium Maser
R. Novick, P. Davidovits, W. Happer, Jr., and W. A. Stern (Columbia U.)
Proc. Physics of Quantum Electronics Conf., McGraw-Hill Book Co., New York, pp. 626–634, 1966 (microwave)

27.95 Interaction of Traveling Waves in a Ring Laser
E. M. Belenov, E. P. Markin, V. N. Morozov, and A. N. Oraevskii (Lebedev Physics Institute, Academy of Sciences, U.S.S.R.)

JETP Letters, Vol. 3, pp. 32–35, January 1, 1966 (3.39-μ He–Ne laser, varied output mirror reflectivity)

27.96 The Generation of Molecular Vibrational Frequencies by Optical Mixing
M. D. Martin and E. L. Thomas (Ministry of Aviation, S.R.D.E., Christchurch, Hants)
Physics Letters, Vol. 19, pp. 651–652, January 1, 1966

27.97 Stresses Developed in Optical Film Coatings
Anthony E. Ennos (Perkin-Elmer)
Appl. Optics, Vol. 5, pp. 51–61, January 1966 (many film materials, single and multilayers, measurements)

27.98 A Recording Sampling System for Measuring Laser Energy
R. C. Williams and Harold A. Mueller (Medical College of Virginia)
Appl. Optics, Vol. 5, pp. 135–138, January 1966 (beam splitter, fiber optic, photodiode)

27.99 Electron Beam Excitation in Laser Crystals
W. W. Anderson (Stanford U.)
Appl. Optics, Vol. 5, pp. 167–168, January 1966 (depth of active region, method proposed)

27.100 Proof of the Manley-Rowe Relations from Quantum Considerations
E. O. Schulz-Dubois and H. Seidel (B.T.L.)
Electronics Letters, Vol. 2, pp. 24–25, January 1966

27.101 A Nonreciprocal Electrooptic Device
John L. Wentz (Westinghouse)
Proc. IEEE, Vol. 54, pp. 97–98, January 1966 (2 Gc/s traveling wave interacting in KDP, no magnetic field)

27.102 Vibrational and Rotational Studies Using Q Switching of Molecular Gas Lasers
G. W. Flynn, M. A. Kovacs, C. K. Rhodes, and A. Javan (M.I.T.)
Appl. Phys. Letters, Vol. 8, pp. 63–64, February 1, 1966

27.103 Locking of Laser Oscillators by Light Injection
H. L. Stover and W. H. Steier (B.T.L.)
Appl. Phys. Letters, Vol. 8, pp. 91–93, February 15, 1966

27.104 Propulsion and Angular Stabilization of Dust Particles in a Laser Cavity
Eric G. Rawson and A. D. May (U. of Toronto, Canada)
Appl. Phys. Letters, Vol. 8, pp. 93–95, February 15, 1966

27.105 Spacers for Fabry-Perot Interferometers
G. D. Saksena (Atomic Energy Estab., Bombay, India)
J. Opt. Soc. Am., Vol. 56, p. 256, February 1966 (three 3-mm diameter Invar rod spacers in perspex ring)

27.106 Energy Echoes
Robert M. White (U. of California)
J. Appl. Phys., Vol. 37, pp. 1693–1696, March 15, 1966 (involve nonlinearity in time domain)

27.107 Detection of Low Angle Optical Scattering by Fabry-Perot Resonance
D. G. Peterson (Lockheed) and Amnon Yariv (Calif. Inst. of Tech.)
Appl. Optics, Vol. 5, pp. 469–470, March 1966

27.108 A Method of Measurement of Thresholds and Fluorescent Lifetimes of Laser Systems
A. M. White (Royal Radar Estab., England)
British J. Appl. Phys., Vol. 17, pp. 429–430, March 1966

27.109 Variable Brewster Angle Flat Used as Gas Laser Gain Control
Stuart A. Schleusener and Alvin A. Read (Iowa State U.)
Rev. Sci. Instr., Vol. 37, pp. 287–289, March 1966

27.110 Self Mode-Locking of Lasers with Saturable Absorbers
A. J. DeMaria, D. A. Stetser, and H. Heynau (United Aircraft)
Appl. Phys. Letters, Vol. 8, pp. 174–176, April 1, 1966 (recirculating loop, Nd glass amplifier, 1 ns pulses at 100 Mc/s rate)

27.111 Generation of Ultrashort Optical Pulses by Mode Locking the YAlG: Nd Laser
M. DiDomenico, Jr., J. E. Geusic, H. M. Marcos, and R. G. Smith (B.T.L.)
Appl. Phys. Letters, Vol. 8, pp. 180–183, April 1, 1966 (acoustic modulator in laser cavity, 8×10^{-11} s pulse length)

27.112 Zapping Paper
Jon H. Myer (Hughes Aircraft)
Proc. 8th Annual Electron and Laser Beam Symposium, April 6–8, 1966. Sponsored by U. of Michigan and IEEE, pp. 105–106 (multilayer coated paper, thermal effects, first layer records at $0.1 \ J/mm^2$)

27.113 Inhomogeneities in Optical Crystals Resulting from Nonplanar Solid-Melt Interface during Growth
James F. Nester (Perkin-Elmer)
J. Appl. Phys., Vol. 37, pp. 2002–2004, April 1966 ($CaWO_4:Nd^{3+}$ crystals, interferogram, radially asymmetric temperature gradient, laser beam 10 times diffraction limit)

27.114 Growth of Laser-Quality Rare-Earth Fluoride Single Crystals in a Dynamic Hydrogen Fluoride Atmosphere
M. Robinson and D. M. Cripe (Hughes)
J. Appl. Phys., Vol. 37, pp. 2072–2074, April 1966

27.115 Optical Pulses Produced by Laser Length Variation
W. C. Henneberger and H. J. Schulte (B.T.L.)
J. Appl. Phys., Vol. 37, p. 2189, April 1966 (0.63-μ laser, 252-cm long cavity, 59.4 Mc/s on moving end mirror, 1 ns pulses)

27.116 Cooler for Semiconductor Light Emitters, Lasers, and Photodetectors
H. K. Kessler (N.B.S., Washington)
Rev. Sci. Instr., Vol. 37, pp. 517–518, April 1966 (use flowing low-temperature gas)

27.117 Scanning Active Interferometer Employing Linear Drive Excitation and Reflectance Monitor
Vern N. Smiley, Adolph L. Lewis, and David K. Forbes (U. S. Navy Electronics Lab.)
Appl. Optics, Vol. 5, pp. 827–829, May 1966 (regenerative gas laser amplifiers)

27.118 Fabrication of High Quality Optical Pinholes
N. K. Sheridon, S. J. Krulikoski, and W. H. Wells (Bendix)
Appl. Optics, Vol. 5, p. 869, May 1966 (100 μ to 0.08 μ diameters, shadow of latex balls, evaporate aluminum)

27.119 Perspective Rendering of the Field Intensity Diffracted at a Circular Aperture
Leo Beiser (CBS Labs)
Appl. Optics, Vol. 5, pp. 869–870, May 1966

27.120 Bowl Feed Technique for Producing Supersmooth Optical Surfaces
Ralph W. Dietz and Jean M. Bennett (Michelson Lab.)
Appl. Optics, Vol. 5, pp. 881–882, May 1966 (roughness down to 3 Å rms)

27.121 Determination of Laser Beam Energy Profile by a Multi-Layer Foil Technique
M. L. Pilcher and B. A. Tozer (Central Electricity Research Labs., England)
British J. Appl. Phys., Vol. 17, pp. 695–696, May 1966 (foil of thin aluminum layer on transparent sheet, Q-sw laser beam measurements)

27.122 Liquid Immersion for Reducing Damaging Effect of Laser Giant Pulses to Dielectric Mirror Coatings

David W. Gregg and Scott J. Thomas (Lawrence Radiation Lab.)
Appl. Phys. Letters, Vol. 8, pp. 316–318, June 15, 1966 (in nitrobenzene, giant pulse lasers)

27.123 Two-Cavity Laser as High-Resolution Spectroscope
N. G. Basov, A. N. Oraevskii, G. M. Strakhovskii, and A. V. Uspenskii (Lebedev Phys. Inst., Acad. of Sci., USSR)
JETP Letters, Vol. 3, pp. 305–306, June 15, 1966

27.124 Trapping of Electrons in a Spatially Inhomogeneous Laser Beam
N. J. Phillips and J. J. Sanderson (English Electric, England)
Physics Letters, Vol. 21, pp. 533–534, June 15, 1966

27.125 Momentum Transfer Produced by Focused Laser Giant Pulses
David W. Gregg and Scott J. Thomas (U. of California)
J. Appl. Phys., Vol. 37, pp. 2787–2789, June 1966 (several metals, maximum momentum transfer at about $10^9 \ W/cm^2$)

27.126 Stability of Fabry-Perot Étalons
F. M. Phelps, III (U. of Michigan) and K. B. Newbound (U. of Alberta, Canada)
J. Opt. Soc. Am., Vol. 56, pp. 831–832, June 1966

27.127 Mode Coupling Due to Backscattering in a He-Ne Traveling-Wave Ring Laser
Frederick Aronowitz (Honeywell), and R. J. Collins (U. of Minnesota)
Appl. Phys. Letters, Vol. 9, pp. 55–58, July 1, 1966

27.128 Behavior of CdS Crystals Under Laser Light Excitation
Keiji Maeda and Seishi Iida (Tokyo Shibaura, Japan)
Appl. Phys. Letters, Vol. 9, pp. 92–94, July 15, 1966 (photoconductivity, luminescence, 5145 Å excitation using argon ion laser)

27.129 Effect of High-Velocity Mirror Translation on Optical Coherence in Laser Interferometers
T. J. Burgess (U. of California)
Appl. Optics, Vol. 5, pp. 1239–1240, July 1966 (reduces coherence)

27.130 Dependence of Laser Output on Air Pressure in a Cavity
Toshiharu Tako and Misao Ohi (Nat'l Research Lab. of Metrology, Japan)
Japanese J. Appl. Phys., Vol. 5, p. 638, July 1966 (separate cell for air within He-Ne laser cavity, hysteresis effect near laser threshold, possible thermal lens effect)

27.131 Discipline Your Laser with an Output-Energy Controller
Herbert Gresser (TRG)
Microwaves, Vol. 5, No. 7, pp. 50–53, July 1966

27.132 A High Time Resolution Polarimeter for Laser Analysis
Sun Lu and T. A. Rabson (Rice U.)
Appl. Optics, Vol. 5, pp. 1293–1296, August 1966 (light beam polarization measured in less than 0.1 μs, spike polarization data on Nd glass laser)

27.133 Automatic Control of the Spacing of Fabry-Perot Interferometers
J. V. Ramsay (Nat'l Stds. Lab., Australia)
Appl. Optics, Vol. 5, pp. 1297–1301, August 1966 (uses wedge-shaped reference interferometer)

27.134 A Tunable Birefringent Fabry-Perot Interferometer
J. V. Ramsay and R. N. Smartt (Nat'l Stds. Lab., Australia)
Appl. Optics, Vol. 5, pp. 1341–1342, August 1966 (uniform thickness mica sheet within cavity)

27.135 A Zero-Field Maser Oscillator
W. E. Hughes (Westinghouse)
IEEE J. Quantum Electronics, Vol. QE-2, pp. 233–235, August 1966 (Fe^{3+} ion, about 9.35 GHz, below 4.2°K)

27.136 Wide Field Active Imaging
        R. A. Myers, H. Wieder, and R. V. Pole (IBM Watson)
        IEEE J. Quantum Electronics, Vol. QE-2, pp. 270–275, August 1966 (pictorial information within laser cavity)

27.137 Two-Photon Spectroscopy of Anthracene
        D. Fröhlich and H. Mahr (Cornell U.)
        IEEE J. Quantum Electronics, Vol. QE-2, p. 294, August 1966 (abstract)

27.138 Exploratory Development to Establish Feasibility of a Millimeter and Submillimeter Wave Imaging System
        H. Jacobs, G. Morris, and R. C. Hofer (US Army Electronics Command)
        IEEE J. Quantum Electronics, Vol. QE-2, p. 294, August 1966 (abstract)

27.139 A Pulsed, Coaxial Transmission Line Gas Laser
        M. Geller, D. E. Altman, and T. A. DeTemple (USN Electronics Lab.)
        J. Appl. Phys., Vol. 37, pp. 3639–3640, August 1966 (low inductance, N_2 laser, 10 ns output pulses)

27.140 Effects of Coherence on Imaging Systems
        Philip S. Considine (Technical Operations)
        J. Opt. Soc. Am., Vol. 56, pp. 1001–1009, August 1966 (edge ringing, transilluminated and reflected coherent imaging, speckling, Hg arc lamp and 0.63-μ He–Ne laser)

27.141 Comparison of Single-Lens and Two-Lens Coherent Imaging of Complex Distributions
        Wade T. Cathey, Jr. (Autonetics)
        J. Opt. Soc. Am., Vol. 56, pp. 1015–1017, August 1966 (two-lens system produces an accurate image)

27.142 Image Comparison by Interference
        Wade T. Cathey, Jr., and J. G. Doidge (Autonetics)
        J. Opt. Soc. Am., Vol. 56, pp.. 1139–1140, August 1966 (He–Ne laser, small object moved, caused interference of images)

27.143 A Diffusely Transmitting, Integrating Sphere for Measuring Laser Output with a Phototransistor
        K. E. Fligsten and M. L. Wolbarsht (USN Med. Research Inst.)
        Proc. IEEE, Vol. 54, pp. 1109–1110, August 1966

27.144 Coupling Effects in a Passive Q-Switched Ruby Laser
        V. Degiorgio and M. Giglio (Lab. CISE, Italy)
        Nuovo Cimento, Vol. 45B, pp. 69–71, September 11, 1966 (vanadium phthalocyanine, two halves of beam decouples with 1.5 mm diam blocking wire)

27.145 A Self-Calibrating Technique Measuring Laser Beam Intensity Distributions
        I. M. Winer (Korad)
        Appl. Optics, Vol. 5, pp. 1437–1439, September 1966 (multiple lens camera, several neutral density filters)

27.146 Nonlinear Effects in an NMR Experiment
        J. Pommier and H. Benoit (CNRS, France)
        IEEE J. Quantum Electronics, Vol. QE-2, pp. 388–390, September 1966 (up to 6th harmonic in a 5-photon transition)

27.147 Relaxation in the Level F = 1 of the Ground State of Hydrogen: Application to the Hydrogen Maser
        J. Vanier and R. F. C. Vessot (Varian)
        IEEE J. Quantum Electronics, Vol. QE-2, pp. 391–398, September 1966

27.148 Modulation Effects in an Ammonia Beam Maser Oscillator
        D. C. Lainé and A. L. S. Smith (U. of Keele, England)
        IEEE J. Quantum Electronics, Vol. QE-2, pp. 399–408, September 1966

27.149 Concentration and Temperature Dependences of Spin-Lattice Relaxation Times in Ruby at Helium Temperatures: Relaxation in a Zero Magnetic Field
        A. A. Manenkov and Yu. K. Danileiko (Lebedev Phys. Inst., Acad. of Sci., USSR)
        IEEE J. Quantum Electronics, Vol. QE-2, pp. 409–412, September 1966 (X-band experiments)

27.150 Laser Pumped Microwave Emission in Excited States of Neon
        T. F. Johnston, Jr., R. L. Abrams, and G. J. Wolga (Cornell U.)
        IEEE J. Quantum Electronics, Vol. QE-2, pp. 412–417, September 1966 (3.39 μ neon laser for pump, 26.4 GHz output)

27.151 A Laser with a Nonresonant Feedback
        R. V. Ambartsumyan, N. G. Basov, P. G. Kryukov, and V. S. Letokhov (Lebedev Phys. Inst., Acad. of Sci., USSR)
        IEEE J. Quantum Electronics, Vol. QE-2, pp. 442–446, September 1966 (feedback from diffuse scattering reflector, output characteristics)

27.152 Laser Action by Enhanced Total Internal Reflection
        Charles J. Koester (American Optical)
        IEEE J. Quantum Electronics, Vol. QE-2, pp. 580–584, September 1966 (high index passive glass fiber core, lower index Nd-doped glass clad, optically pumped, signal in core is amplified)

27.153 Laser Operation with Liquid Semiconductor Mirrors
        M. Birnbaum and T. L. Stocker (Aerospace)
        IEEE J. Quantum Electronics, Vol. QE-2, pp. 632–635, September 1966 (ruby laser, one liquid selenium mirror at 500°C, reflectivity measurements)

27.154 Quenching Effects in Coupled Lasers
        P. W. Pheneger and R. H. Pantell (Stanford U.)
        IEEE J. Quantum Electronics, Vol. QE-2, pp. 644–648, September 1966 (ruby lasers, end coupled)

27.155 High Temperature, Vacuum Tight Windows for Gas Laser Research
        Alfred L. Greilich and James E. Swain (Lawrence Rad. Lab., U. of California)
        Rev. Sci. Instr., Vol. 37, pp. 1257–1258, September 1966 (quartz, 1000°C, procedure)

27.156 Measurement of the Frequency Change of Light Reflected from a Metal Surface
        J. Shamir and R. Fox (Israel Inst. of Tech.)
        Physics Letters, Vol. 23, pp. 314–315, October 31, 1966 (He–Ne laser, aluminum surface, fractional frequency shift of less than 5×10^{-21} per reflection)

27.157 The Antenna Properties of Optical Heterodyne Receivers
        A. E. Siegman (Stanford U.)
        Appl. Optics, Vol. 5, pp. 1588–1594, October 1966 (effects of noise)

27.158 The Measurement of Fast Current Pulses in the Kiloampere Range
        John P. Markiewicz and John L. Emmett (Stanford U.)
        Appl. Optics, Vol. 5, pp. 1687–1688, October 1966 (useful for flash lamps)

27.159 The Antenna Properties of Optical Heterodyne Receivers
        A. E. Siegman (Stanford U.)
        Proc. IEEE, Vol. 54, pp. 1350–1356, October 1966 (effects of noise)

27.160 Theory of the Infrared Generation by Coherently Driven Molecular Vibrations

Francesco De Martini (M.I.T.)
J. Appl. Phys., Vol. 37, pp. 4503–4507, November 1966 (proposal, 10-μ wavelength from calcite, Q-sw ruby pump, Raman effect, 10-kW output)

27.161   Detection of Differences in Real Distributions
James E. Rau (Autonetics)
J. Opt. Soc. Am., Vol. 56, pp. 1490–1494, November 1966 (two spatial distributions, single lens, coherent illumination)

27.162   Coupling of High Peak Power Pulses from He–Ne Lasers
William H. Steier (BTL)
Proc. IEEE, Vol. 54, pp. 1604–1606, November 1966 (KDP modulator within laser cavity, 80 ns 80 mW output pulse, repetition rate up to 100 kHz)

27.163   Pressure-Induced Line Shift and Collisional Narrowing in Hydrogen Gas Determined by Stimulated Raman Emission
P. Lallemand, P. Simova, and G. Bret (Harvard U.)
Phys. Rev. Letters, Vol. 17, pp. 1239–1241, December 19, 1966

27.164   Accuracy and Sensitivity of the Thermal Lens Method for Measuring Absorption
Domenico Solimini (U. of California)
Appl. Optics, Vol. 5, pp. 1931–1939, December 1966 (lenses within optical resonator, spot sizes at mirrors, calculate focal lengths)

27.165   Quantum-Counting Spectroscopy
M. R. Brown (Signals R & D Estab., England), and R. J. Strain (STL, England)
J. Appl. Phys., Vol. 37, pp. 4806–4810, December 1966

AUTHOR INDEX

Aagard, R. L. 3.22, 3.30, 3.56, 26.40
Abella, I. D. 27.89
Abrahams, M. S. 9.188, 9.195
Abrams, R. L. 27.150
Abramson, E. 8.3
Ackerman, J. A. 14.11, 27.25
Adams, A., Jr. 6.18, 6.53, 6.54
Adams, C. M., Jr. 26.89
Adams, I. 3.166
Adams, N. I. 6.17, 6.26, 20.7, 20.13, 20.118
Adams, S. L. 6.152
Adamski, J. A. 27.53
Adamson, M. C. 3.47, 12.32
Adelman, A. H. 23.4, 23.96
Adler, E. 23.39
Adler, R. 17.22, 20.56, 23.48, 26.193, 26.196
Afanas'ev, A. M. 20.59, 20.63
Afanasyev, Yu. V. 23.233
Agranovich, V. M. 20.130
Aharanov, Y. 18.100
Ahearn, W. E. 9.204, 9.211
Ahlstrom, E. R. 3.78
Ahmed, S. A. 6.76, 6.83, 7.233, 14.67, 14.88, 26.55
Aigrain, P. 10.26
Ainslie, N. 9.51
Aisenberg, S. 7.2, 14.66
Akanaev, B. A. 19.40, 23.176
Akhmanov, S. A. 19.40, 19.50, 19.61, 20.60, 20.64, 20.70, 20.75, 20.78, 20.96, 20.104, 20.115, 20.125, 22.17, 23.111, 23.137, 23.141
Akitt, D. P. 7.169
Akselrad, A. 9.190
Alcock, A. J. 21.60, 22.48, 22.51, 23.27
Aldag, H. R. 4.53
Aleksandrov, E. B. 19.113, 23.151

Aleksoff, C. C. 15.129
Alexander, F. B. 10.31
Alkemade, C. Th. J. 6.172, 12.216
Allen, J. W. 3.151
Allen, L. 6.128, 6.173, 24.17, 24.150
Allen, R. B. 7.203
Alley, C. O. 24.212
Allwood, M. J. 26.166
Alpert, S. S. 6.27, 26.128
Alpiner, B. 16.54, 23.48
Altman, D. E. 7.205, 27.139
Altshuler, S. 23.213
Ambartsumyan, R. V. 3.170, 11.44, 11.49, 11.51, 13.82, 16.109, 23.138, 27.151
Ammann, E. O. 6.114, 15.107, 15.116, 18.66, 24.38, 24.260, 24.261, 24.267, 24.307
Anan'ev, Yu. A. 5.57, 11.34, 13.96
Anbe, T. 3.16
Anderson, B. D. 27.68
Anderson, C. H. 5.39
Anderson, J. E. 14.111, 26.73, 26.85
Anderson, L. K. 15.26, 16.12, 16.60, 16.76, 16.90, 16.144, 16.148, 16.151
Anderson, P. H. 26.143, 26.180
Anderson, R. L. 9.1
Anderson, W. W. 2.65, 9.157, 14.115, 27.99
Ando, S. 26.204
Andrade, O. 7.197, 7.223
Andreeva, T. L. 8.30
Andresen, H. G. 3.131, 4.60, 17.35, 20.16
Antes, L. L. 6.44
Antonoff, M. M. 9.100, 10.84
Aoki, Y. 24.303
Aplet, L. J. 4.81, 17.47, 24.37
Arams, F. 11.23

Arbel, A. 5.41
Archer, A. E. 27.28
Archer, R. J. 9.180, 10.36, 10.37
Arecchi, F. T. 2.59, 6.35, 11.31, 11.32, 16.134, 18.32, 18.84, 18.102, 18.103, 18.131, 18.142, 23.106, 26.61
Armstrong, D. E. 24.263
Armstrong, J. A. 9.18, 9.62, 9.70, 9.103, 9.104, 9.134, 9.160, 9.174, 9.191, 13.32, 18.101, 18.122, 18.137, 23.26, 25.7
Arndt, J. 24.308
Arnold, G. W. 9.210
Arnold, K. M. 9.33
Aronowitz, F. 2.62, 27.127
Asami, S. 6.177
Asawa, C. K. 3.34, 4.79, 26.15
Ascoli-Bartoli, U. 13.66
Ashburn, E. V. 1.1, 1.12, 1.36, 1.40, 1.51
Ashby, D. E. T. F. 21.1, 21.19
Ashkin, A. 6.16, 9.4, 9.177, 10.15, 20.6, 20.36, 20.47, 20.98, 20.101, 20.103, 20.117, 23.201
Ashkin, M. 19.161
Ashkinadze, B. M. 23.258
Ashley, E. J. 24.94
Askar'yan, G. A. 14.143, 20.37, 22.16, 22.20, 22.25, 23.109, 23.115, 23.162, 23.199, 23.216, 23.226, 23.255
Astheimer, R. W. 15.140, 24.199
Aten, A. C. 10.65
Atwood, J. G. 2.103, 3.91
Auffret, R. 3.176, 4.67, 4.113, 12.223
Aukerman, L. W. 9.90, 9.182
Auld, B. A. 17.63

Auzel, M. F. 4.8
Avakian, P. 8.3
Aven, M. 10.52, 10.72, 14.86
Avizonis, P. V. 3.165, 11.42, 14.107, 23.122, 23.234, 23.246, 23.252
Axe, J. D. 5.6, 9.21

Babcock, R. V. 9.93
Backlund, P. 23.135, 23.136
Badger, A. S. 3.82
Badoual, R. 24.196
Bagaev, S. N. 6.101, 12.123
Bagaev, V. C. 9.47
Bagaev, V. S. 16.153
Bahr, A. J. 20.49
Bailey, R. L. 6.61, 18.23
Bain, W. C. 26.168
Baird, D. H. 15.14, 15.36
Baird, K. M. 6.15, 6.70, 6.103, 6.130, 12.75, 26.124
Baker, D. A. 21.20, 21.32
Baker, D. J. 14.26, 16.39
Bakos, J. 7.158
Bakumenko, V. L. 5.47
Balashov, I. F. 13.96
Baldwin, G. C. 23.88, 27.10
Baldwin, G. D. 3.150, 17.44
Ballard, G. S. 25.58
Ballard, S. S. 24.302
Ballik, E. A. 7.163, 7.164, 7.238
Ballman, A. A. 5.15, 15.68, 15.110, 15.176, 23.201, 24.282, 27.40
Balzarini, D. 26.128
Band, H. E. 26.33
Barach, J. P. 24.255
Barger, R. L. 19.112
Barker, L. M. 26.121, 26.162
Barker, W. A. 1.30, 2.21
Barkhudarova, T. M. 3.168, 17.48
Barnard, T. W. 26.156
Barnes, C. W. 16.122, 17.23

153

Barnes, F. S. 14.10, 14.23, 14.28
Barnett, N. E. 25.93
Barocchi, F. 19.150, 23.127
Barone, S. R. 3.3, 12.4, 12.69
Barrett, J. J. 19.90, 20.10, 20.118
Barron, P. P. 26.197
Bartell, L. S. 21.37, 21.54
Barynin, V. A. 22.50
Basil, I. T. 3.150, 17.44
Basov, N. G. 1.46, 1.63, 2.57, 2.85, 2.112, 2.113, 3.170, 4.57, 4.73, 7.191, 9.47, 9.135, 9.193, 9.220, 10.27, 10.103, 11.44, 11.49, 11.51, 11.56, 12.196, 13.82, 14.138, 14.146, 16.109, 18.95, 23.18, 23.138, 27.123, 27.151
Bass, J. C. 16.84, 16.119, 16.120, 20.86
Bass, M. 3.127, 23.89
Bates, R. L. 27.6
Bateson, S. 24.48, 26.46
Batulev, V. N. 23.120
Baugh, C. W. 4.18, 14.36
Baumeister, P. W. 24.197, 24.308
Baym, G. 19.71, 20.65, 20.83, 21.52, 23.124
Bazhulin, P. A. 7.112, 7.149, 27.71
Beaubien, M. W. 4.28
Bebb, H. B. 22.8, 22.26, 22.37
Beck, A. C. 24.58, 24.59
Becker, A. G. 5.74
Becker, C. H. 5.4, 27.32
Becker, R. A. 25.27
Beckmann, K. H. 24.116
Bédard, G. 18.112
Bedilov, M. R. 14.101
Beekenkamp, P. 23.192
Beer, R. 24.253
Beiser, L. 27.119
Bekefi, G. 21.38, 21.45, 21.72
Belenov, E. M. 6.134, 12.179, 27.95
Belikova, T. P. 3.117, 23.98, 23.185
Bell, R. L. 10.9, 10.39
Bell, W. E. 6.49, 7.2, 7.45, 7.113, 7.133, 7.134, 7.140, 7.156, 7.180, 16.29
Bellisio, J. A. 6.41
Bel'skii, N. K. 3.133
Ben-Dov, O. 21.44
Benedek, G. B. 19.60, 19.139, 23.146, 24.155, 24.251

Benedetti-Michelangeli, G. 13.66
Benesch, R. 21.15, 21.70
Bennett, G. E. 9.132
Bennett, H. E. 16.136
Bennett, J. M. 12.43, 24.94, 27.120
Bennett, W. R., Jr. 1.20, 2.20, 2.49, 6.10, 6.67, 6.69, 6.88, 7.15, 7.32, 7.55, 7.70, 7.71, 7.130, 7.131, 7.163, 7.164, 7.234, 18.34, 27.15
Benoit à la Guillaume, C. 10.76
Benoit, H. 27.146
Benson, R. C. 3.50, 13.11
Berger, S. B. 24.184
Berger, W. E. E. 6.130, 26.124
Bergstein, L. 12.53, 12.88, 12.163, 12.182, 12.219, 24.193
Berkley, D. A. 3.40, 9.150, 18.13, 27.59
Berkowitz, D. A. 3.156
Berlman, I. B. 27.5
Berman, P. R. 20.73
Bernal, E. 4.29, 15.150, 20.112, 23.150, 23.211, 23.232
Bernard, M. G. A. 10.28, 10.79, 16.1, 24.21
Berné, A. 18.102, 18.131, 18.142
Berozashivili, Yu. N. 16.153
Berreman, D. W. 24.52, 24.53, 24.107, 24.169, 24.310
Berry, M. 22.7
Bershteĭn, I. L. 12.197, 26.145
Bertolotti, M. 18.64
Bertotti, B. 22.55, 23.244
Bespalov, V. I. 23.200
Besson, J. 10.102, 16.145
Besson, J. M. 10.74
Bettinger, R. T. 24.212
Bevacqua, S. F. 9.16
Bevernage, G. 22.36
Bevolo, A. J. 1.30
Beyer, J. B. 12.156
Bhaumik, M. L. 8.10, 8.19
Bhawalkar, D. D. 20.40
Bialecke, E. P. 12.18
Biard, J. R. 9.85, 9.86
Biazzo, M. R. 12.57
Biberman, L. M. 1.26
Bickart, C. J. 3.131, 4.60, 17.35, 17.57
Bicknell, W. E. 23.126
Bielan, C. V. 9.16

Bierstedt, P. E. 4.76, 5.66, 15.139
Billeter, T. R. 17.2
Billings, B. H. 25.53
Bird, V. R. 10.31
Birdsall, C. K. 23.60
Birks, J. B. 19.51
Birks, L. S. 3.143
Birky, M. M. 6.109
Birman, J. L. 19.140
Birnbaum, G. 1.10, 19.33
Birnbaum, M. 3.10, 3.25, 3.27, 3.70, 3.96, 3.202, 12.188, 13.69, 15.96, 18.21, 23.130, 24.97, 24.274, 27.153
Bjorkholm, J. E. 15.61, 20.89
Bjorkstam, J. L. 20.12
Bjørnerud, E. K. 3.7, 12.6, 16.10
Black, J. 9.5
Blakemore, J. S. 10.48
Blanc, J. 8.25, 9.2
Blandin, A. 1.19
Blattner, D. 15.42, 16.4, 16.83
Bleekrode, R. 7.54, 7.64, 8.20, 8.21, 14.75, 27.39
Bletzinger, P. 6.82
Blevin, W. R. 27.22
Blifford, I. H., Jr. 24.201
Bloembergen, N. 1.5, 1.14, 11.24, 16.3, 19.24, 19.32, 19.38, 19.39, 19.69, 19.72, 19.73, 19.86, 19.92, 19.135, 20.4, 20.24, 20.54, 20.80, 20.81, 20.84, 20.87, 20.94, 20.99, 23.20, 23.29, 23.95, 23.113, 23.143, 23.144, 23.152, 23.154, 23.155, 23.170, 23.250
Bloom, A. L. 1.66, 1.72, 6.5, 6.49, 6.146, 7.45, 7.133, 7.134, 7.140, 7.180, 16.29, 18.151, 18.155
Bloom, G. H. 19.149
Bloom, L. R. 15.14, 15.36, 16.8, 17.4, 17.5
Blume, A. E. 3.60, 17.6
Bobrinev, V. I. 25.96
Bobroff, D. L. 6.53, 6.54, 7.1, 12.80, 20.48
Bockasten, K. 7.25, 7.109, 7.197, 7.223
Bodnar, Z. 24.93
Boersch, H. 11.10, 11.54, 12.51, 21.71
Boerschig, B. A. 26.29
Boersma, S. L. 18.138
Bogdankevich, O. V. 9.220, 10.103, 14.138, 14.146

Boĭko, V. A. 23.138
Boivin, A. 23.99, 24.213
Bolgiano, L. P., Jr. 16.113, 18.12, 26.31
Bolmarcich, J. J. 6.13, 9.41
Bolwijn, P. T. 6.87, 6.120, 6.132, 6.170, 6.172, 6.180, 12.216, 18.30, 18.38, 18.51
Bonch-Bruevich, A. M. 19.113, 23.151, 23.195
Bond, W. L. 10.3, 20.29
Bonifacio, R. 2.59, 11.31, 11.32
Boornard, A. 21.49
Boot, H. A. H. 26.133
Borchardt, H. J. 4.76, 5.66, 15.139
Bordogna, J. 15.99, 26.75
Borodulin, V. I. 11.48, 13.48
Borrelli, N. F. 4.58, 14.103, 15.123
Bortfeld, D. P. 3.72, 11.37, 13.20, 19.2, 19.3, 19.4, 19.66, 23.224
Bosomworth, D. R. 17.71, 20.126
Bottka, N. 9.126, 24.112
Bourrabier, C. 23.249
Bouwhius, G. 6.48, 6.126, 6.127, 6.147, 12.70
Bowden, R. A. 2.4
Bowe, P. W. A. 13.53, 13.58
Bowen, D. E. 4.2
Bowen, J. R. 24.273
Bowers, H. C. 23.178, 24.41
Bowness, C. 14.82
Boyd, G. D. 6.16, 12.20, 20.6, 20.29, 20.35, 20.36, 20.47, 20.98, 20.101, 20.102, 20.103, 20.117, 23.201, 24.20
Boyle, W. S. 3.156
Boyne, H. S. 6.109
Brackenridge, J. B. 26.100
Bradbury, R. A. 3.175, 4.3, 17.1
Bradford, J. N. 13.89, 14.56, 17.42
Bradley, D. J. 3.186, 3.195, 3.206, 12.209, 23.131
Bradley, L. C., III 24.28, 27.20
Bradley, L. T. 23.42
Bradsell, R. H. 26.21
Bramley, A. 23.11, 27.41
Brand, F. A. 2.7, 2.34, 3.62, 3.83, 3.167, 11.4, 11.16, 11.43, 16.7
Brand, H. 15.166
Brandewie, R. A. 7.240, 14.15, 16.157
Brändli, H. P. 18.124